46
Topics in Heterocyclic Chemistry

Aims and Scope

The series Topics in Heterocyclic Chemistry presents critical reviews on present and future trends in the research of heterocyclic compounds. Overall the scope is to cover topics dealing with all areas within heterocyclic chemistry, both experimental and theoretical, of interest to the general heterocyclic chemistry community.

The series consists of topic related volumes edited by renowned editors with contributions of experts in the field.

More information about this series at http://www.springer.com/series/7081

Marco Bandini
Editor

Au-Catalyzed Synthesis and Functionalization of Heterocycles

With contributions by

N. Ahlsten · A. Arcadi · X.C. Cambeiro · Z.-Y. Cao · I. Larrosa · F. López · J.L. Mascareñas · D.H. Miles · M.P. Muñoz · G.J.P. Perry · M.T. Quirós · F.D. Toste · Y.-H. Wang · L. Ye · L. Zhang · J. Zhou

Springer

Editor
Marco Bandini
Dipartimento di Chimica "G. Ciamician"
Università di Bologna
Bologna, Italy

ISSN 1861-9282 ISSN 1861-9290 (electronic)
Topics in Heterocyclic Chemistry
ISBN 978-3-319-35142-1 ISBN 978-3-319-35144-5 (eBook)
DOI 10.1007/978-3-319-35144-5

Library of Congress Control Number: 2016943836

Printed on acid-free paper

This Springer imprint is published by Springer Nature
The registered company is Springer International Publishing AG Switzerland

Preface

"Noble," "coinage," "precious"...so many "labels" have been utilized over the years to describe gold. All of them are certainly correct with the exception of the one that probably affected mostly the role of the late-transition metal in chemical reactivity: "inert". As a matter of fact, besides the intrinsic inertness of gold in the elemental form, [Au(I)] and [Au(III)] species have displayed unique physical-chemical properties that led to unexpected applications in homogeneous catalysis with particular regard to electrophilic activations of unactivated unsaturated hydrocarbons. Doubtless, the chemistry of π-systems has faced a revolution in terms of chemical scope, mildness of reaction conditions, and selectivity over the past fifteen years, due to the establishment of this metal in catalysis. A rough search on database dealing with the item "gold catalysis" can better highlight the state-of-the-art impact of organometallic gold species in organic synthesis. Impressively, gold (141 articles in 2014) has already reached longtime used metals such as palladium (125 articles in 2014) and copper (60 articles in 2014) and already overcame other "neighbors" in the periodic table such as silver (30 articles in 2014) and rhodium (38 articles in 2014).

The innate tolerance of gold catalysis toward "hard" hetero moieties contributed to its diffusion on the synthesis of densely functionalized molecular scaffolds including heterocyclic cores. This volume provides an overview of the most efficient and synthetically useful approaches to the construction of heterocyclic motifs by means of homogeneous gold catalysis.

The chapters have been written by leading experts; emphasis has been addressed to both scope and limitation of the methodologies. Additionally, mechanistic insights are provided in order to ensure a proper rational of the chemical outcomes. Main activation modes and reaction machineries (i.e., electrophilic additions), triggered by gold coordination to π-systems, have been highlighted with particular regard to alkynes, alkenes, and allenes. Besides these intriguing C–X bond-forming events, gold catalysis found substantial applications also in the manipulation of C–C triple bonds via rearrangement reactions as well as site-selective oxidative protocols. Recent and leading examples in these realms have been also accounted and described in dedicated chapters.

Additionally, the well-established attitude of gold-based catalytic systems in assisting the activation/functionalization of inert C–H bonds has been documented with the aim of shedding light on the applicability of readily available and unfunctionalized heteroarenes for the synthesis of added value compounds.

The latter two chapters have been dedicated to the impact of gold on asymmetric transformations and total synthesis of natural products. I personally consider these frameworks two hot topics of the nowadays-homogeneous gold catalysis.

Obviously, I feel indebted to the colleagues who agreed to contribute in the realization of this account, and I apologize to the authors left out this time.

I do really hope that the book could contribute in stimulating new perspectives and developments in the blooming research area and could encourage even more practitioners in engaging in this fascinating research topic.

Bologna, Italy
April 2016

Marco Bandini

Contents

Top Heterocycl Chem (2016) 46: 1–52
DOI: 10.1007/7081_2015_5006

Published online: 3 February 2016

Synthesis of Oxygenated Heterocyclic Compounds via Gold-Catalyzed Functionalization of π-Systems

Jose L. Mascareñas and Fernando López

Abstract During the last decade, there has been an impressive amount of new reactions catalyzed by gold (I or III) complexes, many of which lead to oxygen heterocycles in novel, practical, and efficient ways. In this chapter we have selected the most representative examples of the use of gold catalysis for the synthesis of oxygenated heterocycles, with an organization based on the type of reaction. Each section pretends to give an overview of the potential of each particular strategy and its usefulness to prepare synthetically relevant oxygenated rings. Heterocyclic systems that contain other heteroatoms in addition to oxygen have been essentially excluded, as well as those examples in which the oxygen atoms merely acts as a connecting element between two reactive functional groups.

Keywords Cyclization • Cycloaddition • Gold catalysis • Heterocycles • Organometallic catalysis • Oxacycles

Contents

J.L. Mascareñas (✉)
Centro Singular de Investigación en Química Biológica y Materiales Moleculares (CIQUS) and Departamento de Química Orgánica, Universidad de Santiago de Compostela, C/ Jenaro de la Fuente, s/n, 15782 Santiago de Compostela, Spain
e-mail: joseluis.mascarenas@usc.es

F. López (✉)
Centro Singular de Investigación en Química Biológica y Materiales Moleculares (CIQUS) and Departamento de Química Orgánica, Universidad de Santiago de Compostela, C/ Jenaro de la Fuente, s/n, 15782 Santiago de Compostela, Spain

Instituto de Química Orgánica General (CSIC), Juan de la Cierva 3 28006, Madrid, Spain
e-mail: fernando.lopez@csic.es

1 Introduction

Oxygen-based heterocycles are common structural components of a wide range of naturally occurring and biologically active molecules [1–4]. Among then, five- and six-membered oxacycles like tetrahydropyrans, dihydropyrans, tetrahydrofurans, or several types of γ- and δ-lactones are motifs with a high prevalence in medically and biologically relevant natural and nonnatural products. However, smaller (three and four membered) as well as larger oxacycles (from seven membered to macrolides) are also of paramount importance in organic synthesis and in medicinal chemistry. For instance, oxetane-containing building blocks have attracted the attention of medicinal chemists in recent years [5]. On the other hand, and although less common in biologically active natural products, seven-membered oxacycles like oxepanes, oxepins, or benzoxepines as well as their eight- to ten-membered counterparts (e.g., oxocanes, oxonanes, and oxecanes) have also attracted the attention of chemists.

Consequently, the development of new and very efficient methods for the synthesis of all these types of oxygen-based heterocyclic compounds is of paramount importance [6–8]. While most methods for the construction of these systems are based on classical cyclization or cycloaddition methods, there have been also many reports based on the use of transition metal catalysis, particularly those involving metals of the groups 8 to 10. The relatively recent demonstration that metals of group 11, in particular gold, can open novel reactivity paths for C–C unsaturated systems has opened new opportunities to discover new types of transformations which could not be achieved using other transition metal catalysts [9–12]. The unique reactivity of gold complexes can in part be explained in terms of relativistic effects [13, 14], which induce the contraction of 6s and the expansion of 5d orbitals. As a consequence, this metal exhibits singular characteristics, such as a high affinity for π-unsaturated systems like alkynes, alkenes, or allenes, and a low propensity to participate in typical redox processes characteristic of other transition metal catalytic cycles (e.g., oxidative additions and reductive eliminations). Moreover, gold (I) or (III) complexes tend to activate alkynes, alkenes, or allenes in a highly chemoselective manner, activation which opens interesting reaction pathways that usually involve carbocationic intermediates. Also very important is the possibility of modulating the properties of the metal through modification of its ancillary ligands (e.g., phosphines, *N*-heterocyclic carbenes, etc.). This previously unsuspected property has considerably widened the potential and versatility of gold catalysts, in particular of those based on cationic gold (I) complexes (e.g., Ligand-$Au^+ X^-$).

During the last decade, there has been an impressive amount of new reactions reported in this field, many of which allowed to ensemble oxygen heterocycles in novel, practical, and efficient ways. Although the assembly of oxygenated heterocyclic compounds by means of gold catalysis has been previously discussed as part of several reviews devoted either to gold chemistry or to heterocyclic synthesis, in this chapter we will focus explicitly on the use of gold catalysis for the synthesis of oxygenated heterocycles. We have chosen an organization based on the type of reaction. Given that it is extremely difficult to cover comprehensively all the cases, within each section we have selected the most representative examples to give an overview of the potential of each particular strategy and its usefulness to prepare synthetically relevant oxygenated rings. Heterocyclic systems that contain other heteroatoms in addition to oxygen are barely discussed. Importantly, we have specifically selected those examples in which the oxygen atoms play a significant role in the mechanism of the process. Thus, those cases where the oxygen merely acts as a connecting element between two reactive functional groups have been discarded.

2 Oxycyclizations Initiated by a Gold-Promoted Activation of π-Bonds

2.1 Cycloisomerizations Involving Hydroalkoxylations and Hydrocarboxylations of π-Systems

Many examples of homogeneous catalysis employing gold complexes exploit the ability of these metal catalysts to bind chemoselectively to π-unsaturated bonds and promote a subsequent attack of an oxygen-based nucleophile [15–18]. In general these metal carbophilic catalysts can coordinate and activate alkynes, alkenes, dienes, and allenes, but when the reacting precursors contain several unsaturated moieties, nucleophiles seem to have a kinetic preference for attacking activated alkynyl systems, which warrants high chemoselectivities.

2.1.1 Hydroalkoxylation of Allenes

Allenes are stable but also kinetically reactive substrates, with π-electrons that can readily coordinate to appropriate electrophilic transition metals, including gold

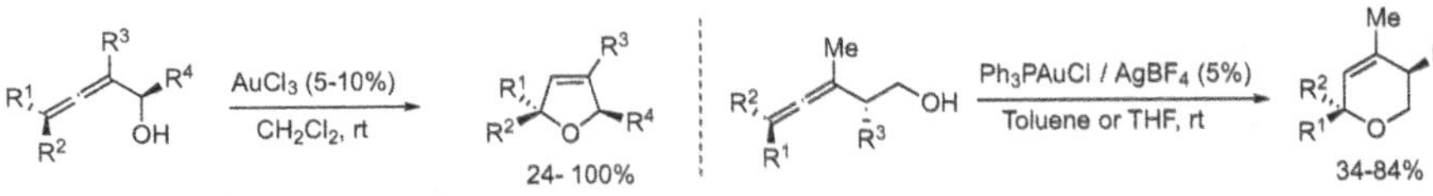

Scheme 1 Hydroalkoxylation of α-hydroxyallenes and β-hydroxyallenes

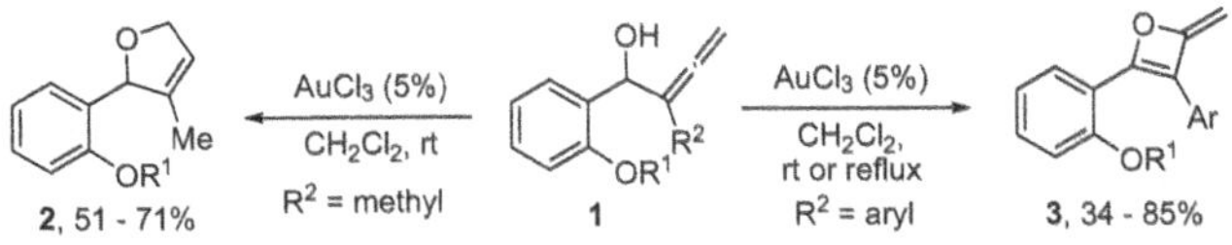

Scheme 2 4-*exo*-dig vs 5-*endo*-trig cyclizations of α-allenols

complexes [19, 20]. In 2001, Krause and coworkers reported the $AuCl_3$-catalyzed cyclization of α-hydroxyallenes that efficiently affords, at room temperature, tri- and tetrasubstituted 2,5-dihydrofurans, in good yields. The reaction involves a complete axis to center chirality transfer (Scheme 1) [21]. This and other groups expanded the scope of this type of *endo-trig* cyclizations by using different α-hydroxyallenes or β-hydroxyallenes as well as alternative gold catalysts like $AuBr_3$, AuCl, and Ph_3PAuBF_4 (Scheme 1) [22–27]. The transformation can also be carried out in ionic liquids which allows the easy recovery of the catalyst [28].

α-Allenols can also undergo 4-*exo*-dig cyclizations, although the transformation is rather unusual. Thus, Almendros and Alcaide showed that, although α-allenols such as **1** derived from salicylaldehyde generally provide the 5-*endo*-trig cyclization products **2** under catalysis with $AuCl_3$, when the allene incorporates an aryl substituent at its internal position (R^2), oxetenes of type **3** resulting from a tandem 4-*exo*-dig cyclization/dehydrogenation process are obtained (Scheme 2) [29].

Widenhoefer and coworkers reported in 2006 that the gold(I) catalyst JohnPhosAuOTs was able to promote a highly efficient and regioselective *exo*-hydroalkoxylation of γ-hydroxy and δ-hydroxyallenes to provide, after just a few minutes at *rt*, the corresponding tetrahydrofuran and tetrahydropyran derivatives of type **5** in good to excellent yields. Interestingly, the nature of the counterion was key to achieve high levels of regioselectivity (*exo* vs *endo* hydroalkoxylation) (Scheme 3) [30].

Alcaide and Almendros examined related hydroalkoxylations of α,-γ-dihydroxyallenes like **6**, incorporated into a β-lactam framework. Curiously, whereas the $AuCl_3$-catalyzed cyclization of the TBS-protected substrate afforded the 5-*exo*-attack product **7**, the analog MOM-protected derivative underwent an in situ deprotection followed by a 5-*endo*-trig cyclization to provide **8**. On the other hand, if the MOM group is placed at the γ-position, the 7-*endo*-trig hydroalkoxylation product **9** is observed (Scheme 4) [31–33].

The same group also described a $AuCl_3$-catalyzed oxycyclization of allenols of type **10** to provide tetrahydrooxepines **11** through a regioselective 7-*endo*-trig-annulation (Scheme 5) [34]. This transformation differs markedly from most of

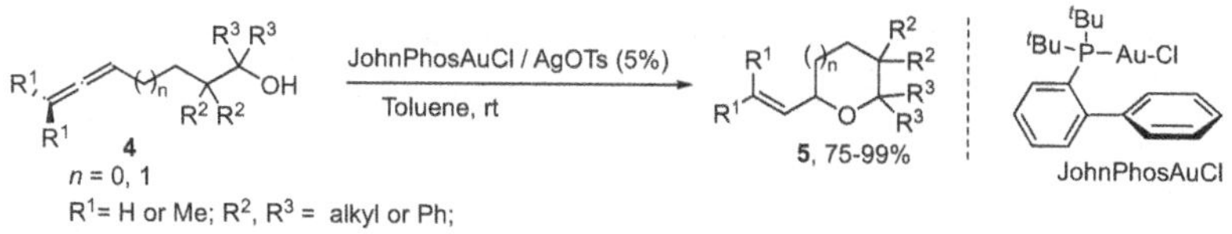

Scheme 3 Hydroalkoxylation of hydroxyallenes

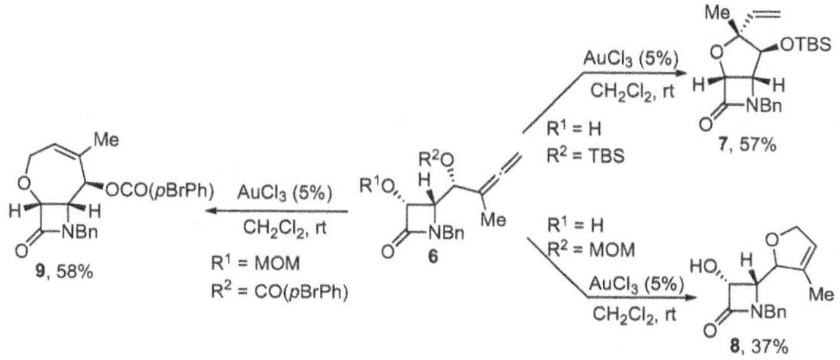

Scheme 4 Hydroalkoxylations of α,γ-dihydroxyallenes **6**

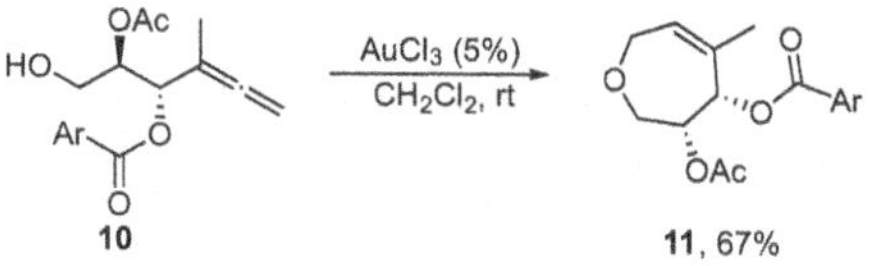

Scheme 5 Hydroalkoxylations of α,γ-dihydroxyallenes **10**

the oxycyclizations of γ-allenols, which proceed via a 5-*exo*-dig cyclization, leading to 2-alkenyltetrahydrofurans.

In 2007, Widenhoefer and coworkers reported the first gold-catalyzed enantioselective hydroalkoxylation of allenols. In particular, the ionic chiral catalyst derived from DTBM-BIPHEP $(AuCl)_2$ (2.5%)/AgOTs (5%) was able to promote the cyclization of a variety of γ- and δ-allenols to provide their corresponding tetrahydrofurans and tetrahydropyrans with good yields and excellent enantioselectivities (Scheme 6) [35]. Interestingly, the transformation was particularly effective with γ-hydroxyallenes incorporating an axially chiral allenyl moiety. When a racemic allenol was used as reactant, a mixture of *E*/*Z* isomers were obtained with the same absolute configuration at the new Csp^3 center. Thus, the absolute configuration of the catalyst determines the configuration of this stereocenter of the product, whereas the stereochemical relationship between the catalyst and the substrate determines the relative configuration of the double bond of the product.

In the same year, Toste and coworkers reported that chiral gold(I) complexes consisting of a racemic ancillary ligand (typically a bisphosphine ligand) and a chiral counterion (chiral binaphthol-derived phosphate) were also excellent

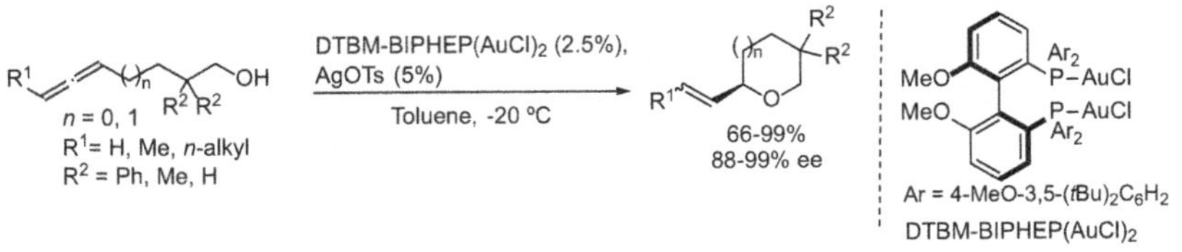

Scheme 6 Enantioselective hydroalkoxylation of allenols with a gold complex featuring a chiral ligand

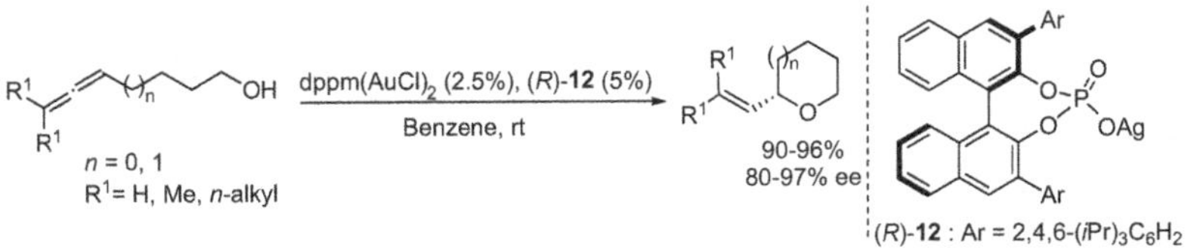

Scheme 7 Enantioselective hydroalkoxylations with gold complexes featuring chiral counterions

catalysts for the *exo*-hydroalkoxylation of γ and δ-allenols, providing the corresponding 2-vinyl tetrahydrofurans and tetrahydropyrans with high yields and excellent enantioselectivities (Scheme 7) [36]. A related chiral catalyst was recently used by the same group for the desymmetrizing hydroalkoxylation of 2-aryl-2-allenyl-1,3-diols to yield multisubstituted tetrahydrofurans with good yields and excellent diastereo and enantioselectivities [37]. On the other hand, Lalic and coworkers reported an *exo*-selective cyclization of enantioenriched trisubstituted δ-allenols to provide the corresponding cyclic ethers with complete chirality transfer. In this case, it was found that the use of the gold complex $^{t}Bu_3PAuOTs$ is key to avoid the epimerization of the allene prior to the cyclization [38].

Importantly, the versatility of the above gold-catalyzed hydroalkoxylations to access to highly substituted dihydro- and tetrahydropyran moieties was successfully illustrated by several groups in the context of the synthesis of biologically relevant products which contain these subunits as key structural components. For instance, Krause and coworkers reported the total synthesis of alkaloids (−)-isocyclocapitelline and (−)-isochrysotricine using a key dihydrofuran intermediate (**13**) which was built by a $AuCl_3$-catalyzed cycloisomerization of an α-allenol precursor, a process that also proceeds with complete axis-to-center chirality transfer [39, 40]. Another elegant example was included in the total synthesis of ionomycin-calcium complex by Kocienski and coworkers, who assembled its 2,5-dihydrofuran subunit by means of an α-allenol cycloisomerization [41]. The gold-catalyzed cycloisomerization of β-allenols to dihydropyrans has also been used in total synthesis, in particular, the cyclization of **14** to yield **15** was employed by Krause and coworkers as key step in the synthesis of the sesquiterpenoid bejarol (Scheme 8) [42].

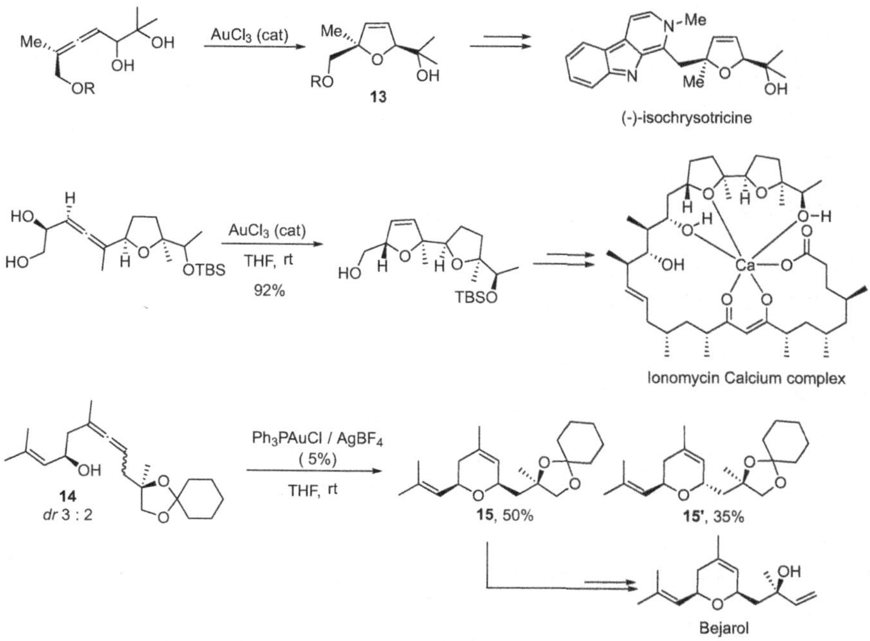

Scheme 8 Applications of gold-catalyzed hydroalkoxylation of allenes in total synthesis

2.1.2 Hydrocarboxylation of Allenes

Carboxylic acids can also be used as nucleophiles for the hydrofunctionalization of allenes, thus providing a direct access to chiral lactones. Toste and coworkers studied the asymmetric cycloisomerization of **16** with several types of chiral gold complexes. Interestingly, catalysts exclusively based on the chirality of the ancillary ligand or on that of the counterion provided good yields but poor *ee*'s in a model reaction (up to 38% *ee*). However, a catalyst composed of a chiral counterion as well as the chiral bisphosphine BINAP was able to provide the product with a remarkable 88% yield and 82% *ee*, demonstrating again the potential of chiral counterions in gold catalysis (Scheme 9) [36].

Carboxylic esters have also been used as substrates in these reactions. For instance, Shin and coworkers demonstrated that $AuCl_3$ efficiently catalyzes the cyclization of *t*-butyl allenoates to γ-butenolides (Scheme 10, Eq. 1) [43]. Interestingly, Hammond and coworkers isolated and characterized the vinyl gold intermediate that is formed in the cyclization process and demonstrated that it can participate in subsequent protodeauration and iododeauration processes [44].[1] On the other hand, Bäckvall reported a related procedure for the synthesis of

[1] In Sect. 2.4 a series of tandem processes, where this vinyl gold species are used as key intermediates, are discussed.

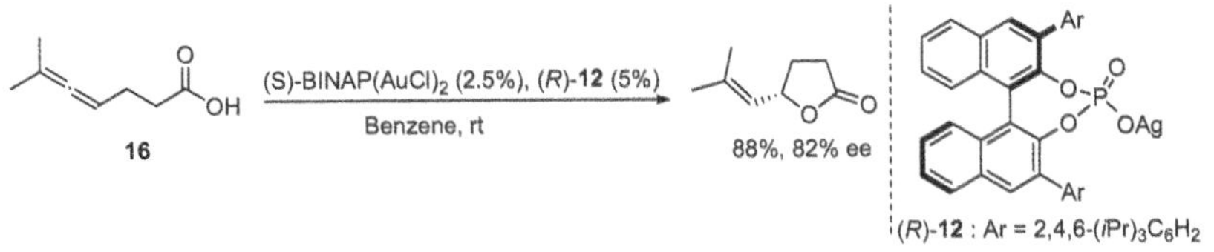

Scheme 9 Enantioselective hydrocarboxylation of allenes

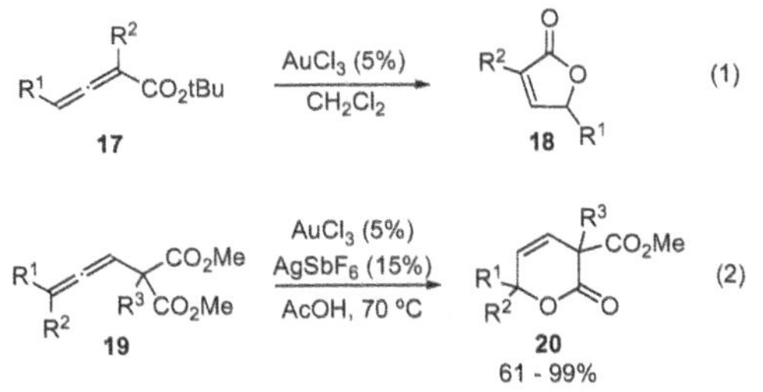

Scheme 10 Cyclizations of allenoates

δ-lactones like **20** using $AuCl_3/AgSbF_6$ as catalyst and allenyl esters **19** as substrates [45].

In summary, the hydroalkoxylation and hydrocarboxylation of allenes can be considered among the most well-developed transformations in gold catalysis. Its expansion not only has allowed the access to several key building blocks for natural product synthesis but has also provided relevant mechanistic information on gold-catalyzed processes.

2.1.3 Hydroalkoxylation of Alkynes

The hydroalkoxylation of alkynes is one of the most convenient methods for the synthesis of oxygen heterocycles such as tetrahydrofurans and pyrans (for reviews covering this topic, see [17, 33, 46–48]). The high alkynophilicity of gold complexes usually allows these reactions to be performed under mild conditions, providing either the *exo*-dig or *endo*-dig products. In this section, we show only examples which involve a simple hydroalkoxylation. Processes where the hydroalkoxylation of the alkyne is followed by a second, or more additional steps, in a tandem sequence, are described in Sect. 2.3.

In 2007, Pale and coworkers reported that ω-acetylenic alcohols could be regio- and stereoselectively converted to α-alkylidene tetrahydropyran or tetrahydrofurans in the presence of catalytic amounts of AuCl and K_2CO_3. The reactions proceed at *rt*, with good or very good yields and complete *exo*-dig selectivity [49] and involve an *anti*auration hydroxylation sequence followed by a base-mediated deprotonation and protodeauration. This methodology has been applied by the same group to the straightforward synthesis of the aurones, a family of natural flavonoids that

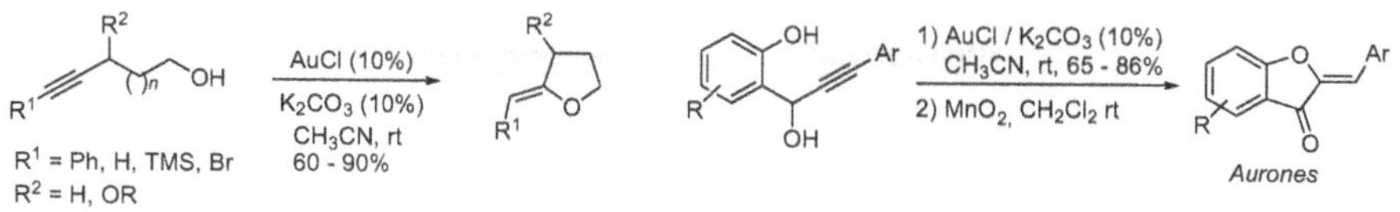

Scheme 11 Examples of hydroalkoxylation of acetylenic alcohols

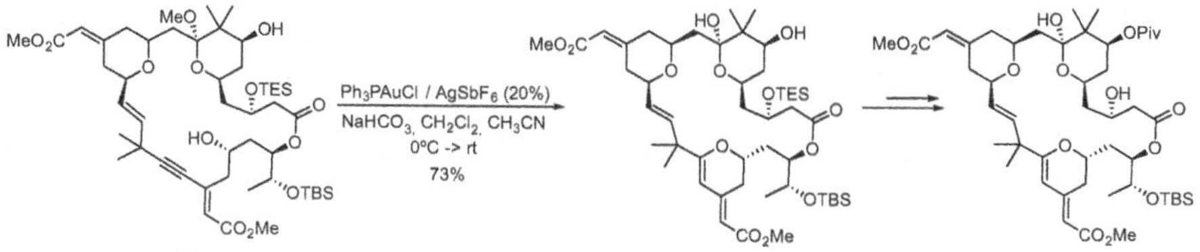

Scheme 12 An intramolecular hydroalkoxylation in the total synthesis of bryostatin 16

exhibit interesting biological activities (Scheme 11) [50, 51] (for mechanistic studies, see [52]).

Moreover, Trost and coworkers demonstrated in their recent total synthesis of bryostatin 16 that these oxycyclizations could be also be performed in complex molecules, showing a high degree of functional group tolerance. Thus, the gold-catalyzed 6-*endo*-dig cyclization of enynol **21** was performed with catalytic amounts of $Ph_3PAuCl/AgSbF_6$ and $NaHCO_3$, providing the tetrahydropyranic product **22** in a very remarkable 73% yield. Curiously, and probably due to the Lewis acidity of the gold catalyst, the methyl ketal moiety was also hydrolyzed under the cyclization conditions (Scheme 12) [53].

More recently, Czekelius and coworkers reported gold-catalyzed hydroalkoxylations of diynols like **23** [54, 55]. In particular, selective *exo* or *endo* cyclizations could be obtained depending on the substituents at the position 3 of the diyne. When an oxygen atom is placed in this position (X=O), only the *endo* adduct was observed. On the contrary, when X=CH_2, the *exo*-cyclization occurred. Based on this and additional experimental data, the authors suggested that stereoelectronic interactions between the σ(C–Au) orbital and the σ*(C–O) orbital favor the *endo*-cyclization when X=O. The higher energy of the σ*(C–C) orbital when X=CH_2 would impede these interactions, promoting the change in regioselectivity (Scheme 13).

Interestingly, Uemura and coworkers reported a desymmetrization of prochiral 1,3-dihydroxymethyl-2-alkynylbenzene chromium complexes (**24**) based on an asymmetric hydroalkoxylation reaction catalyzed by chiral bisphosphinegold complexes like $XylylBINAP(AuCl)_2/AgSbF_6$. The resulting isochromene chromium complexes (**25**) were obtained with excellent enantioselectivities [56] (Scheme 14).

Enynols do also undergo oxycyclization reactions. For example, Liu and coworkers demonstrated that enynols like **26**, exhibiting a tertiary alcohol group at C1, undergo an extremely mild gold-catalyzed *exo-dig* cyclization to provide

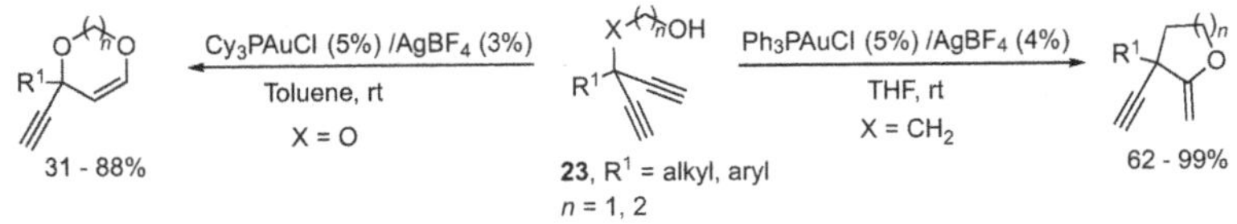

Scheme 13 Gold-catalyzed hydroalkoxylations of diynols **23**

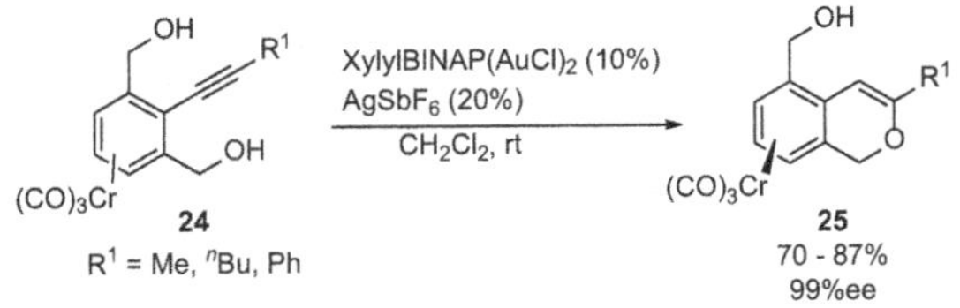

Scheme 14 Desymmetrization 1,3-dihydroxymethyl-2-alkynylbenzene chromium complexes

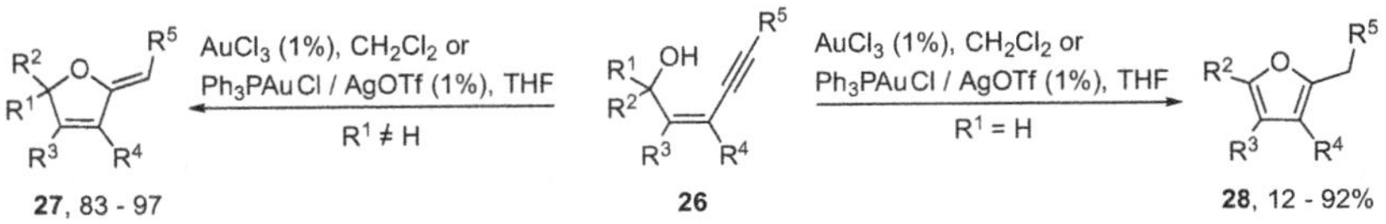

Scheme 15 Oxycyclization reactions of enynols

5-yliden-2,5-dihydrofurans (**27**) with excellent yields. When C1 presents a secondary hydroxyl group, furans of type **28**, resulting from an isomerization reaction, which could take place before or after the protodeauration, are obtained (Scheme 15). A very related gold(I)-catalyzed hydroxycyclization of easily accessible 2-alkynylallyl alcohols to afford 2,4-disubstituted and 2,3,5-trisubstituted furans, under mild conditions, was also independently reported by Hashmi and Lee groups [57, 58].

2.1.4 Hydrocarboxylation of Alkynes

In 2006, Genet, Michelet, and coworkers reported that acetylenic carboxylic acids like **29** can undergo a AuCl *exo*-selective cycloisomerization at room temperature to provide *exo*-methylenic lactones in good yields and with a remarkable reaction scope (Scheme 16) [59–62]. More recently, Conejero, Michelet, Cadierno, and coworkers also demonstrated that water soluble gold complexes such as **31** were also competent catalysts in related oxycyclizations of γ-alkanoic acids in toluene/water mixtures. Strikingly, the potentially competitive hydration of the alkynes was not observed in these examples [63].

Van de Weghe and coworkers showed that the intramolecular $AuCl_3$-catalyzed cyclization of *o*-alkynylbenzoic methyl esters directly proceeded by a 6-*endo-dig* mode to provide isocoumarins (Scheme 17, Eq. 1). The presence of traces of water (2 equiv.) is required hypothetically for both, the hydrolysis of the oxonium

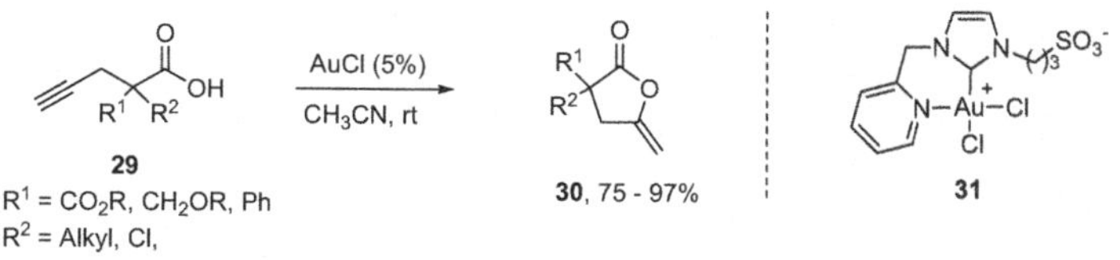

Scheme 16 *Exo*-selective cycloisomerization of acetylenic carboxylic acids

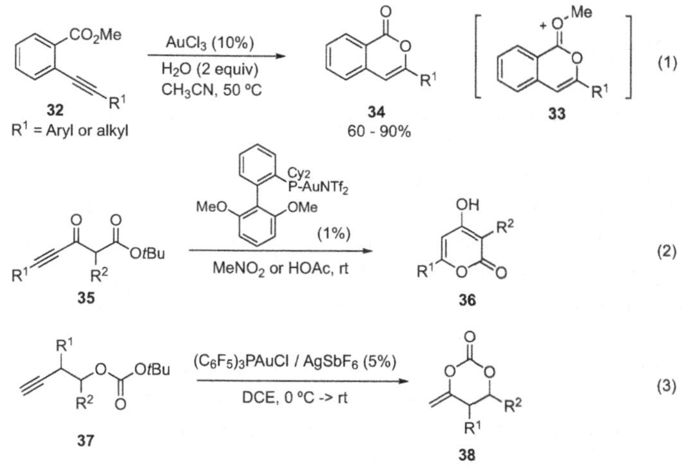

Scheme 17 Cyclizations of alkynyl carboxylic esters and related carbonates

intermediate **33** and for the final protodeauration [64]. A related cycloisomerization using a *t*-butyl carboxylic esters like **35** to yield 4-hydroxy-2-pyrones **36** was developed in 2012 by Fürstner and coworkers (Scheme 17, Eq. 2). Moreover, the method was applied as one of the key steps of the total synthesis of neurymenolide A [65]. Analogously, when a *t*-butyl carbonate derived from a homopropargylic alcohol is used, the corresponding cyclic enol carbonate can be obtained in good yields, as demonstrated by Shin and coworkers (Scheme 17, Eq. 3) [66].

In summary, the hydroalkoxylation and hydrocarboxylation of alkynes are powerful methods for the construction of oxygenated heterocycles, in particular lactones, furans, dihydropyrans, and dihydrofurans. Elegant applications in total synthesis have already been reported, highlighting its versatility and potential in total synthesis.

2.1.5 Hydroalkoxylation of Alkenes

In contrast to allenes or alkynes, alkenes are less susceptible of undergoing the oxycyclization reactions, and therefore examples of these cyclizations are much scarcer (for reviews covering this topic, see [17, 67]). Although as early as in 2004

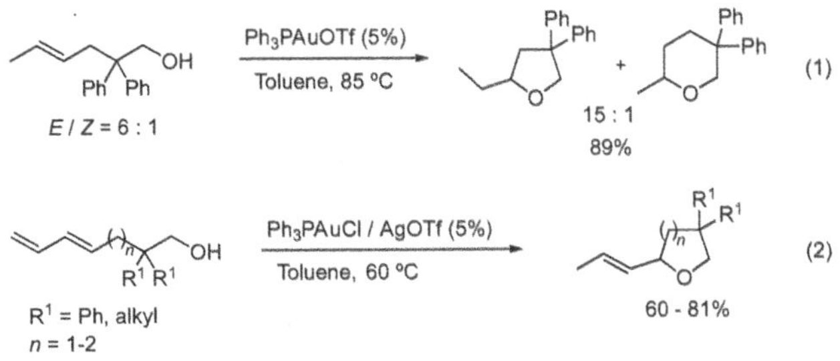

Scheme 18 Intramolecular hydroalkoxylation of alkenes and conjugated dienes

Widenhoefer and coworkers reported intramolecular hydroalkoxylations of γ- and δ-hydroxy olefins using Pt catalysts, similar examples with gold catalysts are very scarce. In 2005 Yang and He reported an isolated example of the intramolecular cyclization of a particular γ-hydroxyl alkene **39**, mediated by a cationic [Au(I)] catalyst. A 15:1 mixture of the corresponding five- and six-membered oxacycles was isolated (Scheme 18, Eq. 1) [68]. From a mechanistic perspective, the reaction involves a nucleophilic addition of the alcohol to the gold-activated double bond to generate a σ-alkylgold intermediate, which undergoes the protodeauration reaction to give the expected saturated oxacycle. Two years later, Sakurai and coworkers demonstrated that gold nanoclusters stabilized by poly(*N*-vinyl-2-pyrrolidone) (Au: PVP) were also competent catalysts for these cyclizations in aqueous media and under aerobic conditions [69]. More recently, Ryu and coworkers reported the first gold-catalyzed intramolecular hydroalkoxylation of a conjugated diene to provide substituted tetrahydropyrans and tetrahydrofurans with moderate to good yields (Scheme 18, Eq. 2) [70]. In Sect. 2.4, we will present examples in which the σ-alkylgold species generated in these cyclizations participate in an oxidative C–C coupling to provide more elaborated oxacycles [71, 72].

It is interesting to note that the presence of an additional hydroxyl group at the allylic position of the alkene can lead to formal allylic substitution reactions. This reactivity stems from the ability of gold complexes to act as σ and π acids [67]. Thus, Aponick and coworkers developed a gold-catalyzed hydroalkoxylation of allylic alcohols which provides 2-vinyl tetrahydropyrans of type **40** in excellent yields [73]. When 2,6-disubstituted tetrahydropyrans are obtained, the reaction is highly diastereoselective towards the *cis* isomer. Based on DFT calculations and experimental data, this group postulated that the most probable mechanism involves a two-step sequence that begins with intramolecular C–O bond formation by *anti*-addition of the non-allylic alcohol to the Au-activated alkene, followed by a concerted hydrogen transfer/*anti*-elimination to release water (Scheme 19). A hydrogen bonding between the two hydroxyl groups, maintained throughout the course of the process, facilitates the reaction and is also responsible for the stereochemical outcome [74–77].

Finally, it is worth to mention that Robertson and coworkers applied this methodology in the context of a short synthesis of isoaltholactone [78].

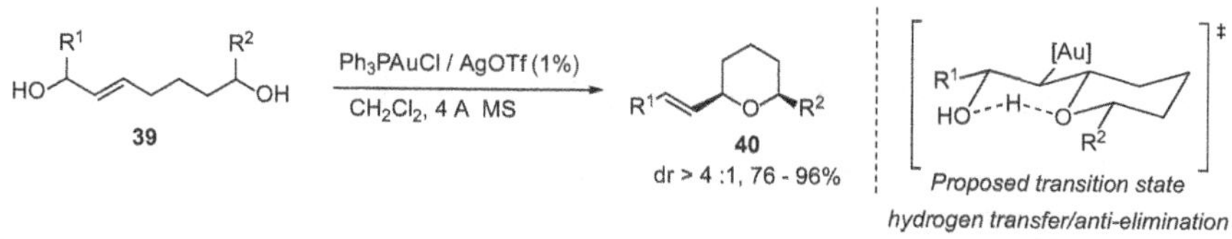

Scheme 19 Hydroalkoxylation of allylic alcohols **39** leading to 2-vinyl tetrahydropyrans

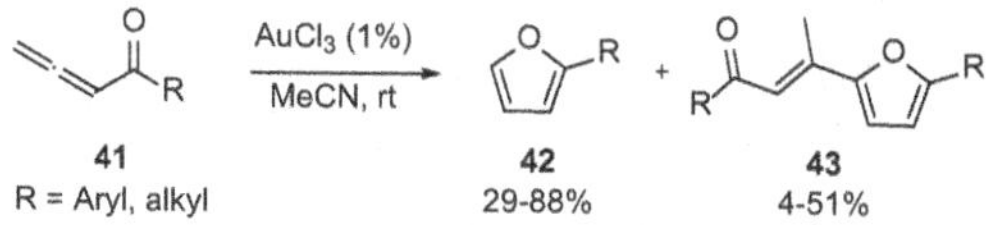

Scheme 20 Cycloisomerizations of allenyl ketones pioneered by Hashmi and coworkers

2.2 *Cycloisomerizations of Allenyl and Alkynyl Ketones, and Alkynyl Epoxides*

As might be expected, carbonyl groups can also be used as nucleophiles in gold-catalyzed oxycyclization processes provided the resulting oxonium species can evolve to stable products. In general keto-allenes are the most studied systems, although keto-alkynes have also been reported to provide diversely substituted furans with great efficiencies. In all these cases, the driving force for the cycloisomerization is the formation of an aromatic heterocycle.

As early as in the year 2000, Hashmi and coworkers reported that α-allenyl ketones like **41** can undergo $AuCl_3$-catalyzed oxycyclizations to yield different types of substituted furans (Scheme 20) [79]. The method constituted the first gold-catalyzed addition of a heteroatom nucleophile to an allenyl moiety.

Che and coworkers employed the cationic gold(III) porphyrin complex **44** for the cyclization of various mono- or disubstituted allenic ketones of type **45** to the corresponding furans [80]. From a mechanistic point of view, the authors proposed that the active gold species coordinates to the "distal" double bond of the allene and induces a nucleophilic attack of the oxygen atom furnishing a cationic intermediate, which evolves to the furan by deprotonation and protodeauration (Scheme 21).

On the other hand, the gold-catalyzed cycloisomerization of bromoallenones **47** was reported by Gevorgyan and coworkers, who demonstrated that the structure of the resulting product is highly dependent on the gold catalyst (Scheme 22, Eq. 1) [81, 82]. Moreover, silyl-, thio-, or seleno-furans can be obtained from the corresponding allenes by a 1,2-Si, 1,2-S, or 1,2-Se shift [83]. Additionally, Waser and coworkers demonstrated that 3-alkynyl furans could be directly obtained from α-allenyl ketones by means of a domino cyclization/alkynylation process employing a hypervalent iodine reagent like **48** (Scheme 22, Eq. 2) [84].

Alk-4-yn-1-ones like **50** can also undergo *5-exo-dig* cycloisomerizations, leading to furans when treated with a gold catalyst such as $Ph_3PAuOTf$ [85, 86]. Alternatively, when the alk-4-yn-1-one is disubstituted at the β-position of the ketone,

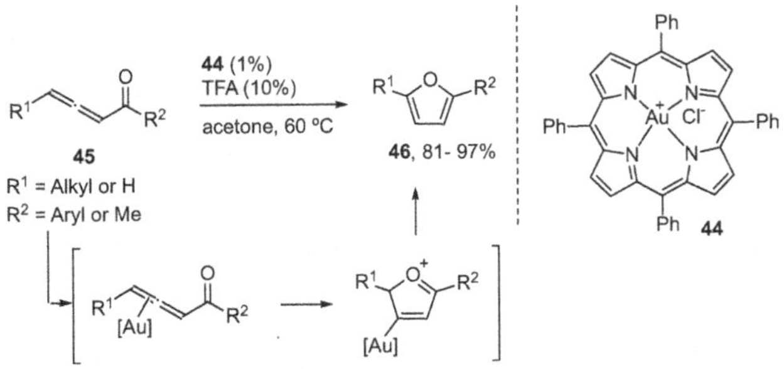

Scheme 21 Cycloisomerizations of allenyl ketones reported by Che and coworkers

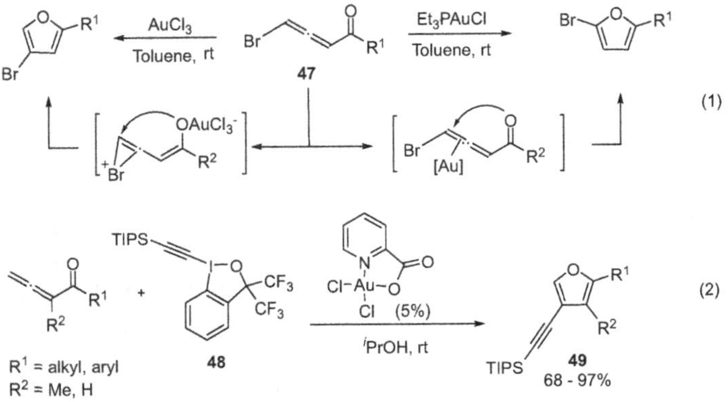

Scheme 22 Cycloisomerization of bromoallenones (Eq. 1) and domino cyclization/alkynylation process (Eq. 2)

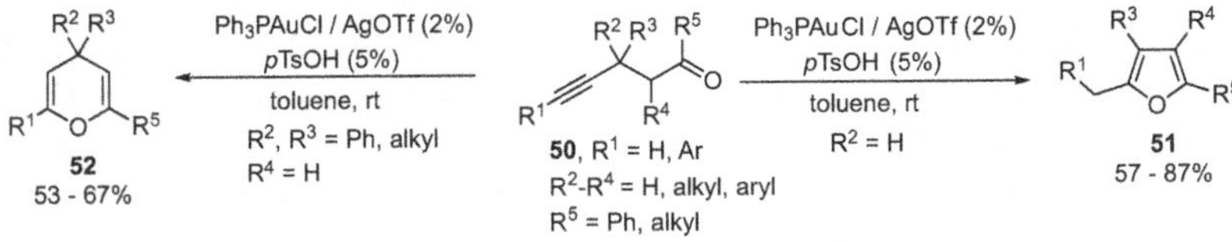

Scheme 23 *5-exo-dig* cycloisomerizations of alk-4-yn-1-ones

pyrans of type **52**, resulting from a *6-endo-dig* cyclization, were obtained (Scheme 23) [85].

Epoxides can also work as nucleophiles in this type of reactions. Hashmi and coworkers reported in 2004 the gold-catalyzed isomerization of epoxyalkynes like **53** to yield substituted furans. After activation of the alkyne by gold(I), the epoxide attacks at its distal position affording the intermediate **54**. Abstraction of a proton followed by protodeauration eventually affords the furan system (Scheme 24) [87–90].

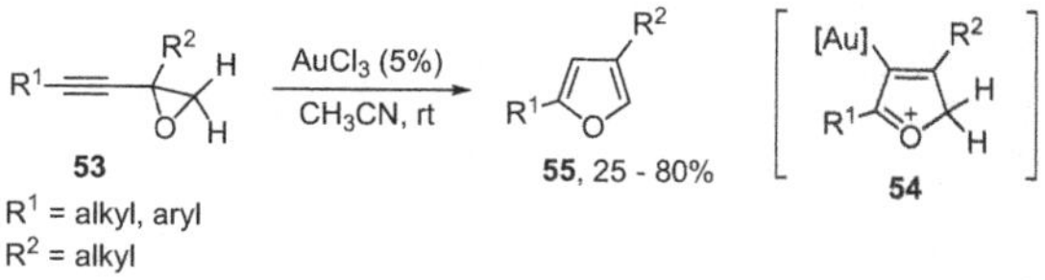

Scheme 24 Cycloisomerizations of alkynyl epoxides leading to furans

2.3 Cycloisomerization Cascades

In this section, we include examples of processes in which the gold-catalyzed addition of oxygen-based nucleophiles to unsaturated C–C bonds is embedded within a more complex cascade cycloisomerization (for reviews partially covering this topic, see [47, 91–93]). The section has been organized based on the π-component which, upon coordination of the Au catalysts, triggers the cascade reaction (e.g., alkyne, allene, or alkene).

2.3.1 Cycloisomerization Cascades Initiated by Activation of Alkynes

In 2005, Kozmin and coworkers reported an efficient gold-catalyzed double cyclization of 1,5-enynols **56** and **57**. The transformation provides several types of oxabridged, fused, and spirocyclic frameworks [94]. The efficiency of the process highlights a highly chemoselective activation of the alkyne by the gold catalyst, which undergoes the 6-*endo*-dig carbocyclization with concomitant intramolecular formation of the C–O bond. Based on their experimental results, the authors indicate that this tandem cyclization is most likely a concerted process which formally results in a formal *anti*-addition of the alkyne and OH groups to the alkene (Scheme 25).

A related stereoselective cycloisomerization/O–H nucleophilic addition was reported by Michelet and coworkers in the context of a gold-catalyzed phenoxycyclization of 1,5-enynes like **60** [95]. The process, which mimics the polyolefin carbocyclizations, involves a 6-*endo* mode of cyclization and directly affords tricyclic systems that are present in hydroquinone-containing sesquiterpenes (Scheme 26).

In 2010, Toste and coworkers reported an enantioselective polycyclization of related 1,6 enynes such as **62**. These processes, initiated by a 6-*exo*-dig cyclization (instead of the previous 6-*endo*), are efficiently catalyzed by MeO-DTBM-BIPHEP $(AuCl)_2/AgSbF_6$ (5 mol%) at *rt*, providing the desired oxa-polycycles in good yields and *ee*'s up to 94% [96]. In addition to phenols as terminating groups, carboxylic acids were also employed, giving rise to the corresponding bicyclic lactones (**65**) in excellent yields and *ee*'s (Scheme 27) (for a racemic variant of this latter polycyclization, see [97]).

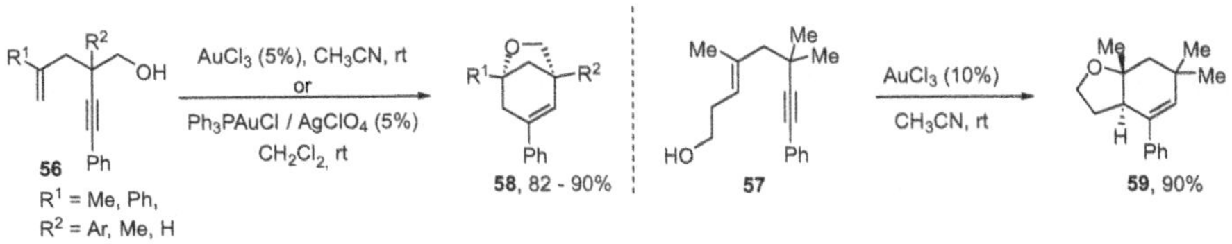

Scheme 25 Cyclization cascade of 1,5-enynols

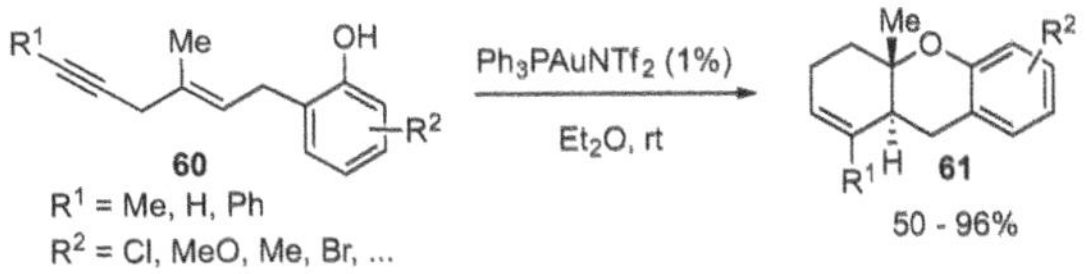

Scheme 26 Phenoxycyclization of 1,5-enynes

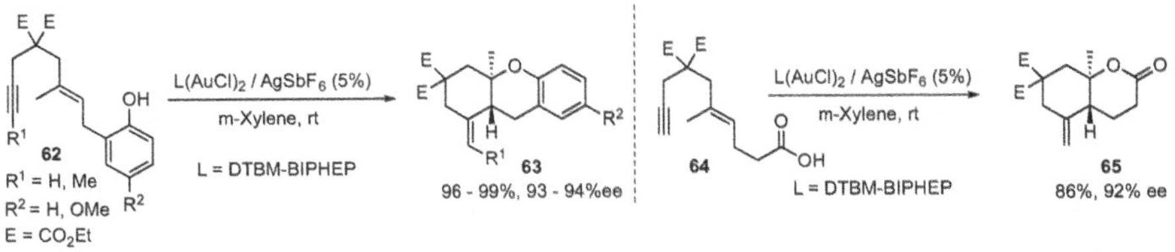

Scheme 27 Enantioselective polycyclization of 1,6 enynes

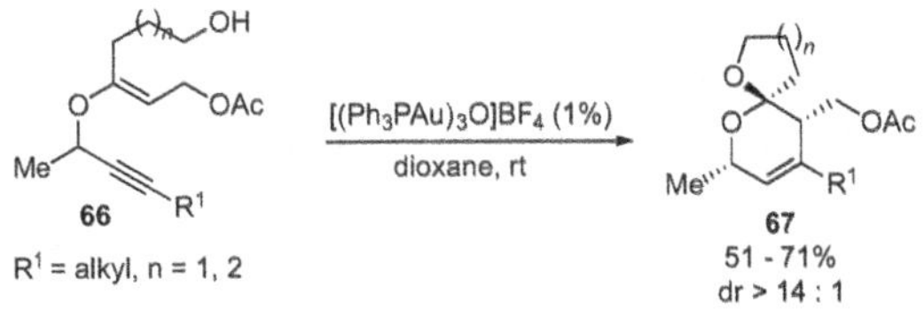

Scheme 28 Intramolecular cyclization of 1,5-enynes to provide spiroketal systems

A related process involving a gold-catalyzed intramolecular reaction of 1,5-enynes like **66** has been reported to provide spiroketals of type **67**, with complete stereocontrol over the three stereogenic centers (Scheme 28). When an enantiomerically enriched propargylic vinyl ether is used, complete chirality transfer was observed [98].[2]

Barluenga and coworkers reported in 2008 an interesting tandem intramolecular hydroalkoxylation/hydroarylation process that leads to benzofused cyclic ethers

[2] For related gold(I)-catalyzed tandem cyclization reactions of alkynylindoles with pendant hydroxyl groups, see [99–101].

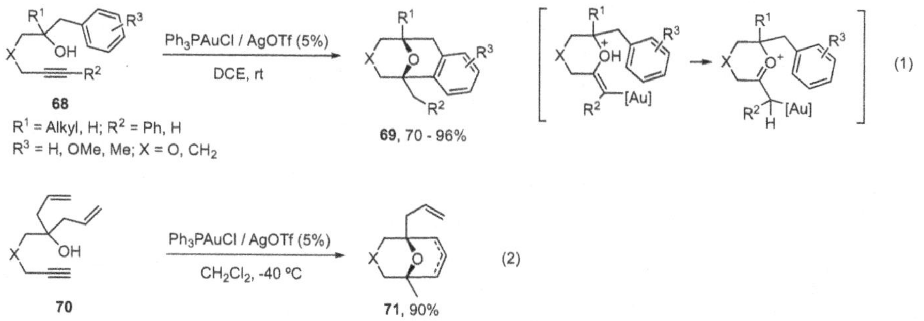

Scheme 29 Tandem hydroalkoxylation/hydroarylation and hydroalkoxylation/Prins-type cyclization processes

from easily accessible benzyl-substituted alkynols. From a mechanistic point of view, the authors proposed that the reaction starts by gold activation of the alkyne, which triggers the hydroalkoxylation (via a *6-exo-dig* cyclization). The resulting cyclic enol ether evolves to an alkylgold oxonium intermediate that undergoes a stereoselective hydroarylation process (Scheme 29, Eq. 1) [102]. A related process, based on a tandem hydroalkoxylation/Prins-type cyclization of readily accessible homoallylic alkynols like **70** was also developed by the same authors (Scheme 29, Eq. 2) [103]. These routes provide a direct approach to bicyclic oxygen-bridged compounds.

The intramolecular reaction of diols with an alkyne has been shown to be an efficient way to construct bicyclic acetals and spiroacetals. Thus, in 2005, Genet, Michelet, and coworkers reported the AuCl-catalyzed formation of bicyclic acetals like **73**, from easily accessible bis-homopropargylic diols **72** [104]. The processes consist of two consecutive hydroalkoxylations, the first one on the alkyne and the second on the resulting enol vinyl gold ether intermediate (Scheme 30) (for an intermolecular variant of the latter example, see [105]).

In a more recent example, Hammond and coworkers described a related cycloisomerization of 2-alkynyl-1,5-diols (**74**) to dioxabicyclo[4.2.1]ketals (**75**) under AuCl catalysis. Curiously, by increasing the catalyst loading and prolonging the reaction time up to 24 h, tetrahydropyrans of type **76**, resulting from a AuCl-catalyzed rearrangement of the initially formed ketals (**75**), were selectively obtained in good yields (Scheme 31) [106].

In a related process, Aponick and coworkers demonstrated that monopropargylic triols like **77** can be transformed into five- and six-membered ring spiroketals **78** when treated with catalytic amounts of JohnPhosAuCl/AgOTf at 0°C (Scheme 32) [107, 108].[3] Interestingly, this chemistry was further applied to natural product synthesis, as elegantly shown by Forsyth and Corey in their respective synthesis of okadaic acid and azaspiracid and platensimycin analogs [115–117].

[3] For related cyclizations, see also [46, 109–114].

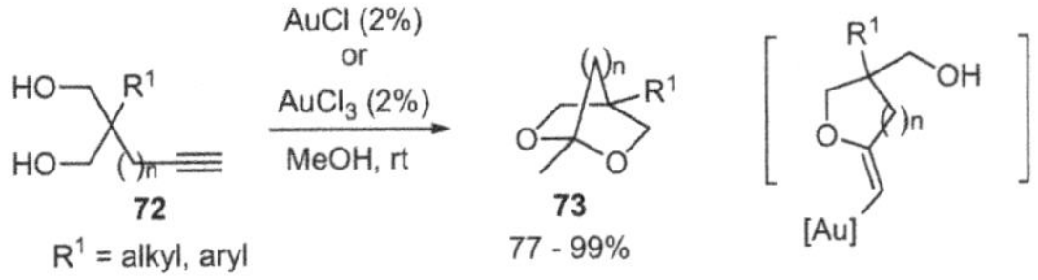

Scheme 30 Intramolecular cyclization cascade of bis-homopropargylic diols

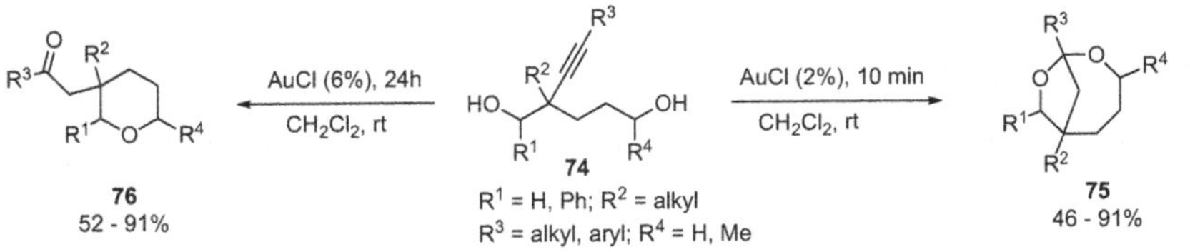

Scheme 31 Cycloisomerization of 2-alkynyl-1,5-diols **74**

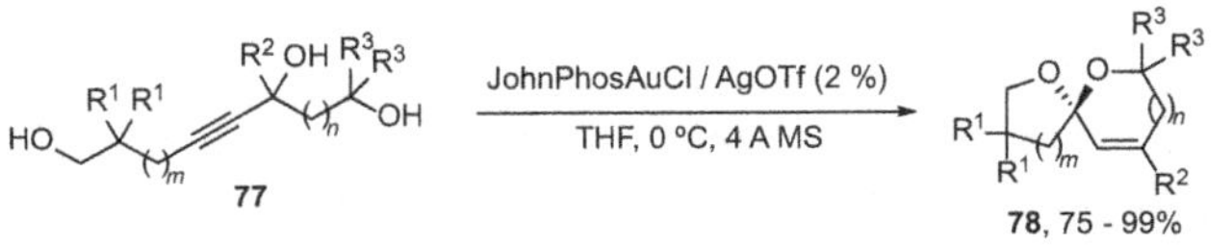

Scheme 32 Synthesis of five- and six-membered ring spiroketals by Aponick and coworkers

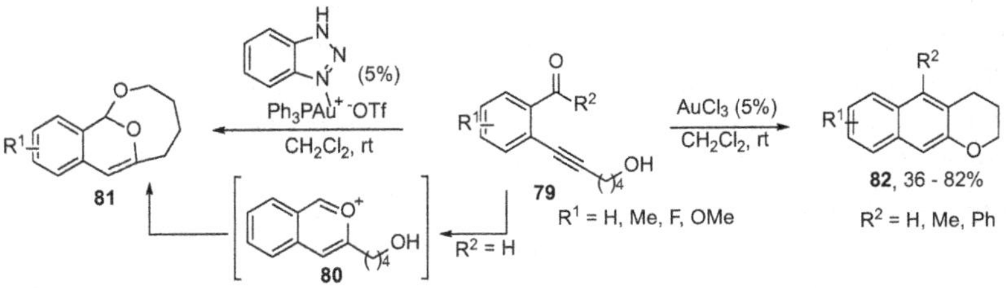

Scheme 33 Cascade annulations of 2-(ynol)aryl aldehydes **79**

Hammond and coworkers studied the gold-catalyzed cascade annulation of 2-(ynol)aryl aldehydes like **79**. Interestingly, they found that when the triazole (TA)-stabilized gold complex Ph_3PAu-(TA)OTf was employed, benzobicyclo [n.3.1]acetals (**81**), resulting from the intramolecular trapping of the pyrylium intermediate **80** by the OH group, are obtained. Alternatively, when $AuCl_3$ was used as catalyst, the reaction produces benzochromanes of type **82** (Scheme 33) [118].

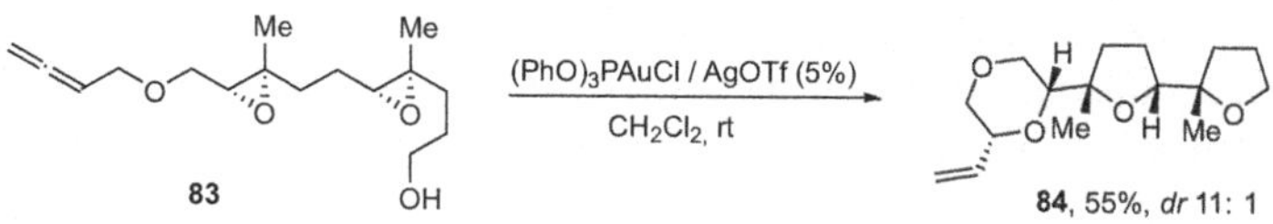

Scheme 34 Cascade cycloisomerizations of allenes tethered to hydroxyepoxides

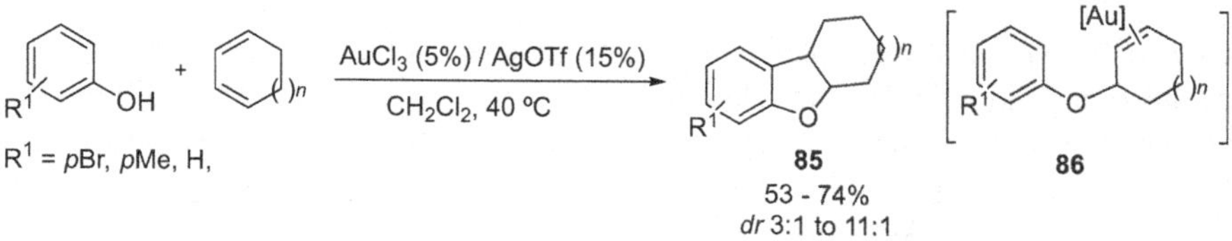

Scheme 35 Annulation of phenols with dienes to generate various dihydrobenzofurans

2.3.2 Cycloisomerization Cascades Initiated by Activation of Allenes

Gagné, Lee, and coworkers demonstrated that cationic gold(I) catalysts promote cascade cycloisomerizations of allenes tethered to hydroxyepoxides [119]. As a result, a variety of polyether skeletons could be obtained depending on the type of allene–hydroxyepoxide used. As can be seen in Scheme 34, the tricyclization cascade of the diepoxide **83** provided the tricyclic ether **84** in a remarkable 55% yield and an excellent 11:1 diastereoselectivity (Scheme 34).

2.3.3 Cycloisomerization Cascades Involving a Gold-Promoted Activation of Alkenes

Li and coworkers reported a highly efficient gold-catalyzed annulation of phenols and naphthols with dienes to generate various dihydrobenzofuran derivatives efficiently. The proposed mechanism involves a hydroalkoxylation of one of the double bonds followed by an intramolecular hydroarylation of the resulting alkene intermediate **86** (Scheme 35) [120].

2.4 Oxycyclizations with the Participation of Additional Exogenous Components

Several variants of the above cycloisomerization reactions which involve external additional components have also been reported. In many cases, the external component is an oxygen-based nucleophile like water, an alcohol, or a carboxylic acid. In some other cases, an oxidant such as selectfluor is used to stoichiometrically oxidize [Au(I)] to [Au(III)] and thus allows novel oxidative couplings to be performed.

Alternatively, diverse pyridine oxides have also been used to generate α-oxo gold–carbenes from alkynes. Some of the most relevant examples, which are classified according to the type of external component, are summarized in the following section.

2.4.1 Incorporation of Oxygen-Based Nucleophiles (H_2O, ROH)

One of the simplest examples to illustrate this kind of processes is the tandem cycloisomerization/hydroalkoxylation of homopropargylic alcohols (**87**) in the presence of an exogenous alcohol (ROH) and catalytic amounts of a protic acid (e.g., *p*TsOH), as reported by Krause and coworkers [121]. The role of the acid is to promote the hydroalkoxylation of the initially formed enol ether. Remarkably, when $HAuCl_4$ is used as catalyst, the presence of *p*TsOH is not required, as this gold complex acts both as carbophilic gold source and as a Brønsted acid. The method provides tetrahydrofuranyl ethers (**88**) in moderate to good yields (Scheme 36, Eq. 1). In a more recent example, Reddy and coworkers reported the gold-catalyzed intramolecular hydroalkoxylation of bromoalkynols **89** in the presence of water to provide γ-butyrolactones (**90**) in good yields (Scheme 36, Eq. 2) [122]. Several other groups have reported alternative examples involving tandem oxycyclizations of alkynol derivatives with a concomitant hydroalkoxylation or hydroxylation process, usually with a Brønsted acid cocatalyst [123, 124].

In the context of hydroalkoxylation cascades, Hashmi and coworkers reported a gold-catalyzed tandem reaction of diyne–diols of type **91** in the presence of external water to provide highly rigid heterocyclic cages of type **93** as the only product. Overall, eight new bonds are formed in the process, which involves two highly selective hydroalkoxylations to yield intermediate **92** (fully identified by X-ray), followed by a monohydroxylation with a molecule of water, and intramolecular hydroalkoxylation of the resulting hemiacetal (Scheme 37) [125].

Sheppard and coworkers described the synthesis of a variety of 3-alkoxyfurans from propargylic alcohols equipped with an acetal moiety (**94**). The authors

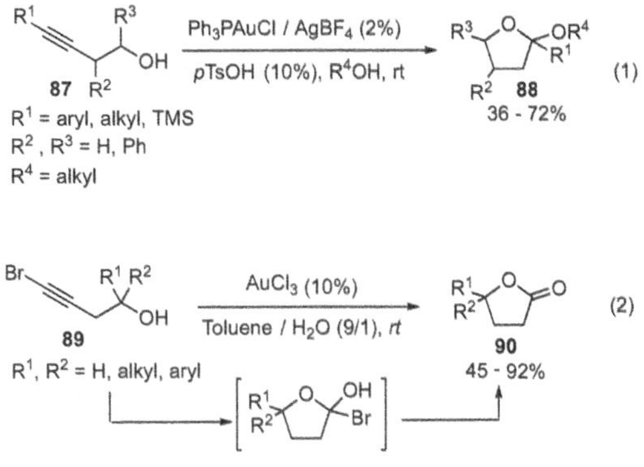

Scheme 36 Cycloisomerization/hydroalkoxylation of homopropargylic alcohols

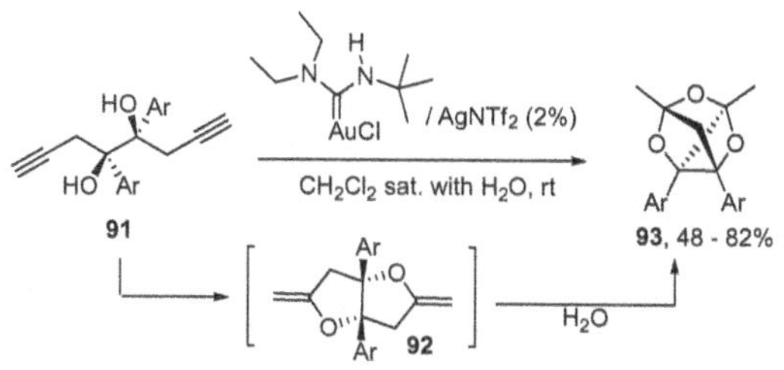

Scheme 37 Tandem reaction of diyne–diols of type **91** in the presence of water

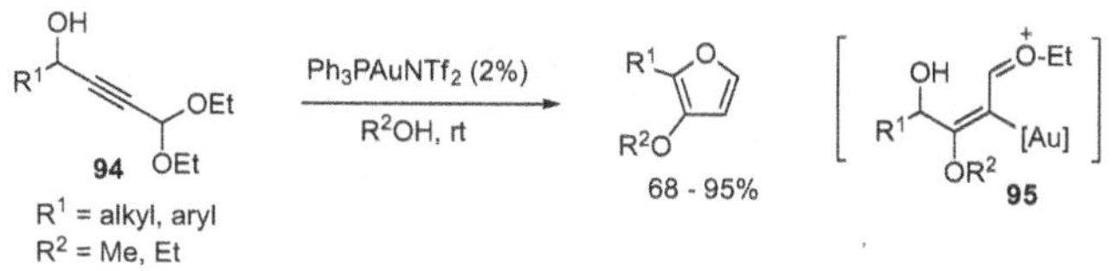

Scheme 38 Synthesis of 3-alkoxyfurans from propargylic alcohols **94**

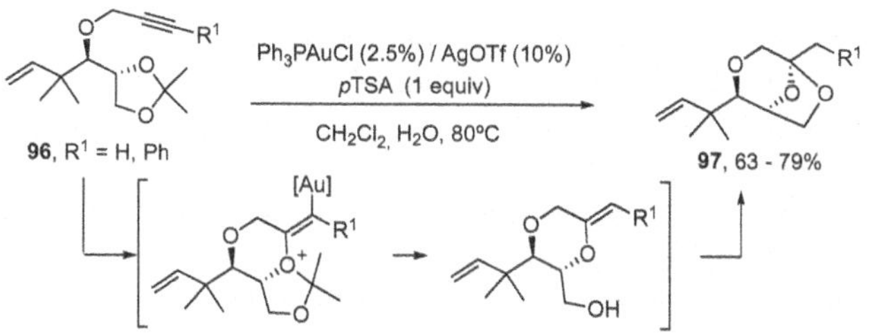

Scheme 39 Alkyne cycloketalization of alkyne tether 1,3-dioxolanes **96**

proposed a regioselective gold-catalyzed addition of the alcohol (R^2OH) to the alkyne to eventually generate a vinyl gold intermediate of type **95**. An intramolecular attack of the OH to the oxonium generates a dihydrofuran that easily evolves to the final furan by protodeauration and loss of ethanol (Scheme 38) [126].

While these latter examples involve the use of free hydroxyl groups, it is also possible to employ ethers like 1,3-dioxolanes as nucleophilic moieties. In particular, Almendros and Alcaide found that the catalyst system composed by Ph_3PAuCl/AgOTf along with a Brønsted acid (PTSA) was competent for the alkyne cycloketalization of alkyne tethered 1,3-dioxolanes (**96**) (Scheme 39) [127].

Wang and coworkers reported a gold-catalyzed reaction of *ortho*-acetylenyl phenyl diazoacetates with water leading to 1H-isochromenes derivatives. The overall process involves a gold-catalyzed decomposition of the diazo with concomitant insertion of water and the subsequent intramolecular hydroalkoxylation (6-*endo*-dig) to yield the isochromene derivatives after the protodeauration. Side products resulting from the 5-*exo*-dig cyclization are also generally observed (Scheme 40) [128].

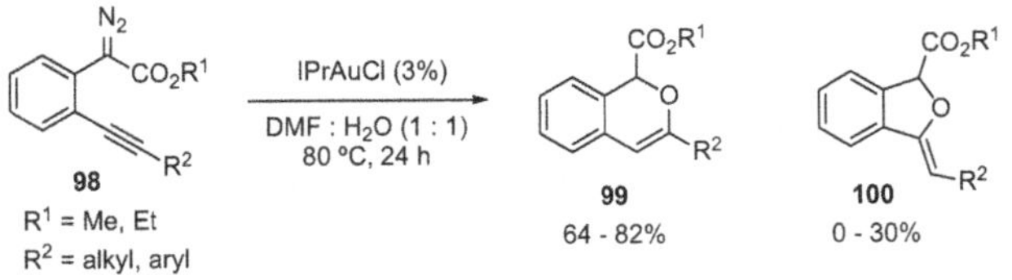

Scheme 40 Reaction of *ortho*-acetylenyl phenyl diazoacetates leading to isochromene derivatives

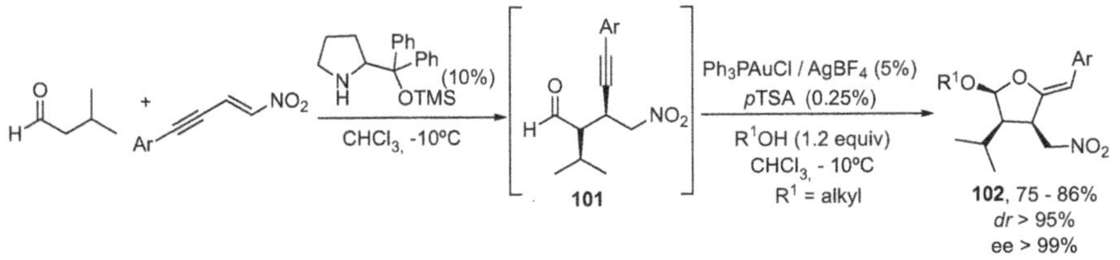

Scheme 41 One-pot Michael addition and gold-catalyzed cycloisomerization leading to nitromethyl substituted dihydrofuranes

Krause, Alexakis, and coworkers demonstrated that enantiomerically enriched β-alkynyl aldehydes (**101**), directly obtained from an organocatalytic Michael addition of aldehydes to nitroenynes, could participate in a one-pot gold-catalyzed cycloisomerization in the presence of an external alcohol (R^1OH) and *p*TsOH to provide nitro-substituted dihydrofuranes like **102** in good yields and excellent diastereo- and enantioselectivities [129] (Scheme 41).

Shi reported a three component intermolecular domino reaction using 2-(arylmethylene)cyclopropilcarbinols of type **103**, terminal alkynes and an alcohol. The resulting 3-oxabicyclo[3.1.0]hexanes are obtained in high yields and moderate diastereoselectivities. The authors proposed a mechanistic scenario consisting of an intermolecular *tandem* hydroalkoxylation of the alkyne followed by a gold-promoted *Prins*-type cyclization on the resulting enol ether intermediate **104**. This process might preferentially occur through a chair-like transition state, thus explaining the observed stereoselectivity (Scheme 42) [130].[4]

Liu and Zhao reported a $AuCl_3$-catalyzed multicomponent domino reaction of 4-acyl-1,6-diynes with water and alkanols that directly affords bicyclic ketals of type **110**. The authors proposed an initial attack of the carbonyl oxygens of **106** to the gold-activated alkynes to yield a bisoxonium intermediate **107**. Subsequent addition of a molecule of water could lead to the dicarbonyl intermediate **108** that eventually could provide the product through an intramolecular cyclization followed by addition of R^2OH, isomerization, and alkanolysis. Despite this mechanistic complexity, the adducts are obtained in good to excellent yields (Scheme 43) [132].

[4] For a related example, see [131].

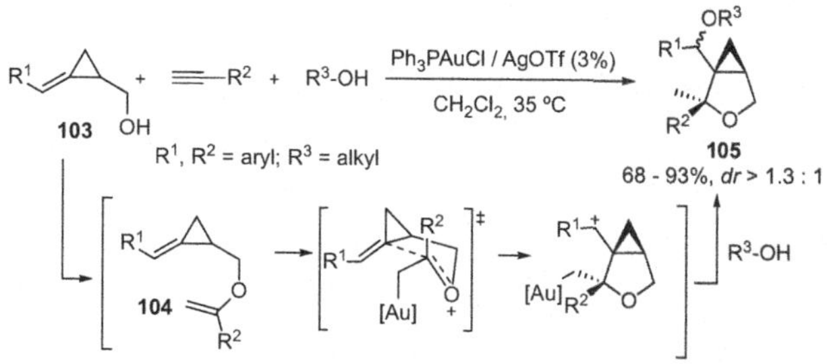

Scheme 42 Three component intermolecular domino reaction of 2-(arylmethylene) cyclopropylcarbinols **103**

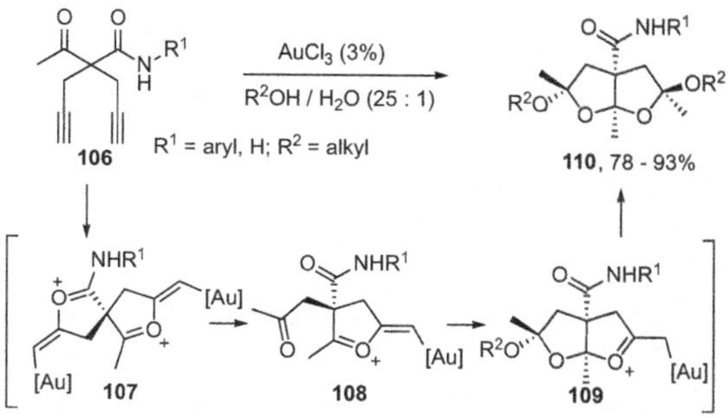

Scheme 43 Multicomponent domino reaction of 4-acyl-1,6-diynes with water and alkanols

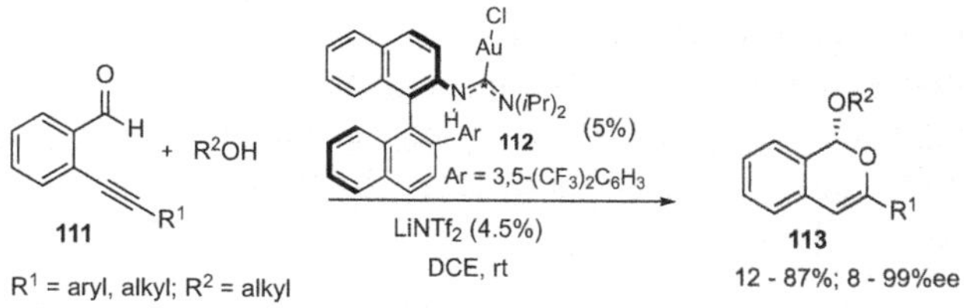

Scheme 44 Enantioselective tandem acetalization/cycloisomerization of *ortho*-alkynylbenzaldehydes

In 2012, Slaughter and coworkers reported an enantioselective tandem acetalization/cycloisomerization of ortho-alkynylbenzaldehydes to yield 1H-isochromenes like **113** with good yields and excellent enantioselectivities [133]. Key for the success of this process is the use of a new class of chiral monodentate acyclic amino carbene gold complex (**112**), specifically developed by the authors (Scheme 44).

In 2013, the groups of Gong [134] and Fañanas and Rodríguez [135] independently reported a three-component stereo- and enantioselective reaction between an alkynol, an aldehyde, and an aniline, under catalysis of a chiral gold binol-phosphate complex JohnPhosAu$^+$[(*R*)-**116**]$^-$. In particular, in the example reported

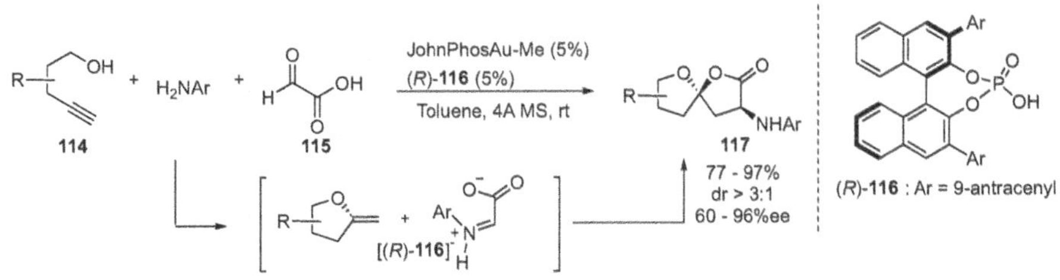

Scheme 45 Three-component stereo- and enantioselective reaction between an alkynol, an aldehyde, and an aniline

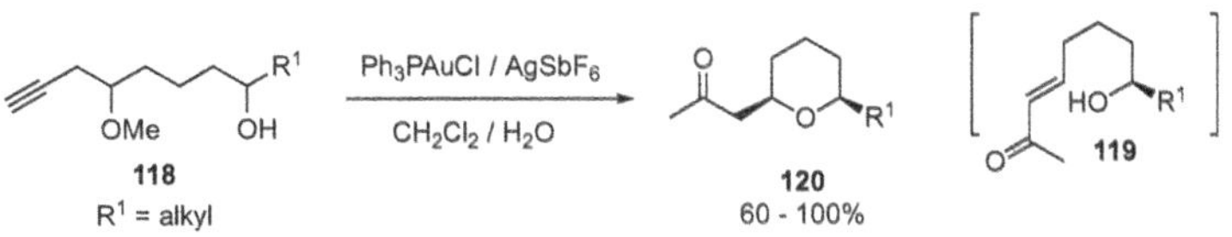

Scheme 46 Transformation of homopropargylic ethers **118** into 2,6-*cis*-disubstituted tetrahydropyrans

by Fañanas and Rodríguez, the authors proposed that the initially formed exocyclic enol ether adds to the in situ generated chiral phosphate iminium complex to generate an oxonium species that undergoes an intramolecular cyclization to deliver the observed spirocyclic product (Scheme 45).

In 2007, Floreancig and coworkers reported the transformation of homopropargylic ethers containing a pendant hydroxyl group into 2,6-*cis* disubstituted tetrahydropyrans **120**. The reaction consists of an initial catalyzed by gold(I)-promoted hydration of the alkyne with concomitant elimination of MeOH to give an α,β-unsaturated ketone of type **119**. A subsequent intramolecular conjugate addition of the alcohol provides the thermodynamically favored 2,6-*cis* tetrahydropyran moiety (Scheme 46) [136].

2.4.2 Participation of Other Type of Exogenous Components

Barluenga and coworkers reported in 2010 a $AuCl_3$-catalyzed cascade cycloisomerization/Diels–Alder reaction using enynols like **121** and an external dienophile such as *N*-phenyl maleimide or tetracyanoethylene. Depending on the length of the alcohol chain, *endo-* or *exo*-cycloisomerization leading to diene intermediates **122** or **123** was observed. Interestingly, the authors demonstrate that the gold catalysts also facilitate the Diels–Alder reaction, allowing the cycloaddition to be performed at relatively low temperatures (Scheme 47) [137].

Li and coworkers reported a tandem alkynylation/cyclization of *ortho*-alkynylaryl aldehydes with terminal alkynes to yield 1-alkynyl-1H-isochromenes (**127**). The reaction is carried out in a water:toluene mixture and is catalyzed by the

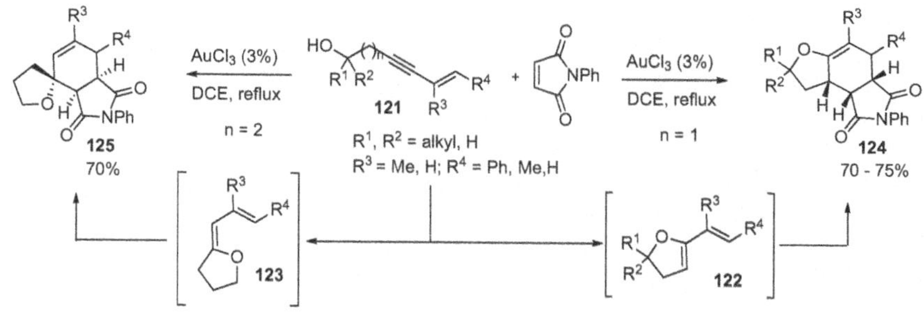

Scheme 47 Cascade cycloisomerization/Diels–Alder reaction of enynols like **121**

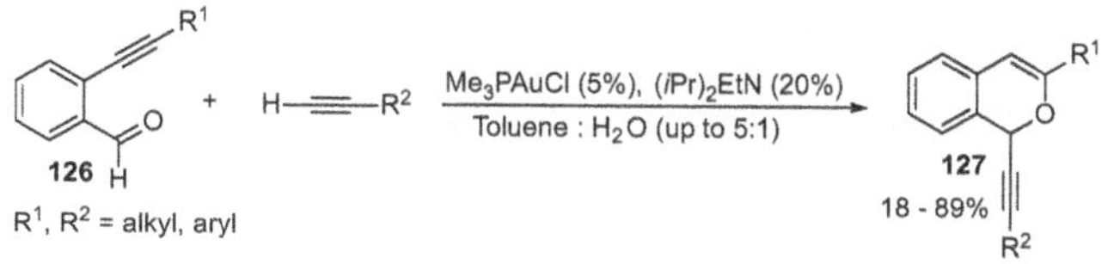

Scheme 48 Tandem alkynylation/cyclization of *ortho*-alkynylaryl aldehydes with terminal alkynes

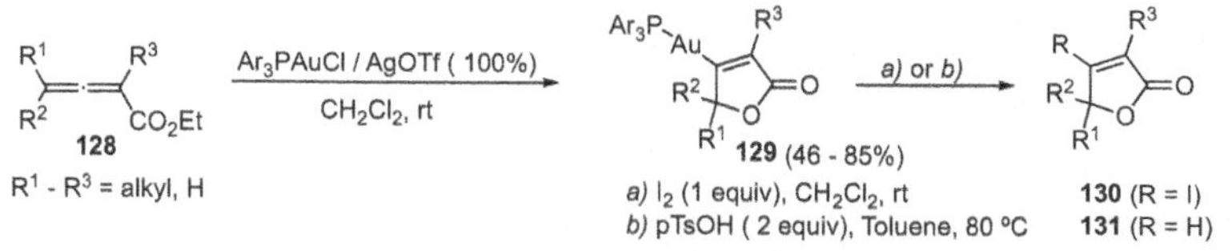

Scheme 49 Reactivity of stable organogold intermediates **129**

phosphine gold complex Me_3PAuCl (5%) in the presence of $^{i}Pr_2EtN$ (20%) (Scheme 48) [138].

Hammond and coworkers studied the reactivity of stable organogold intermediates (**129**) which are generated from the oxycyclization of allenoates **128**. Thus, they have shown that whereas the addition of *p*TsOH-H_2O promotes their protodeauration, other reagents such as I_2 allowed the formation of the corresponding γ-iodolactone **130** (Scheme 49) [44].

The organogold complexes resulting from alkoxylation reactions could also get involved in oxidative C–C coupling reactions with an external reaction component (for a review on this topic, see [139]). A first example of an oxidative coupling was observed by Hashmi and coworkers while studying the $AuCl_3$-catalyzed cyclization of allenyl carbinols **132**, which yielded in addition to the dihydrofurans **133**, the dimers **134** [140]. Similarly, Wegner and coworkers observed the products of cycloisomerization/oxidative coupling upon the treatment of arylpropionic ester **135** with $HAuCl_4$ (5%) and *t*BuOOH (5 equiv.) at 60°C [141].[5] In both cases, the

[5] For a related example, see [142].

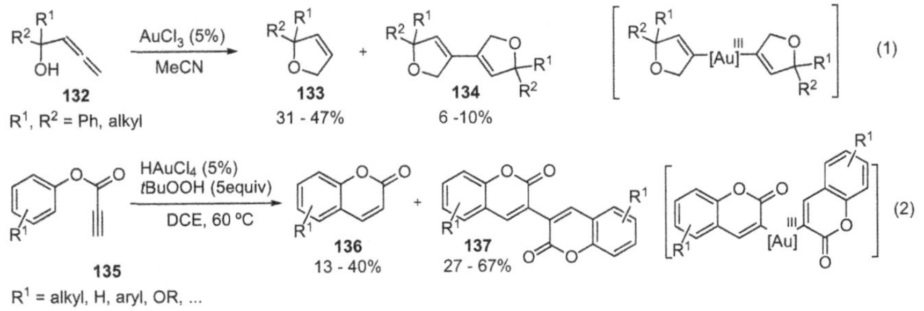

Scheme 50 C–C couplings of organogold complexes resulting from alkoxylation reactions

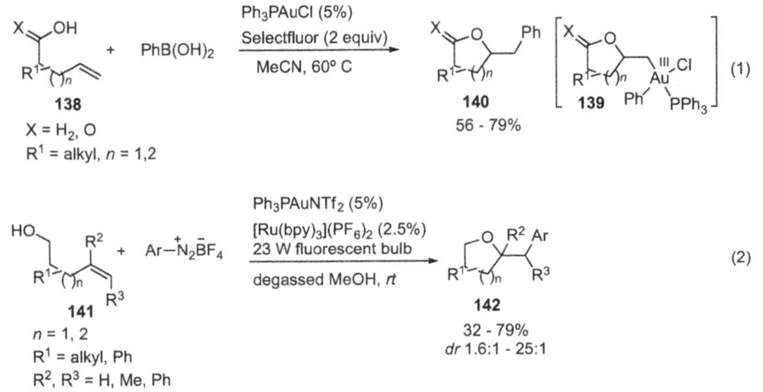

Scheme 51 Intramolecular carboalkoxylations and carbolactonizations of alkenes via oxidative gold catalysis

transformation is best explained assuming the initial formation of a bis-vinyl gold (III) intermediate which after a reductive elimination provides the coupling product (Scheme 50). The role of *t*BuOOH in the example of Wegner is to act as oxidant to regenerate [Au(III)] from the released [Au(I)], while in the case of Hashmi, the maximum yields of dimers (**134**) were up to twice the percentage of $AuCl_3$, since no oxidant is used.

In 2008, Zhang and coworkers reported gold-catalyzed carboalkoxylations and carbolactonizations of alkenes like **138** via oxidative gold catalysis using an exogenous boronic acid and selectfluor as stoichiometric oxidant for gold(I). The use of a neutral gold complex as catalyst suggested that the [Au(I)] oxidation to [Au (III)] takes places prior to the *anti*-attack of the OH moiety to the alkene. The authors proposed the generation alkylgold(III) intermediate like **139** that undergoes reductive elimination to provide the final oxacycle and a gold(I) complex (Scheme 51, Eq. 1) [71]. Also in 2010, Lloyd–Jones and Russell demonstrated that these cyclizations could also be achieved using aryl silanes instead of boronic acids [143]. A related dual-catalytic process, using a visible light-mediated photoredox catalyst and an aryl diazonium salt, was recently reported by Glorius and coworkers [72]. The reaction takes place at room temperature under irradiation

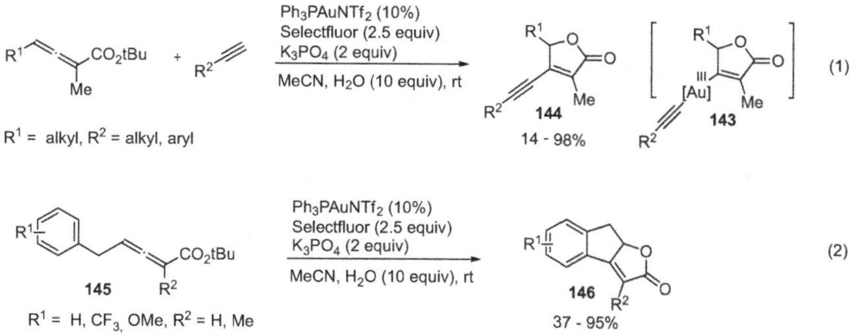

Scheme 52 Oxycyclization/alkynylation and oxycyclization/ C–H bond functionalization cascades

with a simple household light bulb and constitutes the first example that demonstrates the compatibility of gold and photoredox catalysts in a dual-catalyzed transformation (Scheme 51, Eq. 2).

Gouverneur reported in 2010 an oxidative gold-catalyzed oxycyclization/ alkynylation cascade using allenoates as starting materials and a terminal alkyne as external component. In this process, the vinyl gold (I or III) species generated in the allenoate cyclization might react with an in situ generated gold (I or III) alkynylide to render a gold(III) intermediate of type **143** that releases the product after reductive elimination (Scheme 52, Eq. 1) [144]. A similar process but involving a selective intramolecular aryl C–H bond functionalization as the second step of the tandem process was also reported by the same group to provide a direct entry to tricyclic indenofuranones of type **146** from benzyl-substituted butyl allenoates (**145**) (Scheme 52, Eq. 2) [145].[6]

Hashmi and coworkers reported that the vinyl gold species resulting from these allenoate cyclizations can transmetallate to palladium and undergo a C–C coupling with an aryl iodide [146]. The compatibility of gold and palladium was also demonstrated by Blum and coworkers who demonstrated that the oxonium vinyl gold species **148**, resulting from the gold-promoted cyclization of allyl allenoates **147**, undergoes a deallylation reaction in the presence of [Pd(0)] to yield a π-allyl palladium complex that directly cross-couples with the vinyl gold intermediate [147]. Both the gold and the palladium catalysts are regenerated in this latter step (Scheme 53).

Zhang and coworkers reported that homopropargylic alcohols could be directly transformed into dihydrofuranones in the presence of a gold catalyst and stoichiometric amounts of both a protic acid and an oxidant such as a pyridine or quinolone *N*-oxide. Under these conditions, the triple bond is converted into an α-oxo gold–carbene (**152**) that is intramolecularly trapped by the hydroxyl group. The role of the Brønsted acid is to protonate the pyridine product and avoid deactivation of the

[6] For a review summarizing these and other Au(I)/Au(III) oxidative couplings, see [139].

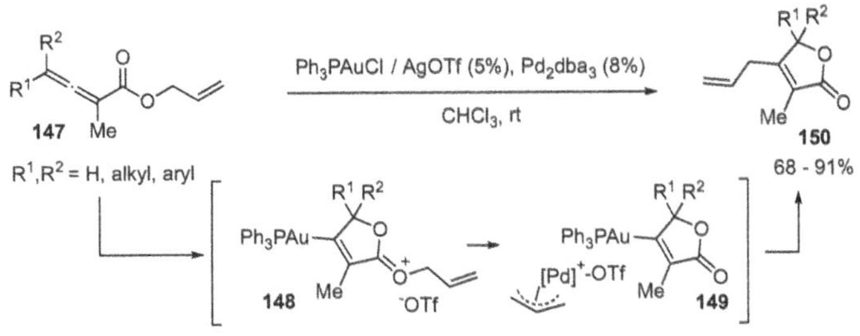

Scheme 53 Dual gold and palladium catalysis by Blum and coworkers

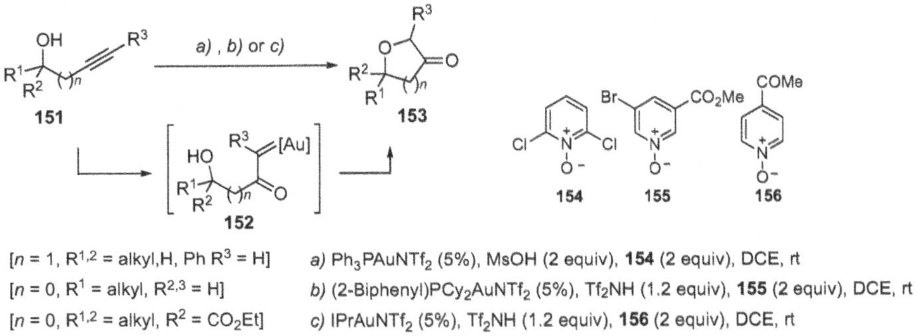

Scheme 54 Generation and reactivity of α-oxo gold–carbenes from alkynes and pyridine oxides

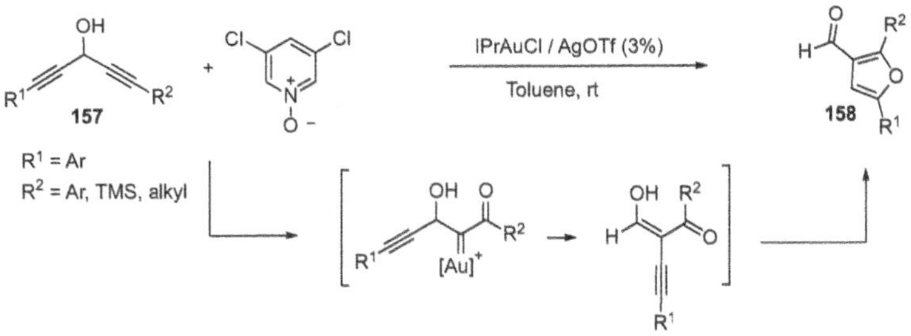

Scheme 55 Cascade transformation of 1,4-diyn-3-ols to 3-formylfurans

cationic gold catalyst. Remarkably, this methodology has also been applied by the same group to the synthesis of oxetan-3-ones from propargylic alcohols as well as to chroman-3-ones from propargyl aryl ethers (Scheme 54) [148–153].

In another interesting example, Hashmi and coworkers reported a gold-catalyzed cascade transformation of 1,4-diyn-3-ols (**157**) to 3-formylfurans (**158**) by using a pyridine oxide to generate the key α-oxo-gold carbenoid. Based on DFT calculations and deuterium isotope labeling, the authors proposed a mechanism involving a 1,2-alkynyl shift on this intermediate followed by a 5-*endo*-dig cyclization and a protodeauration to yield the 3-formyl furan product (Scheme 55).

A variant of the process involving an iododeauration as the final step has also been achieved by the same authors, by using *N*-iodosuccinimide as iodinating reagent [154–156].

3 Cycloisomerizations Involving Rearrangements

In this section we include processes in which the oxacyclic product results from coupling a cycloisomerization with some kind of skeletal rearrangements, including in some cases ring opening or long-range hydrogen migrations. In all the cases discussed, the oxygen plays a key role in the mechanism.

3.1 Oxycyclizations Preceded by Rearrangements of Propargylic Esters

Propargylic esters featuring internal alkynes are frequently used as allene surrogates under gold catalysis, since they can undergo a very facile [1,3]-rearrangement to the corresponding allenyl ester. Usually, this rearrangement is coupled to a subsequent gold-catalyzed transformation of the resulting allene [157, 158]. Following this strategy, Gagosz and coworkers demonstrated in 2006 that the treatment of easily accessible butynediol monobenzoates **159** with $Ph_3PAuNTf_2$ directly affords 2,5 dihydrofurans related to those efficiently obtained from the hydroalkoxylation of related α-allenols (see Sect. 2.1). A mechanistic rational, involving the initial formation of intermediate allene **160**, prior to its hydroalkoxylation, was proposed by the authors [159] (Scheme 56).

Later, in 2008, De Brabander and coworkers demonstrated that this strategy could also be applied to ω-hydroxy propargylic esters, which can be transformed into tetrahydropyranic systems by means of a 6-*exo*-trig hydroalkoxylation of putative allenyl acetate intermediates of type **163**. Remarkably, the reaction is stereoselective, providing the THP 2,6-*cis* isomers (**164**) as major products (Scheme 57) [160].

In 2011, Toste and coworkers reported the enantioselective synthesis of 2-substituted chromenyl pivalates like **168** from racemic chiral β-hydroxy propargylic esters (**165**). The chiral acyclic diaminocarbene gold complex used

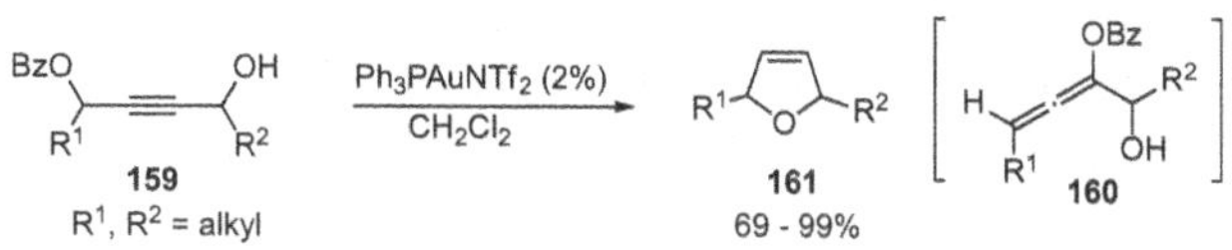

Scheme 56 2,5 Dihydrofurans from butynediol monobenzoates via α-allenol intermediates

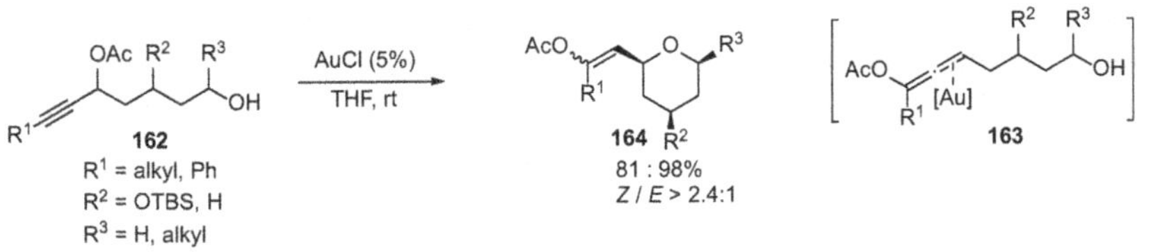

Scheme 57 Tandem cycloisomerization of ω-hydroxy propargylic esters

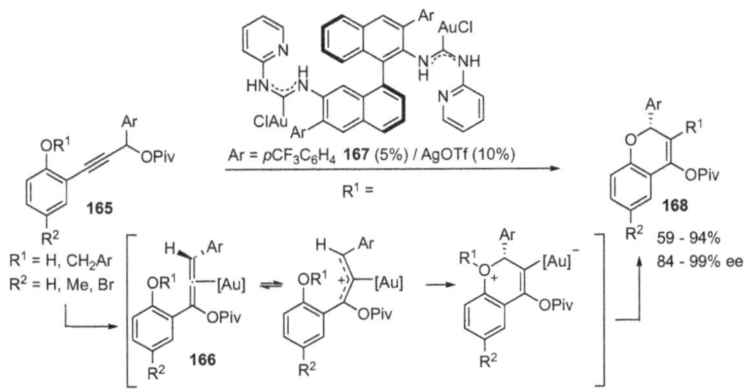

Scheme 58 Enantioselective synthesis of 2-substituted chromenyl pivalates

by the authors catalyzes the dynamic kinetic asymmetric process, involving the initial rearrangement to the allenyl acetate intermediate **166** and its subsequent 6-*endo*-trig hydroalkoxylation to yield the chromenyl adducts (Scheme 58) [161]. A related racemic example catalyzed by $Ph_3PAuCl/AgSbF_6$, to yield dihydropyrans/ 2H-pyrans was also reported by Swamy and coworkers [162].

In all these previous examples, the propargyl ester undergoes a 1,3-migration to provide an allenyl intermediate; however, this is not always the case. When the alkyne does not have substituents at its terminal position, a 1,2-migration usually prevails, thus leading to vinyl gold–carbene intermediates [158]. Based on this reactivity, Toste and coworkers reported a highly enantioselective access to benzopyrans like **173**, via a carboalkoxylation reaction of propargyl esters **169**. The process, catalyzed by MeO-DTBM-BIPHEP$(AuCl)_2$ /$AgSbF_6$, involves the abovementioned 1,2-migration of the propargyl ester to yield the carbene species **170**, which is intercepted by the ether oxygen to generate an oxonium intermediate **171**. The formation of the final benzopyran with a quaternary stereocenter is proposed to proceed through the allyl gold intermediate **172** (Scheme 59) [163].

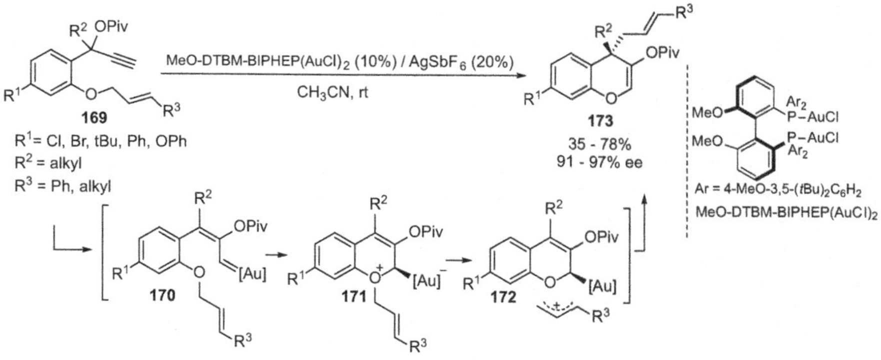

Scheme 59 Enantioselective access to benzopyrans **173**, via carboalkoxylations of propargyl esters **169**

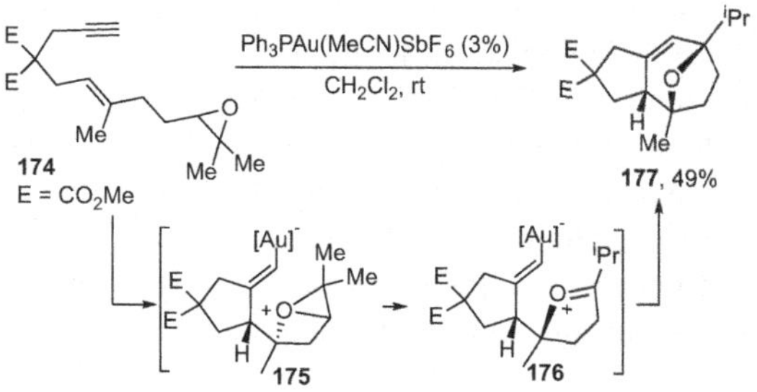

Scheme 60 Cyclization of 1,6-enynes bearing an epoxide tethered to the alkene

3.2 *Oxycyclizations Preceded by Enyne Cycloisomerizations*

Echavarren and coworkers reported a gold-catalyzed cyclization of a 1,6-enyne bearing an epoxide tethered to the alkene that yields highly relevant oxabridged bicyclic products (**177**) in moderate yield. The key intermediate of this process is the oxonium **175**, which undergoes C–O bond cleavage and a subsequent 1,2-hydrogen migration to yield the oxonium **176**. A *Prins*-like cyclization affords the observed oxabridged product (Scheme 60) [164]. This type of product can also be obtained more efficiently if a ketone, instead of an epoxide is attached to the enyne (vide infra) [165, 166].

Alternatively, Echavarren and coworkers reported a gold(I)-catalyzed cycloisomerization of 1,6-enynes **178**, bearing an O-allyl group at its propargylic position. The reaction is proposed to involve an intramolecular 1,5-migration of the O-allyl group (**179**) affording an α,β-unsaturated gold–carbene/allyl gold cation **180**. This species undergoes an intramolecular cyclopropanation to yield the final tricyclic compound **181** (Scheme 61) [167].

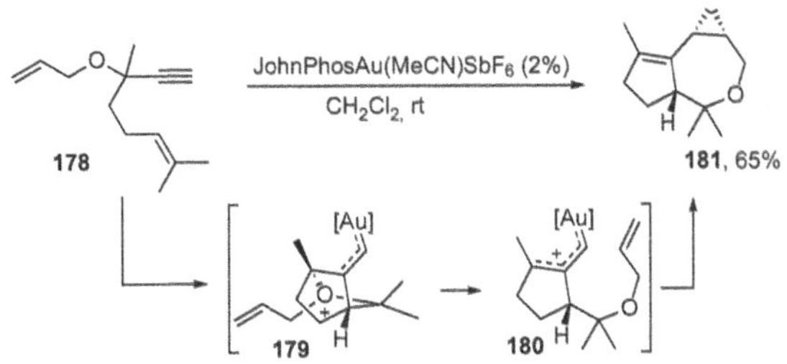

Scheme 61 Cycloisomerization of 1,6-enynes bearing an O-allyl group at its propargylic position

3.3 Oxycyclizations Followed by Rearrangements

Liang and coworkers reported in 2008 a gold-catalyzed tandem cyclization/[1,2]-alkyl migration of epoxyalkynes like **182** to afford spiropyranones of type **184** with good yields and in an stereospecific manner [168]. The reaction, catalyzed by $[NaAuCl_4]_2 \cdot H_2O$ in toluene at 80°C, is proposed to proceed by an initial *endo-dig* cyclization to yield a strained oxonium intermediate (**183**) that eventually undergoes the 1,2-alkyl migration to yield the isolated spiropyranone (Scheme 62).

Also in 2008, Liu and coworkers reported the gold-catalyzed cycloisomerization of epoxy alkynes **185** leading to polycyclic 2H-pyrans (**187**). The reaction is proposed to proceed via a 6-*exo-dig* attack of the epoxide at the activated alkyne to give a vinyl gold species that evolves to the gold–carbene intermediate (**186**). A Nazarov-type cyclization would lead the final product (Scheme 63) [169, 170].

In 2010, Liu and coworkers reported a highly diastereoselective hydrative cyclization of *cis*-1-epoxy-1-alkynylcyclopropanes like **188**. The authors proposed a mechanistic rationale based on an initial 6-*endo*-dig cyclization with concomitant 1,2-migration of the cyclopropyl C–C bond to give the key species **189**. A stereoselective addition of water to this species leads to the final product (**190**) [171]. Interestingly, the authors also investigated the reactivity of these bicyclic allylic alcohols under gold catalysis, demonstrating that they can participate in several types of formal [4 + 2] cycloadditions [171]. On the other hand, if the gold-catalyzed cycloisomerization of **188** is carried out with $PicAuCl_2$ (3%), in the presence of $Ph_3P{=}O$ (10%), with the subsequent addition of *N*-chlorosuccinimide (one-pot process), eight-membered ring ethers like **193**, arising from the ring-opening of the vinyl gold intermediate **192**, are formed (Scheme 64) [172].

Zhang, Tu, and coworkers reported in 2009 the synthesis of pyran-fused indene cores (**196**) by means of a tandem gold-catalyzed oxycyclization–rearrangement of 2-acyl-1-alkynyl-1-aryl cyclopropanes (**194**). The process is initiated by the nucleophilic attack of the carbonyl to the gold-activated alkyne via a 5-*exo*-dig cyclization. The resulting vinyl gold species evolve to the gold–carbene **195** which, after cyclopropane opening and subsequent C–H insertion process, delivers the observed dihydropyran-fused indene product (Scheme 65) [173].

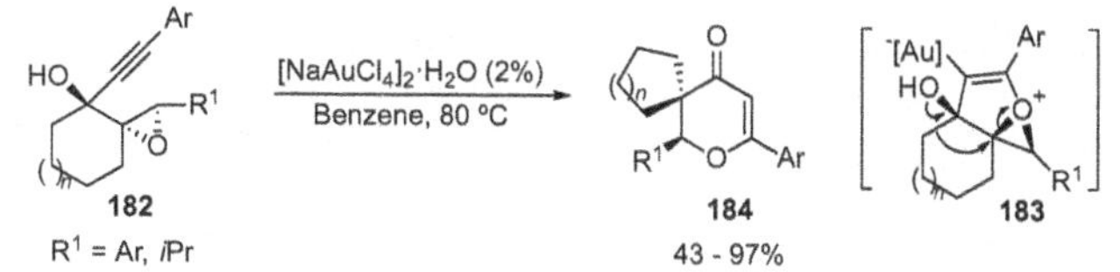

Scheme 62 Tandem cyclization/[1,2]-alkyl migration of epoxyalkynes

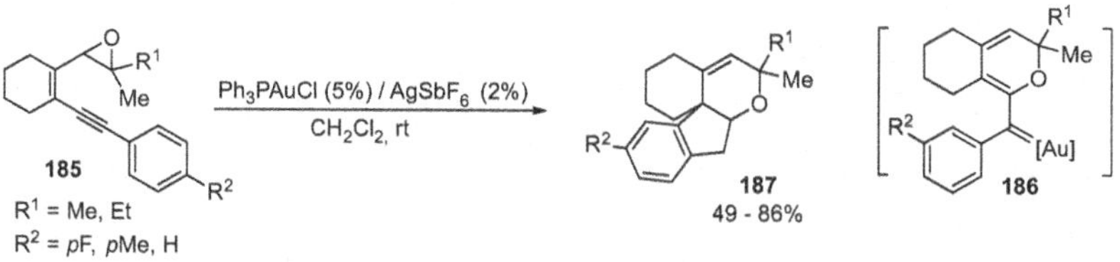

Scheme 63 Cycloisomerization of epoxy alkynes leading to polycyclic 2H-pyrans

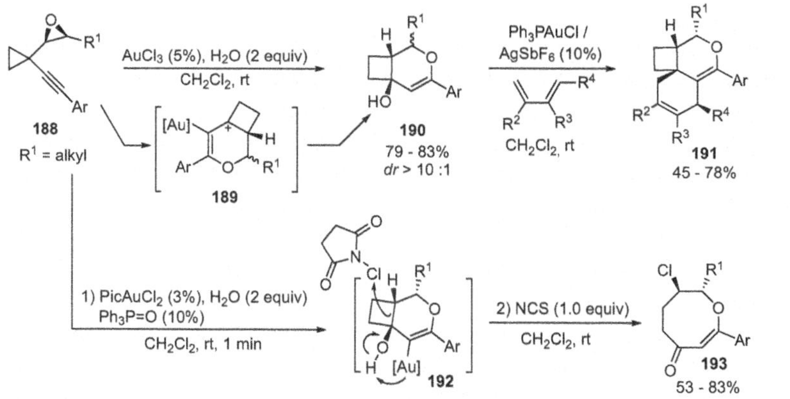

Scheme 64 Diastereoselective hydrative cyclization of *cis*-1-epoxy-1-alkynylcyclopropanes

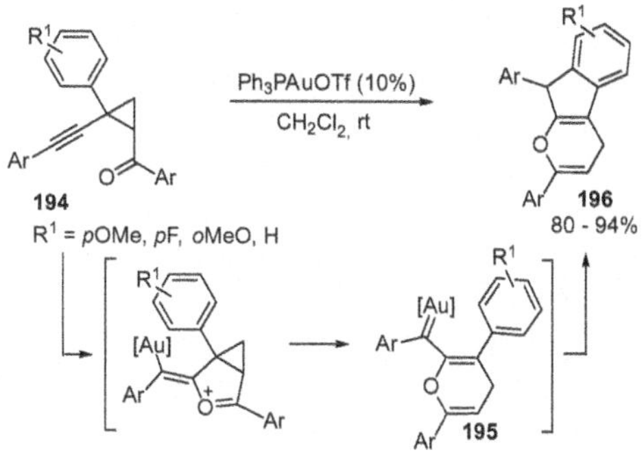

Scheme 65 Tandem oxycyclization–rearrangement of 2-acyl-1-alkynyl-1-aryl cyclopropanes

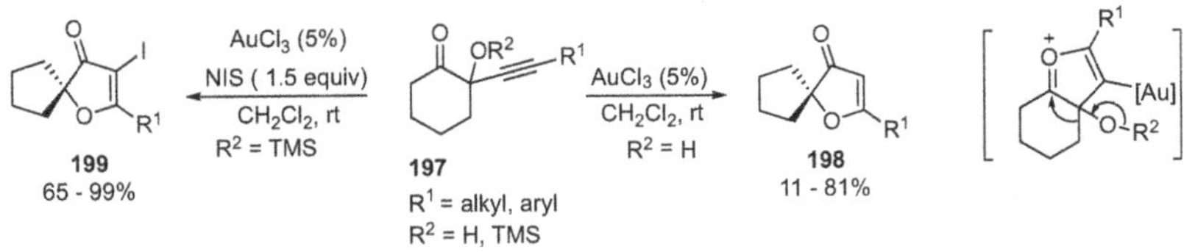

Scheme 66 Tandem cyclization/1,2-migration of 2-alkenyl-2-hydroxy carbonyl compounds

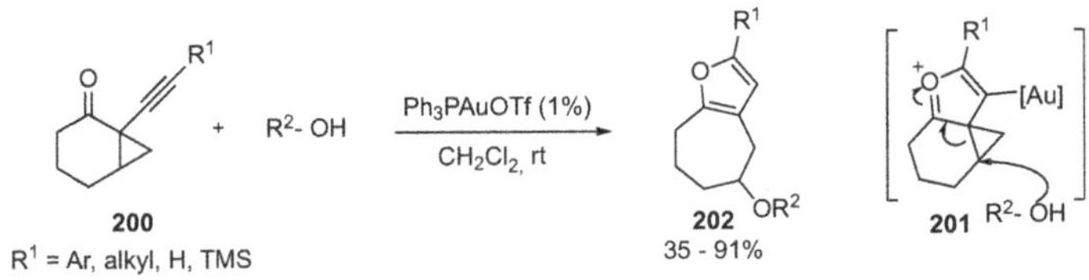

Scheme 67 Fused-substituted furans **202** from 1-(1-alkynyl)-cyclopropyl ketones

Kirsch and coworkers reported a gold-catalyzed tandem cyclization/1,2-migration of 2-alkynyl-2-hydroxy carbonyl compounds (**197**) to yield fully substituted 3(2H)-furanones **198**. After the initial nucleophilic attack of the carbonyl oxygen to the activated alkyne, a 1,2-alkyl shift on the generated oxonium intermediate followed by protodemetalation provides the product (Scheme 66) [174, 175]. On the other hand, when the tandem process is performed with their 2-silyloxy analogs, in the presence of *N*-iodosuccinimide, fully substituted 3(2H)-iodofuranones (**199**), resulting from an iododemetalation, are obtained (Scheme 66) [176].

Zhang and Schmalz reported the synthesis of fused-substituted furans like **202** by subjecting 1-(1-alkynyl)-cyclopropyl ketones (**200**) to gold catalysis in the presence of an external nucleophile such as MeOH [177]. As later demonstrated by Zhao and coworkers through computations, the first step of the catalytic cycle is the cyclization of the carbonyl oxygen onto the triple bond to form a new and stable resonance structure of an oxonium ion and a carbocation intermediate (**201**). The key step is the attack of the nucleophile to the C–C σ-bond of the cyclopropane unit to open the cyclopropane ring and afford, after the protodemetalation, the final product (Scheme 67) [178].

In 2010, Shi and coworkers reported a tandem addition/ring opening reaction of diarylvinylidene cyclopropanols (**203**). The reaction, which is catalyzed by $[(Ph_3PAu)_3O]BF_4/AgOTf$, provides tetrahydropyran derivatives with a pendant allenyl moiety (**204**) [179]. From a mechanistic point of view, the initial activation of the allene by the gold(I) complex most probably weakens the proximal cyclopropyl C–C bond, triggering the attack of the OH moiety and concomitant ring opening. A final protodeauration would lead to the observed tetrahydropyran product (Scheme 68).

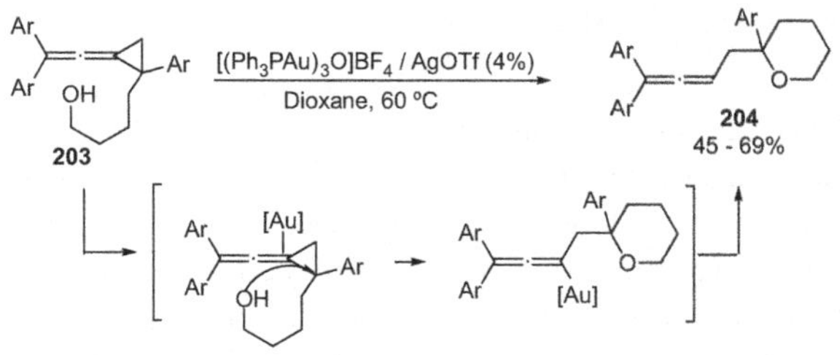

Scheme 68 Tandem addition/ring opening reaction of diarylvinylidene cyclopropanols

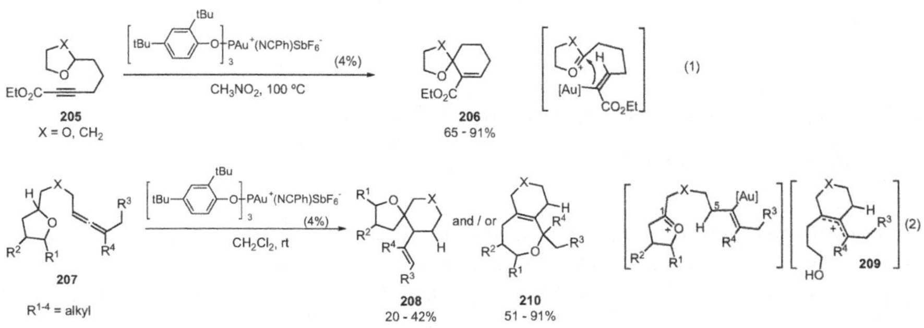

Scheme 69 Cycloisomerization of alkynyl ethers and allenyl ethers involving 1,5-hydride migrations

3.4 1,5-Hydrogen Migrations

In 2010, Gagosz and coworkers described a cycloisomerization of alkynyl ethers like **205** which involves a 1,5-hydride shift/cyclization sequence (for a review, including this topic, see [180]). The process, catalyzed by a highly electrophilic phosphite gold(I) catalyst, has a remarkable scope and provides, depending on the particular type of substrate, five- or six-membered spiro- or fused-dihydrofurans and dihydropyrans. Deuterium-labeling experiments agreed with a mechanistic pathway involving an initial 1,5-hydride transposition onto a gold(I)-activated alkyne followed by the addition of the vinyl gold moiety onto the resulting oxonium ion. Thus, the transformation represents a formal conversion of C(sp3)–H bonds into new C(sp3)–C(sp3) bonds (Scheme 69, Eq. 1) [181, 182]. In 2011, the same authors also demonstrated that a series of allenyl ethers like **207** could also undergo related cascade 1,5-hydride shift/cyclization sequences to yield oxacyclic products. The formation of the oxepanes of type **210** is explained assuming a gold-catalyzed tetrahydrofuran ring opening of the initially formed spiro compound (**208**) to yield an allylic carbocation (**209**) that subsequently ring closes to the oxepane (Scheme 69, Eq. 2) [183].

4 Formal Cycloadditions

While cyclization reactions allow the construction of cycles from acyclic precursors by formally forming a single bond, cycloaddition processes are particularly relevant from the constructive point of view because of the possibility of making at least two bonds and one cycle in a single step. The development of new formal cycloaddition reactions promoted by gold (I and III) catalysts has experienced a remarkable growth during the last decade [184, 185]. In the following section, we summarize some of the most relevant contributions that lead to oxygenated heterocyclic compounds.

4.1 Cycloadditions Based on the Gold-Promoted Generation of Reactive Dipoles

Several groups have demonstrated that certain substrates containing an alkyne or allene and a carbonyl unit can be transformed, under Au (I/III) catalysts, into gold-containing zwitterionic intermediates which undergo different types of intermolecular cycloadditions with external dipolarophiles (e.g., alkyne, alkene, carbonyl. For a review including this topic, see [186]). Although many examples involve nitrogen-containing zwitterionic species, there are also relevant examples exclusively involving oxygenated zwitterionic intermediates. In other cases, the zwitterionic species is an all-carbon gold intermediate that reacts with an external carbonyl component.

Liu and coworkers demonstrated that is possible to produce reactive zwitterionic intermediates of type **212** through an *exo-dig* cyclization of 1-oxo-5-ynes like **211** promoted by the gold complex JohnPhosAuCl/$AgNTf_2$. These dipolar intermediates undergo an intermolecular cycloaddition with a vinyl ether, leading to 9-oxabicyclo[3.3.1]nona-4,7-dienes (**213**). The reaction pathway proposed by the authors involves a [3 + 2] dipolar cycloaddition between the carbonyl ylide **212** and the alkene, followed by ring expansion, and elimination (Scheme 70, right) [187]. Alternatively, when the substrate contains an internal alkyne (R^3=alkyl, Ph), gold activation of the alkyne triggers a 1,3-acyloxy shift that leads to ketone allenic intermediates of type **214**. Then, an intramolecular attack of the carbonyl group on the activated allene generates a benzopyrylium species that undergoes a stereoselective [4 + 2] cycloaddition with the alkene to yield oxycyclic systems **215** with notable diastereoselectivities (Scheme 70, left) [188].

Zhang and coworkers reported in 2008 a novel approach to generate all-carbon 1,3-dipoles (**217**) via a gold-catalyzed migration–fragmentation of ketals of type **216**. These intermediate species participate in a [3 + 2] intermolecular cycloaddition with an aldehyde to yield highly functionalized 2,5-dihydrofurans with excellent diastereoselectivities (Scheme 71) [189].

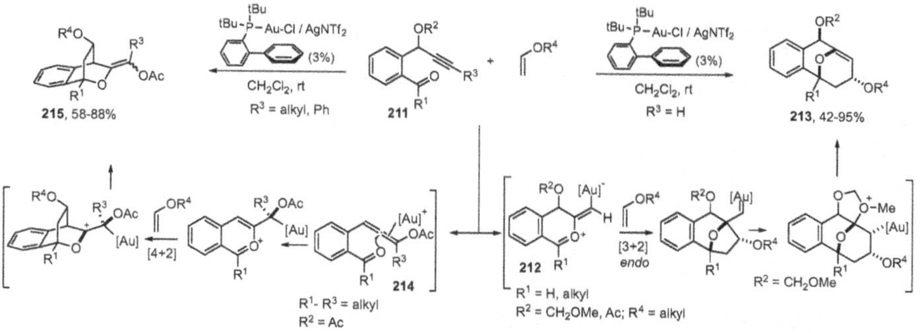

Scheme 70 Cycloadditions of 1-oxo-5-ynes **211** through zwitterionic intermediates

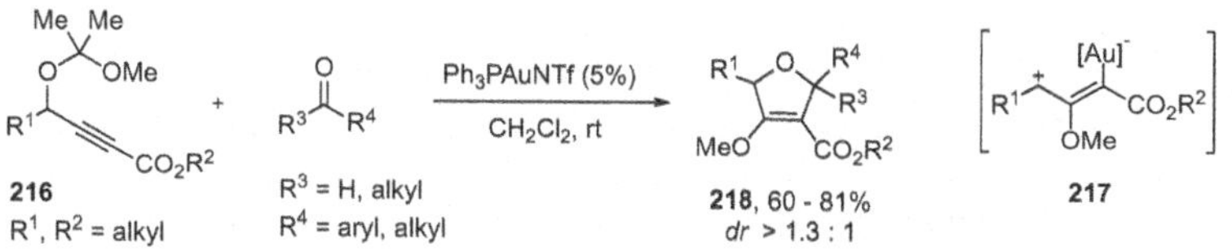

Scheme 71 Gold-catalyzed migration–fragmentation of ketals **216** and their [3 + 2] cycloadditions

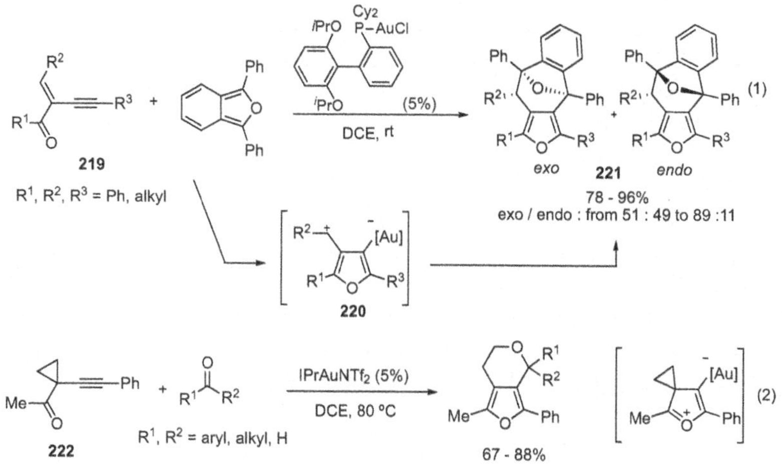

Scheme 72 Cycloadditions of 1,3-furanyl gold dipoles generated from 2-(1-alkynyl)-2-alken-1-ones

On the other hand, Zhang and coworkers have extensively exploited the cycloadditions of 1,3-furanyl gold dipoles (**220**) generated from 2-(1-alkynyl)-2-alken-1-ones (**219**) and a gold catalyst. For instance, these intermediates were used to build oxygen-bridge polyheterocycles in a catalytic heterocyclization/formal [4 + 3] cycloaddition with the 1,3-diphenylisobenzofuran (Scheme 72, Eq. 1). In general, the reaction catalyzed by gold provides the *exo* cycloadducts as major products, whereas a related silver-catalyzed process yields the *endo* counterparts [190]. In another example, L. Zhang and coworkers demonstrated that related dipoles generated from 1-(1-alkynyl)cyclopropyl ketones (**222**) undergo [4 + 2] stepwise cycloadditions with a dipolarophile such as an aldehyde or a ketone (Scheme 72, Eq. 2) [191].

4.2 Other Formal Gold-Catalyzed Cycloadditions Involving Stepwise Processes

In addition to the above annulations involving the generation of dipole intermediates, gold catalysts can also promote formal cycloaddition process by alternative mechanisms. Thus, gold–carbenes generated from a gold-catalyzed skeletal rearrangement of 1,5- or 1,6-enynes (**223**) can react with exogenous aldehydes to provide cyclic ethers **224,** resulting from a formal [2 + 2 + 2] cycloaddition [192]. The formation of the products is explained by the attack of the carbonyl to the cyclopropyl gold–carbene followed by a *Prins*-like cyclization and demetalation. Dienes of type **225**, which result from a metathesis-type reaction of the enyne and the aldehyde, were also observed as side products (Scheme 73, right). In the particular case of 1,6-enynes featuring terminally unsubstituted alkenes, tricyclic compounds of type **226** resulting from the cycloaddition between the gold–carbene intermediate and the carbonyl compound are obtained, as demonstrated by Helmchen and coworkers (Scheme 73, left) [193].

Echavarren and coworkers have developed intramolecular versions of these formal [2 + 2 + 2] cycloadditions of carbonyl-tethered enynes. These reactions proceed via intramolecular attack of the carbonyl to the initially generated cyclopropyl gold–carbenes (**228**) to form oxonium cations of type **229**, which undergo a *Prins*-type cyclization with concomitant catalyst regeneration [164]. The methodology was further optimized and applied to the total synthesis of several natural products such as orientalol F (Scheme 74, Eq. 1) [166] or englerins A and B [165, 194]. Moreover, a partially intermolecular variant using

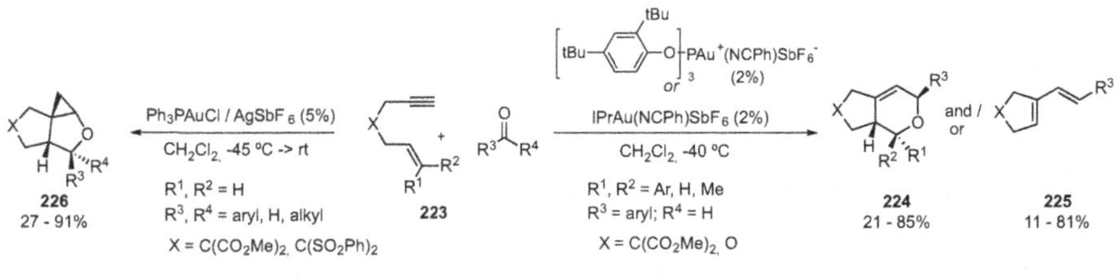

Scheme 73 Cycloadditions of gold–carbenes generated from a skeletal rearrangement of 1,5-enynes

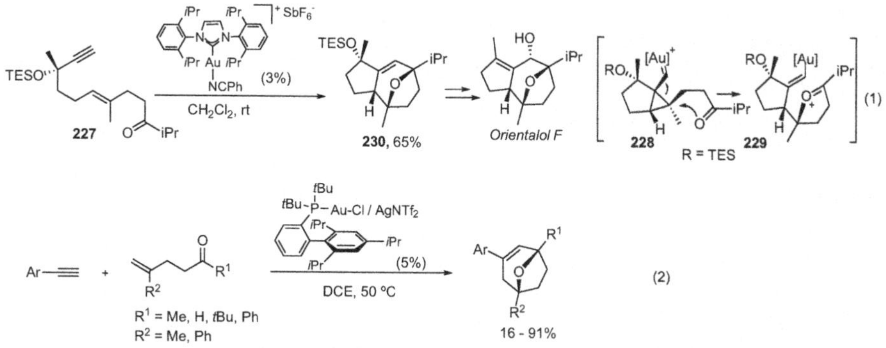

Scheme 74 Intramolecular formal [2 + 2 + 2] cycloadditions of carbonyl-tethered enynes

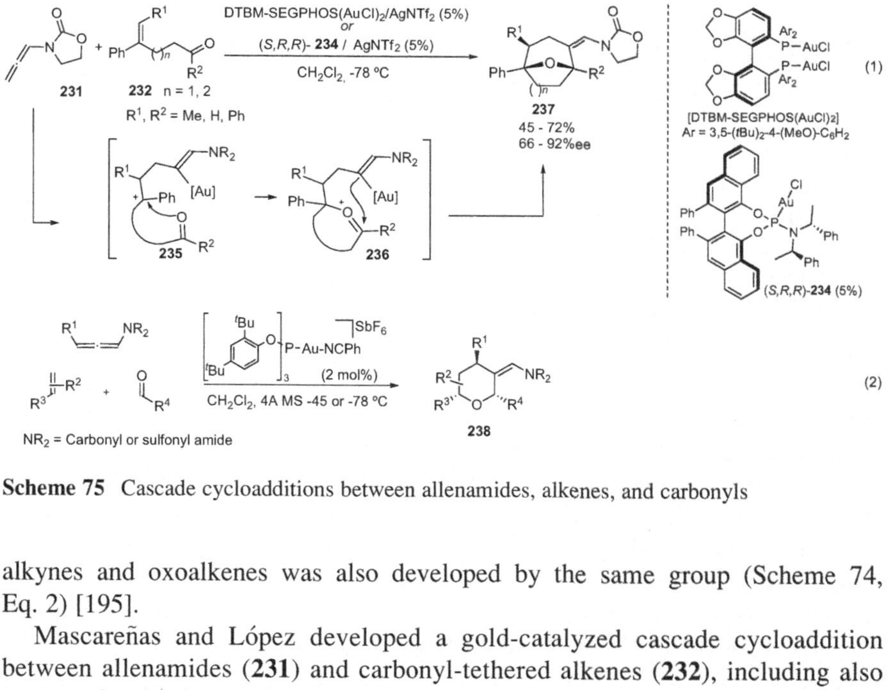

Scheme 75 Cascade cycloadditions between allenamides, alkenes, and carbonyls

alkynes and oxoalkenes was also developed by the same group (Scheme 74, Eq. 2) [195].

Mascareñas and López developed a gold-catalyzed cascade cycloaddition between allenamides (**231**) and carbonyl-tethered alkenes (**232**), including also an enantioselective variant which is catalyzed by DTBM-SEGPHOS$(AuCl)_2$/$AgNTf_2$, or by the phosphoramidite–gold complex (*S*,*R*,*R*)-**234**/$AgNTf_2$. The reaction, which provides synthetically appealing oxabridged seven, eight-, and even nine-membered rings, relies on the interception of intermediates of type **235** by the intramolecular carbonyl group, followed by ring closing on the resulting oxonium intermediate (**236**) (Scheme 75, Eq. 1) [196]. On the other hand, a fully intermolecular variant of these process was recently accomplished by the

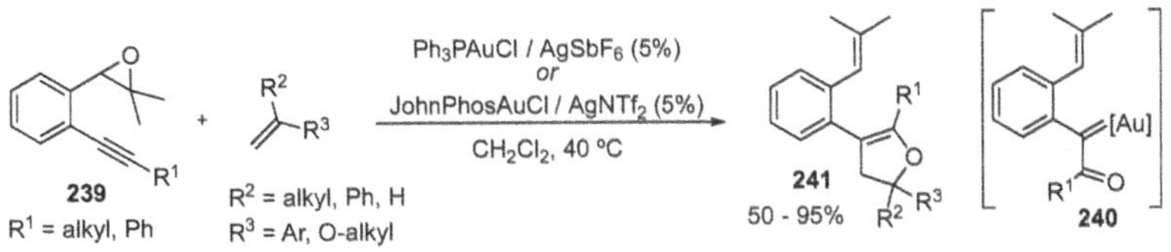

Scheme 76 Cycloaddition of 2-epoxy-1-alkynylbenzenes with electron-rich alkenes

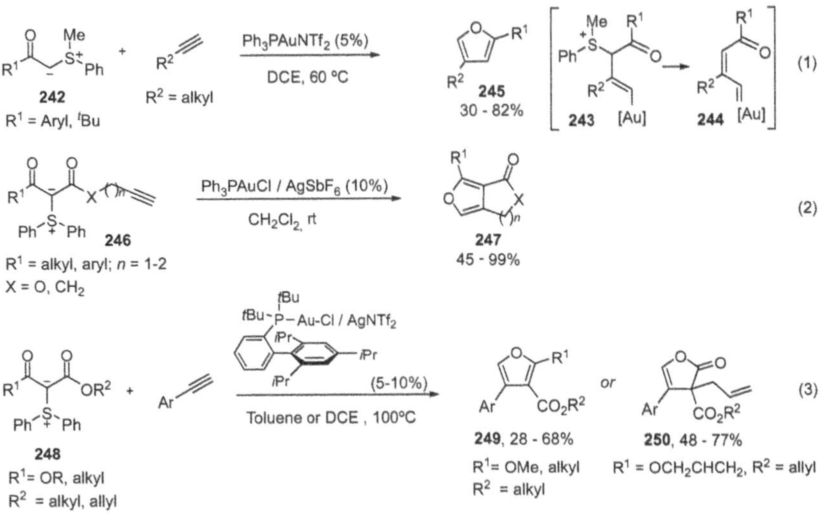

Scheme 77 Cycloadditions involving sulfur ylide precursors

same authors, providing a direct and catalytic entry to 2,6-*cis*-disubstituted tetrahydropyrans of type **238** (Scheme 75, Eq. 2) [197].

Liu and coworkers reported a gold-catalyzed formal cycloaddition of 2-epoxy-1-alkynylbenzenes (**239**) with electron-rich alkenes that directly affords 2-alkenyl-1-(2,3-dihydrofurany-4-yl) benzenes (**241**). The process is initiated by a *endo* cycloisomerization to afford the gold–carbene intermediate **240**, which subsequently undergoes a stepwise [3 + 2] cycloaddition with an electron-rich alkene (i.e., styrene or enol ether) to afford the final dihydrofuran core (Scheme 76) [198, 199].

Sulfur ylides like **242** have also been employed as sources of gold–carbenes that can participate in cycloaddition processes, leading to 2,4-disubstituted furans, as demonstrated by Skrydstrup and coworkers [200]. The reaction presumably starts with the regioselective attack of the ylide at the (η2-alkyne)gold(I) complex to yield intermediate **243**. Back donation of the gold center with concomitant expulsion of the thioether generates an allyl gold–carbene (**244**) that can be intramolecularly intercepted by the carbonyl. A final demetalation generates the 2,4-disubstituted furan (Scheme 77, Eq. 1). Almost simultaneously, Maulide and coworkers also

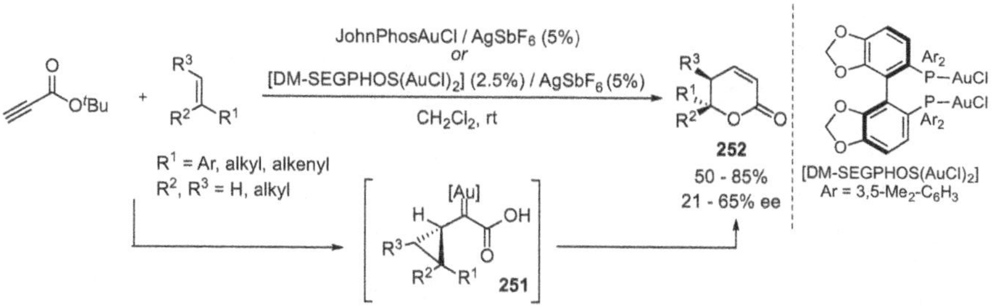

Scheme 78 [4 + 2] Intermolecular cycloaddition of propiolic acids

employed a similar carbene-transfer strategy to synthesize furans and furanones through a formal [3 + 2] cycloaddition process between alkynes and sulfonium ylides like **246** (Scheme 77, Eq. 2) [201]. In the intermolecular version of this reaction, arylacetylenes were found to be the most suitable coupling partners (Scheme 77, Eq. 3). On the other hand, when diallylmalonate-derived ylides were used (**248**, R^1=O-allyl, R^2=allyl), furanones **250** were formed. Assuming the mechanistic rationale shared by Maulide and Skrydstrup, a [3,3]-sigmatropic rearrangement on the initially formed 2-allyloxyfuran **249** would explain the formation of these furanones (Scheme 77, Eq. 3).

Shin and coworkers reported a formal [4 + 2] intermolecular cycloaddition of propiolic acids and alkenes to provide α,β-unsaturated δ-lactones (**252**) which proceeds with complete stereospecificity. A cyclopropyl gold–carbene like **251** is proposed as the key intermediate. Moreover, preliminary enantioselective examples were also reported (up to 65 % *ee*) using DM-SEGPHOS as ancillary ligand (Scheme 78) [202].

Although not directly involving the formation of a C–O bond, in the context of the development of gold-catalyzed allenediene cycloadditions, Mascareñas and López reported some examples of intramolecular [4 + 3] cycloadditions between allenes and furans, catalyzed by $IPrAuCl/AgSbF_6$ catalyst (Scheme 79, Eq. 1) [203]. In a related study, Gung and coworkers reported a transannular version of this process using $JohnPhosAuCl/AgSbF_6$ as the catalyst (Scheme 79, Eq. 2) [204]. Moreover, the same group has also developed the analog transannular and intramolecular [4 + 3] cycloadditions employing propargyl acetates as allene surrogates (Scheme 79, Eq. 3) [205, 206]. Finally, isolated examples of intramolecular [4 + 2] cycloadditions between furans and allenes can also been achieved under phosphite–gold catalysis, provided that the allene is disubstituted at its distal position (Scheme 79, Eq. 4) [207, 208].

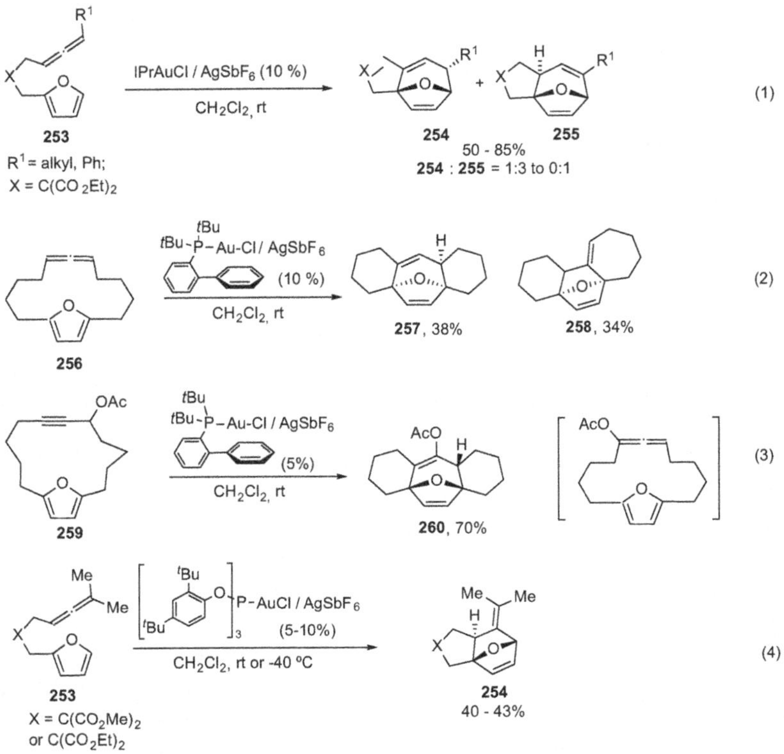

Scheme 79 [4 + 3] and [4 + 2] cycloadditions of allene–furan systems

5 Conclusions

All the above examples come to remark the enormous progress in the assembly of oxygen heterocycles using gold catalysts, which provide for transformations that otherwise could not be achievable. The possibility of tuning the characteristics of the gold complexes with different ligands further extended the enormous potential of these catalysts. Further work to develop efficient processes that rely on readily available and inexpensive starting materials and exploit not only the carbophilic properties of gold but also the recently discovered redox or photoredox-based possibilities is warranted. There is also a need to develop efficient enantioselective variants and further demonstrate the synthetic possibilities of this chemistry to make complex oxacyclic products.

References

1. Elliott MC (2000) Saturated oxygen heterocycles. J Chem Soc Perkin Trans 1:1291–1318
2. Elliot MC, Williams E (2001) J Chem Soc Perkin Trans 1 2303
3. Alali FQ, Liu XX, McLaughlin JL (1999) J Nat Prod 62:504
4. Boivin TLB (1987) Tetrahedron 43:3309
5. Wuitschik G, Carreira EM, Wagner B, Fischer H, Parrilla I, Schuler F, Rogers-Evans M, Müller K (2010) Oxetanes in drug discovery: structural and synthetic insights. J Med Chem 53:3227–3246
6. Maes BUW, Cossy J (eds) (2014) Synthesis of saturated oxygenated heterocycles I, vol 36, Topics in heterocyclic chemistry. Springer, Berlin
7. Cardillo G, Orena M (1990) Stereocontrolled cyclofunctionalizations of double bonds through heterocyclic intermediates. Tetrahedron 46:3321–3408
8. Maes BUW, Cossy J (eds) (2014) Synthesis of saturated oxygenated heterocycles II, vol 36, Topics in heterocyclic chemistry. Springer, New York
9. Toste FD, Michelet V (eds) (2015) Gold catalysis: an homogeneous approach (catalytic science series). Imperial College Press, London
10. Hashmi ASK, Toste FD (eds) (2012) Modern gold catalyzed synthesis. Wiley, Weinheim
11. Mohr F (ed) (2009) Gold chemistry. Wiley, Weinheim
12. Slaughter LM (ed) (2015) Homogeneous gold catalysis, Topics in current chemistry, vol 357. Springer, Cham
13. Gorin DJ, Toste FD (2007) Relativistic effects in homogeneous gold catalysis. Nature 446: 395–403
14. Fürstner A, Davies PW (2007) Catalytic carbophilic activation: catalysis by platinum and gold pi acids. Angew Chem Int Ed 46:3410–3449
15. Huguet N, Echavarren AM (2013) Gold-catalyzed O-H bond addition to unsaturated organic molecules. In: Ananikov VP, Tanaka M (eds) Hydrofunctionalization, vol 43, Topics in organometallic chemistry., pp 291–324
16. Hashmi ASK, Buehrle M (2010) Gold-catalyzed addition of X-H bonds to C-C multiple bonds. Aldrichimica Acta 43:27–33
17. Corma A, Leyva-Pérez A, Sabater MJ (2011) Gold-catalyzed carbon-heteroatom bond-forming reactions. Chem Rev 111:1657–1712
18. Patil NT, Kavthe RD, Shinde VS (2012) Transition metal-catalyzed addition of C-, N- and O-nucleophiles to unactivated C-C multiple bonds. Tetrahedron 68:8079–8146
19. Krause N, Winter C (2011) Gold-catalyzed nucleophilic cyclization of functionalized allenes: a powerful access to carbo- and heterocycles. Chem Rev 111:1994–2009
20. Yang W, Hashmi ASK (2014) Mechanistic insights into the gold chemistry of allenes. Chem Soc Rev 43:2941–2955
21. Hoffmann-Roder A, Krause N (2001) Gold(III) chloride catalyzed cyclization of alpha-hydroxyallenes to 2,5-dihydrofurans. Org Lett 3:2537–2538
22. Winter C, Krause N (2009) Towards sustainable homogeneous gold catalysis: cycloisomerization of functionalized allenes in water. Green Chem 11:1309–1312
23. Gockel B, Krause N (2006) Golden times for allenes: gold-catalyzed cycloisomerization of beta-hydroxyallenes to dihydropyrans. Org Lett 8:4485–4488
24. Gockel B, Krause N (2010) Synthesis of bicyclic ethers by a gold/palladium/gold-catalyzed cyclization/cross coupling sequence. Eur J Org Chem 2010:311–316
25. Zriba R, Gandon V, Aubert C, Fensterbank L, Malacria M (2008) Alkyne versus allene activation in platinum- and gold-catalyzed cycloisomerization of hydroxylated 1,5-allenynes. Chem Eur J 14:1482–1491
26. Eom D, Kang D, Lee PH (2010) Synthesis of 2-alkyl- and ary1-3-ethoxycarbony1-2,5-dihydrofurans through gold-catalyzed intramolecular hydroalkoxylation. J Org Chem 75: 7447–7450

27. Kim S, Lee PH (2008) Cyclization of allenyne-1,6-diols catalyzed by gold and silver salts: an efficient selective synthesis of dihydrofuran and furan derivatives. Adv Synth Catal 350: 547–551
28. Aksın Ö, Krause N (2008) Gold-catalyzed synthesis of 2,5-dihydrofurans in ionic liquids. Adv Synth Catal 350:1106–1112
29. Alcaide B, Almendros P, Martinez del Campo T, Fernandez I (2011) Fascinating reactivity in gold catalysis: synthesis of oxetenes through rare 4-exo-dig allene cyclization and infrequent beta-hydride elimination. Chem Commun 47:9054–9056
30. Zhang ZB, Liu C, Kinder RE, Han XQ, Qian H, Widenhoefer RA (2006) Highly active Au (I) catalyst for the intramolecular exo-hydrofunctionalization of allenes with carbon, nitrogen, and oxygen nucleophiles. J Am Chem Soc 128:9066–9073
31. Alcaide B, Almendros P, del Campo TM, Soriano E, Marco-Contelles JL (2009) Regioselectivity control in the metal-catalyzed o-c functionalization of gamma-allenols, part 1: experimental study. Chem Eur J 15:1901–1908
32. Alcaide B, Almendros P (2011) Gold-catalyzed heterocyclizations in alkynyl- and allenyl-beta-lactams. Beilstein J Org Chem 7:622–630
33. Alcaide B, Almendros P (2014) Gold-catalyzed cyclization reactions of allenol and alkynol derivatives. Acc Chem Res 47:939–952
34. Alcaide B, Almendros P, del Campo TM, Soriano E, Marco-Contelles JL (2009) Metal-catalyzed cyclization of β- and γ-allenols derived from D-glyceraldehyde-synthesis of enantiopure dihydropyrans and tetrahydrooxepines: an experimental and theoretical study. Chem Eur J 15:9127–9138
35. Zhang Z, Widenhoefer RA (2007) Gold(I)-catalyzed intramolecular enantioselective hydroalkoxylation of allenes. Angew Chem Int Ed 46:283–285
36. Hamilton GL, Kang EJ, Mba M, Toste FD (2007) A powerful chiral counterion strategy for asymmetric transition metal catalysis. Science 317:496–499
37. Zi W, Toste FD (2015) Gold(I)-catalyzed enantioselective desymmetrization of 1,3-diols through intramolecular hydroalkoxylation of allenes. Angew Chem Int Ed. 10.1002/anie.201508331
38. Cox N, Uehling MR, Haelsig KT, Lalic G (2013) Catalytic asymmetric synthesis of cyclic ethers containing an alpha-tetrasubstituted stereocenter. Angew Chem Int Ed 52:4878–4882
39. Volz F, Krause N (2007) Golden opportunities in natural product synthesis: first total synthesis of (-)-isocyclocapitelline and (-)-isochrysotricine by gold-catalyzed allene cycloisomerization. Org Biomol Chem 5:1519–1521
40. Volz F, Wadman SH, Hoffmann-Roder A, Krause N (2009) Gold catalysis in stereoselective natural product synthesis: (+)-linalool oxide, (-)- isocyclocapitelline, and (-)-isochrysotricine. Tetrahedron 65:1902–1910
41. Gao ZH, Li YF, Cooksey JP, Snaddon TN, Schunk S, Viseux EME, McAteer SM, Kocienski PJ (2009) A synthesis of an ionomycin calcium complex. Angew Chem Int Ed 48:5022–5025
42. Sawama Y, Sawama Y, Krause N (2008) First total synthesis of (R, R, R)- and (3R,5S,9R)-bejarol by gold-catalyzed allene cycloisomerization and determination of absolute configuration of the natural product. Org Biomol Chem 6:3573–3579
43. Kang J-E, Lee E-S, Park S-I, Shin S (2005) Gold-catalyzed cyclization of tert-butyl allenoate: general synthesis of 2,4-functionalized butenolides. Tetrahedron Lett 46:7431–7433
44. Liu L-P, Xu B, Mashuta MS, Hammond GB (2008) Synthesis and structural characterization of stable organogold(I) compounds. Evidence for the mechanism of gold-catalyzed cyclizations. J Am Chem Soc 130:17642–17643
45. Piera J, Krumlinde P, Strübing D, Bäckvall J-E (2007) Gold-catalyzed cyclization of allene-substituted malonate esters: synthesis of β, γ-unsaturated δ-lactones. Org Lett 9:2235–2237
46. Alcaide B, Almendros P, Miguel Alonso J (2011) Gold catalyzed oxycyclizations of alkynols and alkyndiols. Org Biomol Chem 9:4405–4416
47. Goodwin JA, Aponick A (2015) Regioselectivity in the Au-catalyzed hydration and hydroalkoxylation of alkynes. Chem Commun 51:8730–8741

48. Dorel R, Echavarren AM (2015) Gold(I)-catalyzed activation of alkynes for the construction of molecular complexity. Chem Rev 115:9028–9072
49. Harkat H, Weibel JM, Pale P (2007) Synthesis of functionalized THF and THP through Au-catalyzed cyclization of acetylenic alcohols. Tetrahedron Lett 48:1439–1442
50. Harkat H, Blanc A, Weibel J-M, Pale P (2008) Versatile and expeditious synthesis of aurones via aul-catalyzed cyclization. J Org Chem 73:1620–1623
51. Zhdanko A, Maier ME (2015) Explanation of "silver effects" in gold(I)-catalyzed hydroalkoxylation of alkynes. ACS Catal 5:5994–6004
52. Zhdanko A, Maier ME (2014) Explanation of counterion effects in gold(I)-catalyzed hydroalkoxylation of alkynes. ACS Catal 4:2770–2775
53. Trost BM, Dong GB (2008) Total synthesis of bryostatin 16 using atom-economical and chemoselective approaches. Nature 456:485–488
54. Rüttinger R, Leutzow J, Wilsdorf M, Wilckens K, Czekelius C (2011) Reversal of selectivity in gold-catalyzed cyclizations of 3,3-disubstituted 1,4-diynes. Org Lett 13:224–227
55. Wilckens K, Uhlemann M, Czekelius C (2009) Gold-catalyzed *endo*-cyclizations of 1,4-diynes to seven-membered ring heterocycles. Chem Eur J 15:13323–13326
56. Murai M, Uenishi J, Uemura M (2010) Gold(I)-catalyzed asymmetric synthesis of planar chiral arene chromium complexes. Org Lett 12:4788–4791
57. Hashmi ASK, Häffner T, Rudolph M, Rominger F (2011) Cyclization of 2-alkynylallyl alcohols to highly substituted furans by gold(I)–carbene complexes. Eur J Org Chem 2011: 667–671
58. Kim S, Kang D, Shin S, Lee PH (2010) Gold-catalyzed cyclization of enyne-1,6-diols to substituted furans. Tetrahedron Lett 51:1899–1901
59. Genin E, Toullec PY, Antoniotti S, Brancour C, Genet JP, Michelet W (2006) Room temperature Au(I)-catalyzed exo-selective cycloisomerization of acetylenic acids: an entry to functionalized gamma-lactones. J Am Chem Soc 128:3112–3113
60. Harkat H, Weibel JM, Pale P (2006) A mild access to gamma- or delta-alkylidene lactones through gold catalysis. Tetrahedron Lett 47:6273–6276
61. Harkat H, Dembele AY, Weibel JM, Blanc A, Pale P (2009) Cyclization of alkynoic acids with gold catalysts: a surprising dichotomy between Au-I and Au-III. Tetrahedron 65: 1871–1879
62. Michelet V, Toullec PY, Genin E, Antoniotti S, Genet JP (2008) Au2O3 as a stable and efficient catalyst for the selective cycloisomerization of gamma-acetylenic carboxylic acids to gamma-alkylidene-gamma-butyrolactones. Synlett 707–711
63. Tomas-Mendivil E, Toullec PY, Borge J, Conejero S, Michelet V, Cadierno V (2013) Water-soluble gold(I) and gold(III) complexes with sulfonated N-heterocyclic carbene ligands: synthesis, characterization, and application in the catalytic cycloisomerization of gamma-alkynoic acids into enol-lactones. ACS Catal 3:3086–3098
64. Marchal E, Uriac P, Legouin B, Toupet L, van de Weghe P (2007) Cycloisomerization of γ- and δ-acetylenic acids catalyzed by gold(I) chloride. Tetrahedron 63:9979–9990
65. Chaladaj W, Corbet M, Fürstner A (2012) Total synthesis of neurymenolide a based on a gold-catalyzed synthesis of 4-hydroxy-2-pyrones. Angew Chem Int Ed 51:6929–6933
66. Kang J-E, Shin S (2006) Au(I)-catalyzed cyclization of tert-butyl carbonates derived from - homopropargyl alcohols: a catalytic alternative to cyclic enol carbonates. Synlett 2006: 0717–0720
67. Chiarucci M, Bandini M (2013) New developments in gold-catalyzed manipulation of inactivated alkenes. Beilstein J Org Chem 9:2586–2614
68. Yang C-G, He C (2005) Gold(I)-catalyzed intermolecular addition of phenols and carboxylic acids to olefins. J Am Chem Soc 127:6966–6967
69. Kamiya I, Tsunoyama H, Tsukuda T, Sakurai H (2007) Lewis acid character of zero-valent gold nanoclusters under aerobic conditions: intramolecular hydroalkoxylation of alkenes. Chem Lett 36:646–647

70. Chandrasekhar B, Ryu J-S (2012) Gold-catalyzed intramolecular hydroalkoxylation/cyclization of conjugated dienyl alcohols. Tetrahedron 68:4805–4812
71. Zhang G, Cui L, Wang Y, Zhang L (2010) Homogeneous gold-catalyzed oxidative carboheterofunctionalization of alkenes. J Am Chem Soc 132:1474–1475
72. Sahoo B, Hopkinson MN, Glorius F (2013) Combining gold and photoredox catalysis: visible light-mediated oxy- and aminoarylation of alkenes. J Am Chem Soc 135:5505–5508
73. Aponick A, Li C-Y, Biannic B (2008) Au-catalyzed cyclization of monoallylic diols. Org Lett 10:669–671
74. Ghebreghiorgis T, Biannic B, Kirk BH, Ess DH, Aponick A (2012) The importance of hydrogen bonding to stereoselectivity and catalyst turnover in gold-catalyzed cyclization of monoallylic diols. J Am Chem Soc 134:16307–16318
75. Aponick A, Biannic B (2011) Chirality transfer in Au-catalyzed cyclization reactions of monoallylic diols: selective access to specific enantiomers based on olefin geometry. Org Lett 13:1330–1333
76. Biannic B, Aponick A (2011) Gold-catalyzed dehydrative transformations of unsaturated alcohols. Eur J Org Chem 2011:6605–6617
77. Aponick A, Biannic B, Jong MR (2010) A highly adaptable catalyst/substrate system for the synthesis of substituted chromenes. Chem Commun 46:6849–6851
78. Unsworth WP, Stevens K, Lamont SG, Robertson J (2011) Stereospecificity in the Au-catalysed cyclisation of monoallylic diols. Synthesis of (+)-isoaltholactone. Chem Commun 47:7659–7661
79. Hashmi ASK, Schwarz L, Choi JH, Frost TM (2000) A new gold-catalyzed C-C bond formation. Angew Chem Int Ed 39:2285–2288
80. Zhou CY, Chan PWH, Che CM (2006) Gold(III) porphyrin-catalyzed cycloisomerization of allenones. Org Lett 8:325–328
81. Sromek AW, Rubina M, Gevorgyan V (2005) 1,2-Halogen migration in haloallenyl ketones: regiodivergent synthesis of halofurans. J Am Chem Soc 127:10500–10501
82. Xia YZ, Dudnik AS, Gevorgyan V, Li YH (2008) Mechanistic insights into the gold-catalyzed cycloisomerization of bromoallenyl ketones: ligand-controlled regioselectivity. J Am Chem Soc 130:6940–6941
83. Dudnik AS, Sromek AW, Rubina M, Kim JT, Kel'in AV, Gevorgyan V (2008) Metal-catalyzed 1,2-shift of diverse migrating groups in allenyl systems as a new paradigm toward densely functionalized heterocycles. J Am Chem Soc 130:1440–1452
84. Li Y, Brand JP, Waser J (2013) Gold-catalyzed regioselective synthesis of 2- and 3-alkynyl furans. Angew Chem Int Ed 52:6743–6747
85. Belting V, Krause N (2009) Gold-catalyzed cycloisomerization of alk-4-yn-1-ones. Org Biomol Chem 7:1221–1225
86. Arcadi A, Alfonsi M, Chiarini M, Marinelli F (2009) Sequential gold-catalyzed reactions of 1-phenylprop-2-yn-1-ol with 1,3-dicarbonyl compounds. J Organomet Chem 694:576–582
87. Hashmi ASK, Sinha P (2004) Gold catalysis: mild conditions for the transformation of alkynyl epoxides to furans. Adv Synth Catal 346:432–438
88. Blanc A, Tenbrink K, Weibel JM, Pale P (2009) Mechanistic studies and improvement of coinage metal-catalyzed transformation of alkynyloxiranes to furans: an alcohol addition-cyclization-elimination cascade. J Org Chem 74:5342–5348
89. Jacques B, Hueber D, Hameury S, Braunstein P, Pale P, Blanc A, de Frémont P (2014) Synthesis, characterization, and catalytic activity of alcohol-functionalized nhc gold(i/iii) complexes. Organometallics 33:2326–2335
90. Ji KG, Shu XZ, Chen J, Zhao SC, Zheng ZJ, Liu XY, Liang YM (2009) Gold-catalyzed tandem cyclization/friedel-crafts type reactions toward furan derivatives. Org Biomol Chem 7:2501–2505
91. Michelet V (2015) Gold-catalyzed domino reactions. In: Slaughter LM (ed) Homogeneous gold catalysis, vol 357, Topics in current chemistry., pp 95–132

92. Inamdar SM, Konala A, Patil NT (2014) When gold meets chiral brønsted acid catalysts: extending the boundaries of enantioselective gold catalysis. Chem Commun 50:15124–15135
93. Kirsch SF (2008) Construction of heterocycles by the strategic use of alkyne pi-activation in catalyzed cascade reactions. Synthesis 3183–3204
94. Zhang LM, Kozmin SA (2005) Gold-catalyzed assembly of heterobicyclic systems. J Am Chem Soc 127:6962–6963
95. Toullec PY, Blarre T, Michelet V (2009) Mimicking polyolefin carbocyclization reactions: gold-catalyzed intramolecular phenoxycyclization of 1,5-enynes. Org Lett 11:2888–2891
96. Sethofer SG, Mayer T, Toste FD (2010) Gold(I)-catalyzed enantioselective polycyclization reactions. J Am Chem Soc 132:8276–8277
97. Fürstner A, Morency L (2008) On the nature of the reactive intermediates in gold-catalyzed cycloisomerization reactions. Angew Chem Int Ed 47:5030–5033
98. Sherry BD, Maus L, Laforteza BN, Toste FD (2006) Gold(I)-catalyzed synthesis of dihydropyrans. J Am Chem Soc 128:8132–8133
99. Noey EL, Wang X, Houk KN (2011) Selective gold(I)-catalyzed formation of tetracyclic indolines: a single transition structure and bifurcations lead to multiple products. J Org Chem 76:3477–3483
100. Cera G, Crispino P, Monari M, Bandini M (2011) Stereoselective synthesis of tetracyclic indolines via gold-catalyzed cascade cyclization reactions. Chem Commun 47:7803–7805
101. Liu Y, Xu W, Wang X (2010) Gold(I)-catalyzed tandem cyclization approach to tetracyclic indolines. Org Lett 12:1448–1451
102. Barluenga J, Fernández A, Satrustegui A, Dieguez A, Rodríguez F, Fañanás FJ (2008) Tandem intramolecular hydroalkoxylation-hydroarylation reactions: synthesis of enantiopure benzofused cyclic ethers from the chiral pool. Chem Eur J 14:4153–4156
103. Barluenga J, Fernández A, Dieguez A, Rodríguez F, Fañanás FJ (2009) Gold- or platinum-catalyzed cascade processes of alkynol derivatives involving hydroalkoxylation reactions followed by Prins-type cyclizations. Chem Eur J 15:11660–11667
104. Antoniotti S, Genin E, Michelet V, Genet JP (2005) Highly efficient access to strained bicyclic ketals via gold-catalyzed cycloisomerization of bis-homopropargylic diols. J Am Chem Soc 127:9976–9977
105. Santos LL, Ruiz VR, Sabater MJ, Corma A (2008) Regioselective transformation of alkynes into cyclic acetals and thioacetals with a gold(I) catalyst: comparison with brønsted acid catalysts. Tetrahedron 64:7902–7909
106. Liu L-P, Hammond GB (2009) Highly efficient and tunable synthesis of dioxabicyclo[4.2.1] ketals and tetrahydropyrans via gold-catalyzed cycloisomerization of 2-alkynyl-1,5-diols. Org Lett 11:5090–5092
107. Aponick A, Li CY, Palmes JA (2009) Au-catalyzed cyclization of monopropargylic triols: an expedient synthesis of monounsaturated spiroketals. Org Lett 11:121–124
108. Paioti PHS, Ketcham JM, Aponick A (2014) Controlling regiochemistry in the gold-catalyzed synthesis of unsaturated spiroketals. Org Lett 16:5320–5323
109. Minkler SR, Isley NA, Lippincott DJ, Krause N, Lipshutz BH (2014) Leveraging the micellar effect: gold-catalyzed dehydrative cyclizations in water at room temperature. Org Lett 16:724–726
110. Liu B, De Brabander JK (2006) Metal-catalyzed regioselective oxy-functionalization of internal alkynes: an entry into ketones, acetals, and spiroketals. Org Lett 8:4907–4910
111. Zhdanko A, Maier ME (2014) Gold(I)-, palladium(II)-, platinum(II)-, and Mercury(II)-catalysed spirocyclization of 1,3-enynediols: reaction scope. Eur J Org Chem 2014:3411–3422
112. Jaimes MCB, Bohling CRN, Serrano-Becerra JM, Hashmi ASK (2013) Highly active mono-nuclear NAC-gold(I) catalysts. Angew Chem Int Ed 52:7963–7966
113. Jaimes MCB, Rominger F, Pereira MM, Carrilho RMB, Carabineiro SAC, Hashmi ASK (2014) Highly active phosphite gold(I) catalysts for intramolecular hydroalkoxylation, enyne cyclization and furanyne cyclization. Chem Commun 50:4937–4940

114. Quach R, Furkert DP, Brimble MA (2013) Synthesis of benzannulated spiroacetals using chiral gold-phosphine complexes and chiral anions. Tetrahedron Lett 54:5865–5868
115. Yeung YY, Corey EJ (2008) A simple, efficient, and enantiocontrolled synthesis of a near-structural mimic of platensimycin. Org Lett 10:3877–3878
116. Fang C, Pang YC, Forsyth CJ (2010) Formal total synthesis of okadaic acid via regiocontrolled gold(I)-catalyzed spiroketalizations. Org Lett 12:4528–4531
117. Li YF, Zhou F, Forsyth CJ (2007) Gold(I)-catalyzed bis-spiroketalization: synthesis of the trioxadispiroketal-containing A-D rings of azaspiracid. Angew Chem Int Ed 46:279–282
118. Liu LP, Hammond GB (2010) Gold-catalyzed cascade annulations of 2-(ynol)aryl aldehydes: facile synthesis of benzochromanes and benzobicyclo[n.3.1]acetals. Org Lett 12:4640–4643
119. Tarselli MA, Zuccarello JL, Lee SJ, Gagné MR (2009) Gold(I)-catalyzed cascade cyclization of allenyl epoxides. Org Lett 11:3490–3492
120. Nguyen R-V, Yao X, Li C-J (2006) Highly efficient gold-catalyzed atom-economical annulation of phenols with dienes. Org Lett 8:2397–2399
121. Belting V, Krause N (2006) Gold-catalyzed tandem cycloisomerization-hydroalkoxylation of homopropargylic alcohols. Org Lett 8:4489–4492
122. Reddy MS, Kumar YK, Thirupathi N (2012) A new synthesis of gamma-butyrolactones via $AuCl_3$- or Hg(II)-catalyzed intramolecular hydroalkoxylation o 4-bromo-3-yn-1ols. Org Lett 14:824–827
123. Alcaide B, Almendros P, del Campo TM, Carrascosa R (2010) Metal-catalyzed cycloisomerization and tandem oxycyclization/hydroxylation of alkynols: synthesis of nonfused, spiranic and fused oxabicyclic beta-Lactams. Eur J Org Chem 2010:4912–4919
124. Diéguez-Vázquez A, Tzschucke CC, Crecente-Campo J, McGrath S, Ley SV (2009) AuCl3-Catalyzed hydroalkoxylation of conjugated alkynoates: synthesis of five- and six-membered cyclic acetals. Eur J Org Chem 2009:1698–1706
125. Hashmi ASK, Buhrle M, Wolfle M, Rudolph M, Wieteck M, Rominger F, Frey W (2010) Gold catalysis: tandem reactions of diyne-diols and external nucleophiles as an easy access to tricyclic cage-like structures. Chem Eur J 16:9846–9854
126. Pennell MN, Foster RW, Turner PG, Hailes HC, Tame CJ, Sheppard TD (2014) Gold catalysed synthesis of 3-alkoxyfurans at room temperature. Chem Commun 50:1302–1304
127. Alcaide B, Almendros P, Carrascosa R, Torres MR (2010) Gold/acid-cocatalyzed regiodivergent preparation of bridged ketals via direct bis-oxycyclization of alkynic acetonides. Adv Synth Catal 352:1277–1283
128. Zhou L, Liu YZ, Zhang Y, Wang JB (2011) Sequential Au(I)-catalyzed reaction of water with o-acetylenyl-substituted phenyldiazoacetates. Beilstein J Org Chem 7:631–637
129. Belot S, Vogt KA, Besnard C, Krause N, Alexakis A (2009) Enantioselective one-pot organocatalytic Michael addition/gold-catalyzed tandem acetalization/cyclization. Angew Chem Int Ed 48:8923–8926
130. Tian GQ, Shi M (2007) Gold(I)-catalyzed additions of three-component 2-(arylmethylene) cyclopropylcarbinols, terminal arynes, and alcohols: an efficient access to 3-oxabicyclo [3.1.0]hexanes. Org Lett 9:4917–4920
131. Lu BL, Dai L, Shi M (2012) Strained small rings in gold-catalyzed rapid chemical transformations. Chem Soc Rev 41:3318–3339
132. Meng J, Zhao YL, Ren CQ, Li Y, Li Z, Liu Q (2009) Highly efficient access to bi- and tricyclic ketals through gold-catalyzed tandem reactions of 4-acyl-1,6-diynes. Chem Eur J 15:1830–1834
133. Handa S, Slaughter LM (2012) Enantioselective alkynylbenzaldehyde cyclizations catalyzed by chiral gold(I) acyclic diaminocarbene complexes containing weak Au–arene interactions. Angew Chem Int Ed 51:2912–2915
134. Wu H, He YP, Gong LZ (2013) Direct access to enantioenriched spiroacetals through asymmetric Relay catalytic three-component reaction. Org Lett 15:460–463
135. Cala L, Mendoza A, Fañanás FJ, Rodríguez F (2013) A catalytic multicomponent coupling reaction for the enantioselective synthesis of spiroacetals. Chem Commun 49:2715–2717

136. Jung HH, Floreancig PE (2007) Gold-catalyzed synthesis of oxygen- and nitrogen-containing heterocycles from alkynyl ethers: application to the total synthesis of andrachcinidine. J Org Chem 72:7359–7366
137. Barluenga J, Calleja J, Mendoza A, Rodríguez F, Fañanás FJ (2010) Synthesis of polycyclic compounds by a cascade cycloisomerisation/Diels-Alder reaction. Chem Eur J 16:7110–7112
138. Yao X, Li C-J (2006) Water-triggered and gold(I)-catalyzed cascade addition/cyclization of terminal alkynes with ortho-alkynylaryl aldehyde. Org Lett 8:1953–1955
139. Hopkinson MN, Gee AD, Gouverneur V (2011) Au(I) /Au(III) catalysis: an alternative approach for C-C oxidative coupling. Chem Eur J 17:8248–8262
140. Hashmi ASK, Blanco MC, Fischer D, Bats JW (2006) Gold catalysis: evidence for the in-situ reduction of gold(III) during the cyclization of allenyl carbinols. Eur J Org Chem 1387–1389
141. Wegner HA, Ahles S, Neuburger M (2008) A new gold-catalyzed domino cyclization and oxidative coupling reaction. Chem Eur J 14:11310–11313
142. Zhang R, Xu Q, Chen K, Gu P, Shi M (2013) Gold-catalyzed cascade oxidative cyclization and arylation of allenoates. Eur J Org Chem 2013:7366–7371
143. Ball LT, Green M, Lloyd-Jones GC, Russell CA (2010) Arylsilanes: application to gold-catalyzed oxyarylation of alkenes. Org Lett 12:4724–4727
144. Hopkinson MN, Ross JE, Giuffredi GT, Gee AD, Gouverneur V (2010) Gold-catalyzed cascade cyclization – oxidative alkynylation of allenoates. Org Lett 12:4904–4907
145. Hopkinson MN, Tessier A, Salisbury A, Giuffredi GT, Combettes LE, Gee AD, Gouverneur V (2010) Gold-catalyzed intramolecular oxidative cross-coupling of nonactivated arenes. Chem Eur J 16:4739–4743
146. Hashmi ASK, Lothschütz C, Döpp R, Rudolph M, Ramamurthi TD, Rominger F (2009) Gold and palladium combined for cross-coupling. Angew Chem Int Ed 48:8243–8246
147. Shi Y, Roth KE, Ramgren SD, Blum SA (2009) Catalyzed catalysis using carbophilic Lewis acidic gold and Lewis basic palladium: synthesis of substituted butenolides and isocoumarins. J Am Chem Soc 131:18022–18023
148. Zhang LM (2014) A non-diazo approach to alpha-oxo gold carbenes via gold-catalyzed alkyne oxidation. Acc Chem Res 47:877–888
149. Ye L, Cui L, Zhang G, Zhang L (2010) Alkynes as equivalents of α-diazo ketones in generating α-oxo metal carbenes: a gold-catalyzed expedient synthesis of dihydrofuran-3-ones. J Am Chem Soc 132:3258–3259
150. Ye L, He W, Zhang L (2010) Gold-catalyzed one-step practical synthesis of oxetan-3-ones from readily available propargylic alcohols. J Am Chem Soc 132:8550–8551
151. Wang Y, Ji K, Lan S, Zhang L (2012) Rapid access to chroman-3-ones through gold-catalyzed oxidation of propargyl aryl ethers. Angew Chem Int Ed 51:1915–1918
152. Yeom H-S, Shin S (2014) Catalytic access to alpha-oxo gold carbenes by n-o bond oxidants. Acc Chem Res 47:966–977
153. Shu C, Liu R, Liu S, Li J-Q, Yu Y-F, He Q, Lu X, Ye L-W (2015) Practical, modular, and general synthesis of 3-coumaranones through gold-catalyzed intermolecular alkyne oxidation strategy. Chem Asian J 10:91–95
154. Wang T, Shi S, Rudolph M, Hashmi ASK (2014) Synthesis of fully substituted 3-formyl-4-iodofurans via a gold(i)-catalyzed oxidation/1,2-alkynyl migration/cyclization/iodination cascade. Adv Synth Catal 356:2337–2342
155. Wang T, Huang L, Shi S, Rudolph M, Hashmi ASK (2014) Synthesis of highly substituted n-(furan-3-ylmethylene)benzenesulfonamides by a gold(i)-catalyzed oxidation/1,2-alkynyl migration/cyclization cascade. Chem Eur J 20:14868–14871
156. Wang T, Shi S, Hansmann MM, Rettenmeier E, Rudolph M, Hashmi ASK (2014) Synthesis of highly substituted 3-formylfurans by a gold(I)-catalyzed oxidation/1,2-alkynyl migration/cyclization cascade. Angew Chem Int Ed 53:3715–3719
157. Marion N, Nolan SP (2007) Propargylic esters in gold catalysis: access to diversity. Angew Chem Int Ed 46:2750–2752

158. Correa A, Marion N, Fensterbank L, Malacria M, Nolan SP, Cavallo L (2008) Golden carousel in catalysis: the cationic gold/propargylic ester cycle. Angew Chem Int Ed 47: 718–721
159. Buzas A, Istrate F, Gagosz F (2006) Gold(I)-catalyzed stereoselective formation of functionalized 2,5-dihydrofurans. Org Lett 8:1957–1959
160. De Brabander JK, Liu B, Qian MX (2008) Au(I)- and Pt(II)-catalyzed cyclotherification of omega-hydroxy propargylic esters. Org Lett 10:2533–2536
161. Wang YM, Kuzniewski CN, Rauniyar V, Hoong C, Toste FD (2011) Chiral (acyclic diaminocarbene)gold(I)-catalyzed dynamic kinetic asymmetric transformation of propargyl esters. J Am Chem Soc 133:12972–12975
162. Kotikalapudi R, Swamy KCK (2013) Gold(I) catalysed cycloisomerisation of beta-hydroxy propargylic esters to dihydropyrans/2H-pyrans via allene intermediates. Tetrahedron 69:8002–8012
163. Uemura M, Watson IDG, Katsukawa M, Toste FD (2009) Gold(I)-catalyzed enantioselective synthesis of benzopyrans via rearrangement of allylic oxonium intermediates. J Am Chem Soc 131:3464–3465
164. Jiménez-Nuñez E, Claverie CK, Nieto-Oberhuber C, Echavarren AM (2006) Prins cyclizations in Au-catalyzed reactions of enynes. Angew Chem Int Ed 45:5452–5455
165. Molawi K, Delpont N, Echavarren AM (2010) Enantioselective synthesis of (-)-englerins A and B. Angew Chem Int Ed 49:3517–3519
166. Jiménez-Nuñez E, Molawi K, Echavarren AM (2009) Stereoselective gold-catalyzed cycloaddition of functionalized ketoenynes: synthesis of (+)-orientalol F. Chem Commun 7327–7329
167. Jiménez-Nuñez E, Raducan M, Lauterbach T, Molawi K, Solorio CR, Echavarren AM (2009) Evolution of propargyl ethers into allylgold cations in the cyclization of enynes. Angew Chem Int Ed 48:6152–6155
168. Shu X-Z, Liu X-Y, Ji K-G, Xiao H-Q, Liang Y-M (2008) Au-catalyzed tandem cyclization/[1,2]-alkyl migration reaction of epoxyalkynes: synthesis of spiropyranones. Chem Eur J 14: 5282–5289
169. Lin GY, Li CW, Hung SH, Liu RS (2008) Diversity in gold- and silver-catalyzed cycloisomerization of epoxide-alkyne functionalities. Org Lett 10:5059–5062
170. Lin CC, Teng TM, Tsai CC, Liao HY, Liu RS (2008) Gold-catalyzed deoxygenative Nazarov cyclization of 2,4-dien-1-als for stereoselective synthesis of highly substituted cyclopentenes. J Am Chem Soc 130:16417–16423
171. Yang CY, Lin MS, Liao HH, Liu RS (2010) Diversity of products in the gold-catalyzed cyclization of 1-epoxy-1-alkynylcyclopropanes by using 1-oxyallyl cations. Chem Eur J 16: 2696–2699
172. Liao HH, Liu RS (2011) Effects of haloniums on gold-catalyzed ring expansion of 1-oxiranyl-1-alkynylcyclopropanes. Chem Commun 47:1339–1341
173. Zhang XM, Tu YQ, Jiang YJ, Zhang YQ, Fan CA, Zhang FM (2009) Tandem reactions of cis-2-acyl-1-alkynyl-1-aryl cyclopropanes tuned by gold(I) and silver(I) catalysts: efficient synthesis of pyran-fused indene cores and 2,4,6-trisubstituted phenols. Chem Commun 4726–4728
174. Binder JT, Crone B, Kirsch SF, Liébert C, Menz H (2007) Synthesis of heterocyclic systems by transition-metal-catalyzed cyclization-migration reactions – a diversity-oriented strategy for the construction of spirocyclic 3(2H)-furanones and 3-pyrrolones. Eur J Org Chem 2007: 1636–1647
175. Wegener M, Huber F, Bolli C, Jenne C, Kirsch SF (2015) Silver-free activation of ligated gold(I) chlorides: the use of $[Me_3NB_{12}Cl_{11}]^-$ as a weakly coordinating anion in homogeneous gold catalysis. Chem Eur J 21:1328–1336
176. Crone B, Kirsch SF (2007) Synthesis of 4-iodo-3-furanones utilizing electrophile-induced tandem cyclization/1,2-migration reactions. J Org Chem 72:5435–5438

177. Zhang J, Schmalz HG (2006) Gold(I)-catalyzed reaction of 1-(1-alkynyl)-cyclopropyl ketones with nucleophiles: a modular entry to highly substituted furans. Angew Chem Int Ed 45:6704–6707
178. Fang R, Su CY, Zhao CY, Phillips DL (2009) DFT study on the mechanism and regioselectivity of gold(I)-catalyzed synthesis of highly substituted furans based on 1-(1-alkynyl)cyclopropyl ketones with nucleophiles. Organometallics 28:741–748
179. Li W, Yuan W, Pindi S, Shi M, Li GG (2010) Au/Ag-catalyzed intramolecular ring-opening of vinylidene-cyclopropanes (VDCPs): an easy Access to functional tetrahydropyrans. Org Lett 12:920–923
180. Peng B, Maulide N (2013) The redox-neutral approach to C-H functionalization. Chem Eur J 19:13274–13287
181. Jurberg ID, Odabachian Y, Gagosz F (2010) Hydroalkylation of alkynyl ethers via a gold(I)-catalyzed 1,5-hydride shift/cyclization sequence. J Am Chem Soc 132:3543–3552
182. Vadola PA, Sames D (2009) C-H bond functionalization via hydride transfer: direct coupling of unactivated alkynes and sp(3) C-H bonds catalyzed by platinum tetraiodide. J Am Chem Soc 131:16525–16528
183. Bolte B, Gagosz F (2011) Gold and brønsted acid catalyzed hydride shift onto allenes: divergence in product selectivity. J Am Chem Soc 133:7696–7699
184. López F, Mascareñas JL (2011) Recent developments in gold-catalyzed cycloaddition reactions. Beilstein J Org Chem 7:1075–1094
185. López F, Mascareñas JL (2013) Gold(I)-catalyzed enantioselective cycloaddition reactions. Beilstein J Org Chem 9:2250–2264
186. Garayalde D, Nevado C (2012) Gold-containing and gold-generated 1, n-dipoles as useful platforms toward cycloadditions and cyclizations. ACS Catal 2:1462–1479
187. Teng TM, Das A, Huple DB, Liu RS (2010) Gold-catalyzed stereoselective synthesis of 9-oxabicyclo[3.3.1]nona-4,7-dienes from diverse 1-oxo-4-oxy-5-ynes: a viable formal [4 + 2] cycloaddition on an s-trans-heterodiene framework. J Am Chem Soc 132:12565–12567
188. Teng TM, Liu RS (2010) Gold-catalyzed stereocontrolled oxacyclization/[4 + 2]-cycloaddition cascade of ketone-allene substrates. J Am Chem Soc 132:9298–9300
189. Zhang G, Zhang L (2008) Au-containing all-carbon 1,3-dipoles: generation and [3 + 2] cycloaddition reactions. J Am Chem Soc 130:12598–12599
190. Gao H, Wu X, Zhang J (2010) *Exo*/*endo* selectivity-control in Lewis-acid catalyzed tandem heterocyclization/formal [4 + 3] cycloaddition: synthesis of polyheterocycles from 2-(1-alkynyl)-2-alken-1-ones and 1,3-diphenylisobenzofuran. Chem Commun 46:8764–8766
191. Zhang G, Huang X, Li G, Zhang L (2008) Au-containing all-carbon 1,4-dipoles: generation and [4 + 2] annulation in the formation of carbo-/heterocycles. J Am Chem Soc 130: 1814–1815
192. Escribano-Cuesta A, Lopez-Carrillo V, Janssen D, Echavarren AM (2009) Gold-catalyzed reactions of 1,5-and 1,6-enynes with carbonyl compounds: cycloaddition vs. metathesis. Chem Eur J 15:5646–5650
193. Schelwies M, Dempwolff AL, Rominger F, Helmchen G (2007) Gold-catalyzed intermolecular addition of carbonyl compounds to 1,6-enynes. Angew Chem Int Ed 46: 5598–5601
194. Zhou Q, Chen X, Ma D (2010) Asymmetric, protecting-group-free total synthesis of (-)-englerin A. Angew Chem Int Ed 49:3513–3516
195. Obradors C, Echavarren AM (2013) Intermolecular gold-catalyzed cycloaddition of alkynes with oxoalkenes. Chem Eur J 19:3547–3551
196. Faustino H, Alonso I, Mascareñas JL, López F (2013) Gold(I)-catalyzed cascade cycloadditions between allenamides and carbonyl-tethered alkenes: an enantioselective approach to oxa-bridged medium-sized carbocycles. Angew Chem Int Ed 52:6526–6530
197. Faustino H, Varela I, Mascareñas JL, López F (2015) Gold(I)-catalyzed 2 + 2 + 2 cycloaddition of allenamides, alkenes and aldehydes: a straightforward approach to tetrahydropyrans. Chem Sci 6:2903–2908

198. Li CW, Lin GY, Liu RS (2010) Construction of 2,3-dihydrofuran cores through the [3 + 2] cycloaddition of gold alpha-carbonylcarbenoids with alkenes. Chem Eur J 16:5803–5811
199. Teng TM, Lin MS, Vasu D, Bhunia S, Liu TA, Liu RS (2010) Gold-catalyzed dealkoxylative carbocyclization/[3 + 3] annulation cascade of acetal-allene or ketal-allene substrates. Chem Eur J 16:4744–4748
200. Kramer S, Skrydstrup T (2012) Gold-catalyzed carbene transfer to alkynes: access to 2,4-disubstituted furans. Angew Chem Int Ed 51:4681–4684
201. Huang XL, Peng B, Luparia M, Gomes LFR, Veiros LF, Maulide N (2012) Gold-catalyzed synthesis of furans and furanones from sulfur ylides. Angew Chem Int Ed 51:8886–8890
202. Yeom H-S, Koo J, Park H-S, Wang Y, Liang Y, Yu Z-X, Shin S (2012) Gold-catalyzed intermolecular reactions of propiolic acids with alkenes: 4 + 2 annulation and enyne cross metathesis. J Am Chem Soc 134:208–211
203. Trillo B, López F, Montserrat S, Ujaque G, Castedo L, Lledós A, Mascareñas JL (2009) Gold-catalyzed [4C + 3C] intramolecular cycloaddition of allenedienes: synthetic potential and mechanistic implications. Chem Eur J 15:3336–3339
204. Gung BW, Craft DT (2009) Gold-catalyzed transannular [4 + 3] cycloaddition reactions. Tetrahedron Lett 50:2685–2687
205. Gung BW, Craft DT, Bailey LN, Kirschbaum K (2010) Gold-catalyzed transannular [4 + 3] cycloaddition reactions. Chem Eur J 16:639–644
206. Yang JM, Tang XY, Shi M (2015) Gold(I)-catalyzed intramolecular cycloisomerization of propargylic esters with furan rings. Chem Eur J 21:4534–4540
207. Alonso I, Trillo B, López F, Montserrat S, Ujaque G, Castedo L, Lledós A, Mascareñas JL (2009) Gold-catalyzed 4C + 2C cycloadditions of allenedienes, including an enantioselective version with New phosphoramidite-based catalysts: mechanistic aspects of the divergence between 4c + 3c and 4c + 2c pathways. J Am Chem Soc 131:13020–13030
208. Mauleón P, Zeldin RM, González AZ, Toste FD (2009) Ligand-controlled access to [4 + 2] and [4 + 3] cycloadditions in gold-catalyzed reactions of allene-dienes. J Am Chem Soc 131: 6348–6349

Top Heterocycl Chem (2016) 46: 53–86
DOI: 10.1007/7081_2015_5001

Published online: 21 November 2015

Gold-Catalyzed Synthesis of Nitrogen Heterocyclic Compounds via Hydroamination Reactions

Antonio Arcadi

Abstract Gold catalysis is a very attractive tool for the synthesis of nitrogen heterocyclic compounds by means of hydroamination reactions. This chapter intends to present selected examples of more recent developments in this field which highlight relevant progresses in terms of control of the chemo-, regio-, and stereoselectivity of hydroamination processes in their applications to the synthesis of the corresponding heterocyclic derivatives by means of gold catalysis.

Keywords Asymmetric catalysis • Gold • Hydroamination • Nitrogen heterocyclic compounds • Sequential processes

Contents

A. Arcadi (✉)
Dipartimento di Scienze Fisiche e Chimiche, Università di L'Aquila, Via Vetoio, Coppito, AQ 67010, Italy
e-mail: antonio.arcadi@univaq.it

1 Introduction

The development of general and convenient methods for the addition of N–H bond across a C–C multiple bond (hydroamination [HA] reaction) represents a wide area of interest in organic synthesis because of the relevance of nitrogen-containing compounds, particularly nitrogen heterocycles, as common components of pharmaceuticals and important intermediates in a number of industrial processes (for general reviews about the hydroamination reaction of multiple bonds, see [1, 2]). Despite its simplicity, the HA reaction suffers from a high activation barrier due to electrostatic repulsions between the electron density of the multiple bond and the nitrogen lone pair. Commonly, catalytic methods based on transition metal complexes are used to overcome the activation energy. A myriad of new transition metal-catalyzed HA processes appeared in the literature (for general reviews about the transition metal-catalyzed hydroamination reaction of multiple bonds, see [3–8]). Among them, HA reactions catalyzed by gold are of growing potential in organic synthesis (for general reviews about gold-catalyzed hydroamination reaction of multiple bonds, see [3, 9–14]). Recently, the impact of gold catalysis in organic synthesis has been very impressive [15–42]. This booming of gold catalysis, also, truly revolutioned the area of heterocyclic chemistry (for reviews on gold catalysis applied to the synthesis of heterocycles, see [43–48]). Many research groups have focused on the achievement of better functional group compatibilities and greater levels of molecular complexity for the synthesis of densely functionalized heterocyclic cores under mild conditions trough new developments of gold-catalyzed HA reactions. Although the addition of N–H bond across a C–C multiple bond shows the inherent benefit of optimal atom economy, the control of the chemo-, regio-, and stereoselectivity of the process still encounters many challenges. This chapter examines a selection of more recent developments that put in evidence relevant progress in the application of the gold catalysis to the synthesis of nitrogen heterocyclic compounds by means of HA reactions.

2 Synthesis of Nitrogen Heterocyclic Compounds Through Gold-Catalyzed Intramolecular Hydroamination of Alkynes

Since the first report by Utimoto in 1987 intramolecular HA of 5-yn-1-amines delivering tetrahydropyridines [49, 50], investigation of gold-catalyzed alkyne HA triggered cyclizations as a powerful tool for the construction of biologically important structural motifs has been growing rapidly [51]. The formation of five- and six-membered rings occurs efficiently, whereas the construction of seven-membered rings is more challenging. Intensive research activities have been directed to address the selective *5-endo-dig* gold-catalyzed cyclization of alkyne–amine derivatives to give pyrroles, pyrrolidines, and pyrrolines, important

Scheme 1 Gold-catalyzed dehydrative HA of acetylenic vicinal amino alcohols

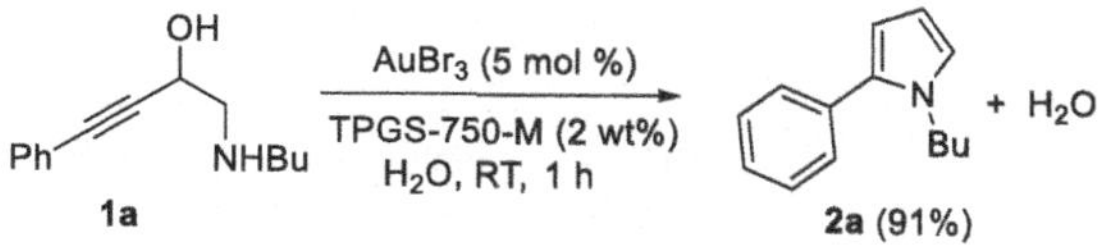

Scheme 2 Enantio- and diastereoselective synthesis of 4-hydroxy-2-pyrrolines

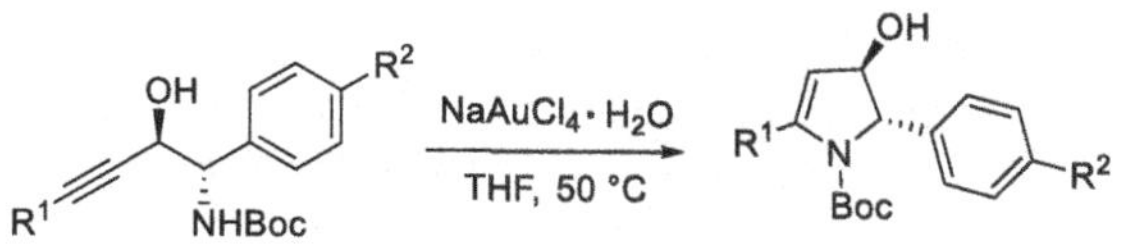

structural motifs in a wide array of natural products and pharmaceutical relevant molecules. Advantageously, gold-catalyzed dehydrative cyclizations of acetylenic amino alcohols **1** to the corresponding pyrroles **2** could take place by means of nanotechnology in an aqueous medium. Indeed, nanomicelles composed of the commercial available surfactant TPGS-750-M enabled the gold-catalyzed annulation reaction to proceed at ambient temperatures in high isolated yields (Scheme 1) [52].

The dehydration reaction was conveniently circumvented in the *5-endo-dig* gold-catalyzed cyclization of acetylene-containing amino alcohols **3** which achieved a facile enantio- and diastereoselective route to aryl-substituted 4-hydroxy-2-pyrrolines **4** in excellent yields (Scheme 2) [53].

Ye developed an enantioselective approach to various 2,3-dihydropyrroles **6** via a *5-endo-dig* gold-catalyzed cycloisomerization of chiral homopropargyl sulfonamides **5** [54]. The *5-endo-dig* intramolecular HA of terminal alkynes is more challenging because it involves an anti-Markovnikov addition. A Markovnikov regioselectivity is more usual for the gold-catalyzed addition to a terminal alkyne. The cyclization proceeded in an open flask by using as additives two catalytic bases, which completely inhibits the formation of dimer by-products (Scheme 3).

Gold catalysis enables the nucleophilic attack not only of primary or secondary amines and amides to the C–C triple bond but also of various other nitrogen nucleophiles. The $IPrAuNTf_2/HNTf_2$ co-catalyzed cyclization of *N*-(2-perfluoroalkyl-3-alkynyl)hydroxylamines **7** produced pyrroles **8** in moderate to excellent yield [55]. Interestingly, by a suitable choice of the catalytic system, a divergent process led to cyclic nitrones **9** in high yield. The coordination of the triple bond of **7** to the metal allowed the *5-endo-dig* closure from the nitrogen atom attack rather than the oxygen to afford the intermediate **11**. In the case of $IPrAuNTf_2$ as the catalyst and $HNTf_2$ as the co-catalyst, the gold-/$HNTf_2$-mediated dehydration of the intermediate **11** led to the iminium species **13**. Subsequent deprotonation/protodeauration delivered the pyrrole product and regenerated the catalyst. Only the protonation/protodemetalation process of **11** occurred in the presence of the silver catalyst to give the cyclic nitrone after the isomerization of the *N*-hydroxyl enamine intermediate **12** (Scheme 4).

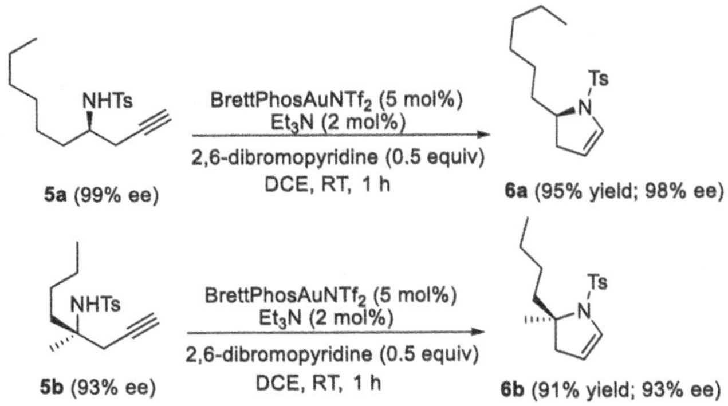

Scheme 3 Gold-catalyzed *5-endo-dig* cycloisomerization of homopropargyl sulfonamides

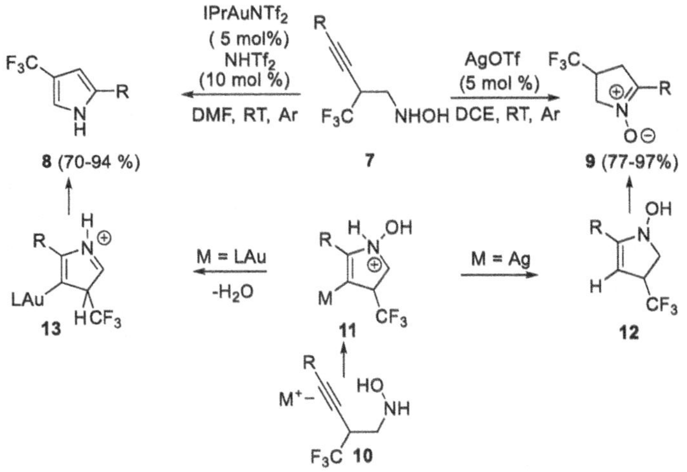

Scheme 4 Catalyst-controlled cyclization of *N*-(2-perfluoroalkyl-3-alkynyl) hydroxylamines

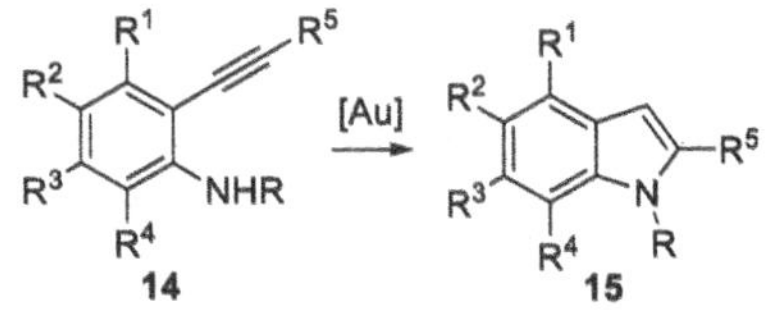

Scheme 5 Gold-catalyzed cycloisomerization of 2-alkynylaniline derivatives

The *5-endo-dig* gold-catalyzed intramolecular HA of propargyl hydrazides into pyrazoles was also described [56]. The 2-alkynylaniline derivatives **14** have been extensively employed to prepare indole **15** by means of their *5-endo-dig* gold-catalyzed cycloisomerization reaction (Scheme 5) [57, 58]. Both inorganic gold salts and organic gold coordination complexes have been used as the catalysts in this transformation. Mechanistic investigation of the activation modes gives

Scheme 6 Gold-catalyzed cycloisomerization of *N*-benzyl-3-alkynyl-5-arylpyridin(or pyrazin)-2-yl amines

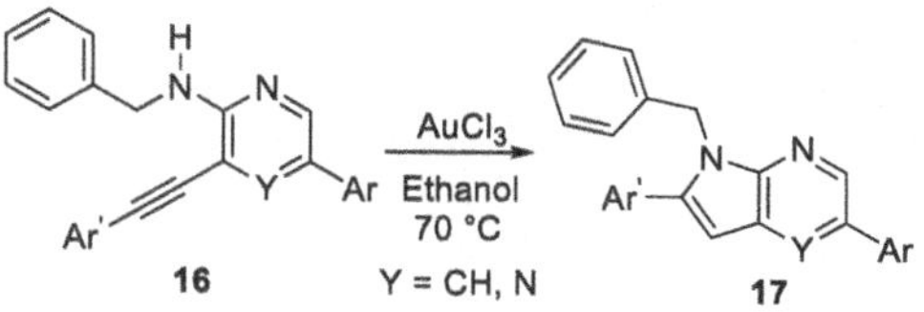

Fig. 1 Two possible limit structures for the substrate–catalyst RC

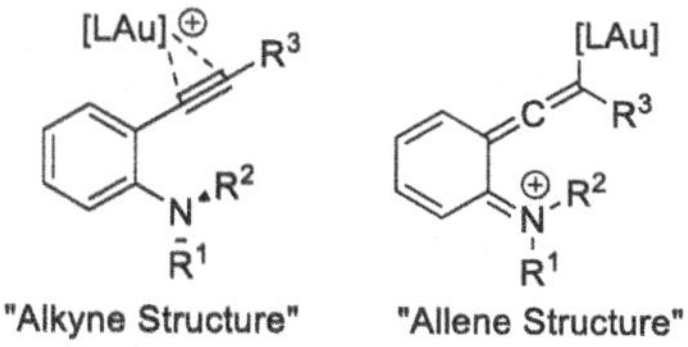

support to the activation of the internal alkyne moiety of the 2-alkynylanilines by π-coordination with the gold catalyst [59, 60].

More recently, the extension to the synthesis of pyrrolo[2,3-*b*]pyridine and pyrrolo[2,3-*b*]pyrazine scaffolds **17** has been achieved by the gold-catalyzed cycloisomerization of the corresponding *N*-benzyl-3-alkynyl-5-arylpyridin (or pyrazin)-2-yl amines **16** with $AuCl_3$ (Scheme 6) [61].

Moreover, the ligand electronic effect in the gold(I)-catalyzed intramolecular alkyne HA was analyzed through a DFT and charge-displacement function (CDF) study [62]. The CDF approach provided a very interesting tool to quantify the π-acidity of the gold catalyst. The detailed CDF analysis of the electronic effect of the ligand on the $L[Au]^+$-catalyzed HA of *N,N*-dimethyl-2-(methylethynyl)aniline indicates that the charge transfer from the alkyne to the $[LAu]^+$ fragment depends on the electron donating power of L. A quantitative relationship has been reported between the acidity of $[LAu]^+$ and the geometry of the reactant complex (RC) which varies along a continuum between two limit structures which are referred as an "alkyne structure (η^2 complex)" and an "allene structure (η^1 complex)" for different ligands (Fig. 1).

The enhancement of the metal fragment acidity by decreasing the electron donating power of L determines an increase of substrate activation power; conversely, the effect of the progressive shift of the reactant complex toward the allene structure produces its deactivation; the reactivity both of the nucleophile and of the site of attack is diminished in the η^1 gold complex. Indeed, carbene-type ligands have been reported to achieve a compromise between the two conflicting effects on the reaction barrier and thus on the catalyst effectiveness [63]. Furthermore, "expanded-ring" six- and seven-membered-ring carbenes (er-NHC_s) have been supposed to be superior to the five-membered-ring carbene complexes. Then, a series of (er-NHC_s) gold(I) complexes have been synthesized, and a comparative study of the catalytic activities of a variety of five-, six-, and seven-membered-ring carbene complexes [(NHC)AuX], [$(PPh_3)AuX$], [$(Me_2S)AuX$], and inorganic compounds of gold in model reactions of indole and benzofuran synthesis was carried out. Quantitative yields of the cyclized products were attained at room temperature at 1 mol% catalyst loadings by using the cationic [(1,3-bis(2,6-diisopropylphenyl)-

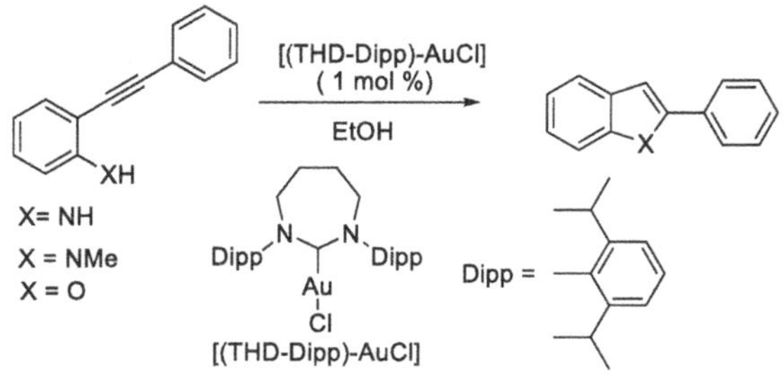

Scheme 7 High activity in π-acid-catalyzed cyclizations of expanded-ring *N*-heterocyclic carbene (er-NHC) gold(I) complexes

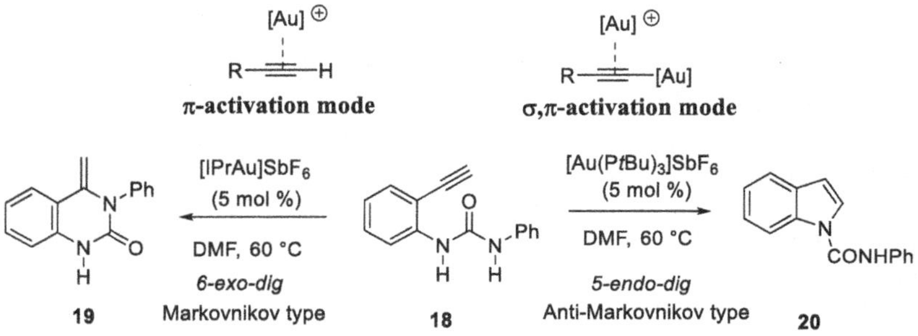

Scheme 8 Selectivity control of the intramolecular HA reaction of urea derivatives

3,4,5,6-tetrahydrodiazepin-2-ylidene gold(I))]tetrafluoborate [(THD)-Dipp)Au] BF_4 (Scheme 7). The implementation of the procedure on a multi-gram scale virtually in quantitative yields in less than 1 h supports the assumption that (er-NHC_s) gold(I) complexes can be particularly suitable for the intramolecular HA and hydrophenoxylation of alkynes under mild, aerobic conditions [64].

Clear evidences that binding of two gold atoms to unsaturated substrates plays an important role in catalysis may offer powerful tools for the chemo- and the regioselective control of intramolecular HA reactions [65–68]. The activation of terminal alkynes by formation of σ–π complexes is well established [69]. The study by Asensio of competitive gold-activation π and σ–π modes of 1(o-ethynyl)urea derivatives **18** showed the significant influence of the ancillary ligand on the selectivity of the intramolecular HA reaction (Scheme 8) [70]. The [Au(IPr)]SbF_6 (IPr = 2,6-bis(diisopropylphenyl)imidazol-2-ylidene) catalyst achieved exclusively the formation of the *6-exo-dig N*-3-attack product (>95%) **19**, whereas [Au(P*t*Bu)$_3$] SbF_6 led to the indole **20** in 80% yield. The formation of an equimolecular amount of **19** and **20** was observed with the more common [Au(PPh)$_3$]SbF_6.

The reversal of the selectivity in the presence of 10 mol% of Et_3N for the [Au (IPr)]SbF_6-catalyzed heterocyclization process suggested the possibility of a gold acetylide species as an intermediate in the *5-endo-dig* ring closure of **18**. Therefore, the σ-gold complex **21** was prepared (Scheme 9). A complete conversion of **21** into the two-metaled indole derivative **22** occurred in the presence of a catalytic amount

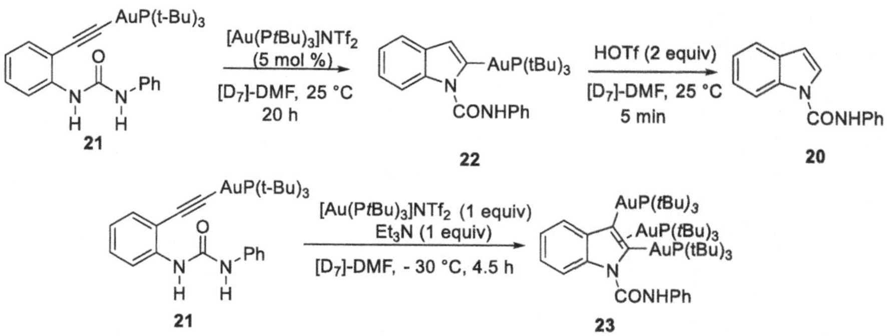

Scheme 9 Synthesis of 2-aurated- and polyaurated indole species

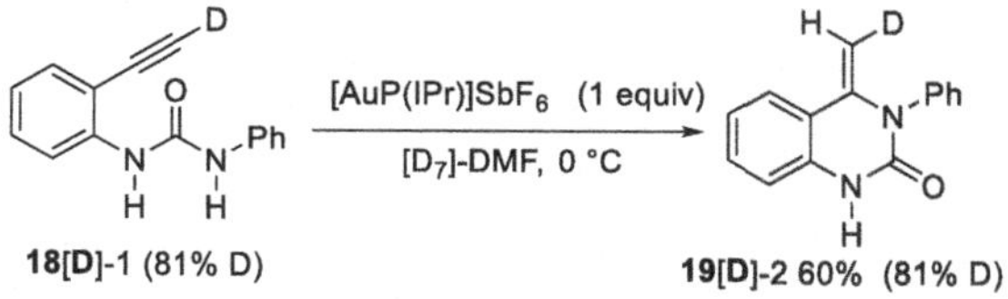

Scheme 10 Deuterium-labeling experiment in the *6-exo-dig* cyclization of urea **18**

of $[Au(PtBu)_3]NTf_2$. Of note, a low-temperature (−30°C) reaction involving acetylide **21** with $[Au(PtBu)_3]NTf_2$ in the presence of Et_3N to prevent protodeauration led to the polyaurated species **23** identified by NMR spectroscopy and MALDI spectrometry.

Labeling experiments ascertained that the formation of the *6-exo-dig* product **19** derived by the faster π-activation of the starting alkyne by the $[Au(IPr)]^+$ cation (Scheme 10). The role of highly hindered ligands with high % V_{buried} value on the selective outcome of the reaction was elucidated. Studies of dispersion and metal effects in the binding of gold cations to alkynyl ligands are emerging at the forefront of current research [71].

Alkynyl compounds bearing proximate ambident nucleophiles are very suitable candidates to explore the different cyclization pathways and activation modes involved in gold catalysis. Triphenylphosphine, low temperature, and the presence of base favored the σ,π activation mode to give selectively the hydroamination products **25** in the heterocyclization reaction of propargylic thioureas **24** (Scheme 11). The selective formation of the S-cyclization products **26** by the classical π-activation mode occurred by the choice of $[Me_4tBuXPhosAuNTf_2]$ as the catalyst [72].

Unsymmetrical ammonium salt-tagged NHC–gold(I) complexes have been applied as recyclable catalysts in water in *5-exo-dig* gold-catalyzed cyclization reactions of acetylenic amides to lactams [73]. Selective *5-exo-dig* gold-catalyzed cyclizations found application for the synthesis of 2-fluoroalkyl-5-methyl imidazoles from propargyl amidines [74]. Moreover, *5-exo-dig* cyclization of chiral α-hydrazino esters **27** was favored over *6-endo-dig* process, and aza-proline derivatives **28** were isolated in good yields without epimerization at the stereogenic center (Scheme 12) [75].

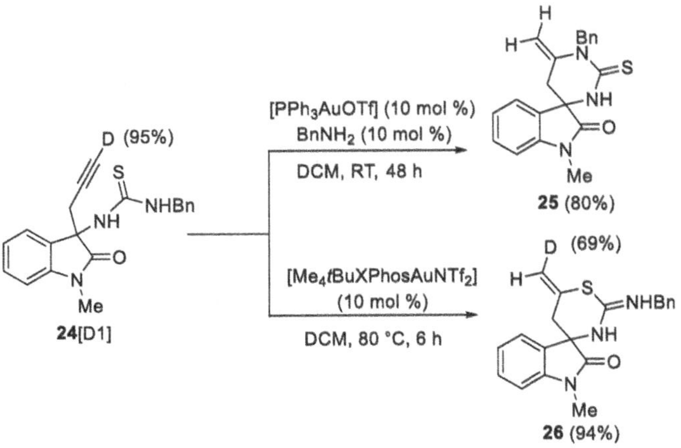

Scheme 11 Selective gold(I)-catalyzed intramolecular heterocyclization of propargylic thioureas

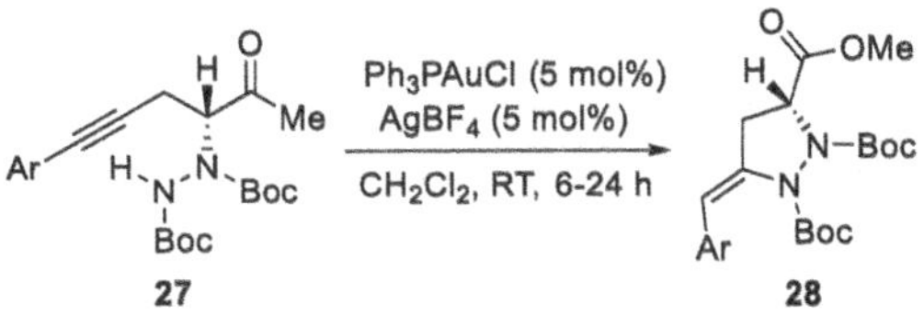

Scheme 12 Synthesis of chiral aza-proline derivatives

Unexpected regio- and chemoselective *6-endo-dig* intramolecular HA of the in situ generated phenyl isocyanate-derived propargyl ureas **29** gave access to tetrasubstituted 3,4-dihydropyrimidin-2(1*H*)-ones **30** (Scheme 13) [76].

This investigation complemented previous studies aimed to determine the main effects which direct the chemo- and regioselective outcome of Ag- versus Au-catalyzed transformations [77, 78]. Thermodynamic/kinetic factors more than the catalyst features might determine the nature of the reaction products. Related example of chemo- and regioselective *6-endo N-* versus *O*-cyclization reactions of the enyne–urea system **31** for the synthesis of annulated indole derivatives **32** versus **33** has been explained in terms of counterion effect on Au-/Ag-catalyzed regiodivergent pathways (Scheme 14) [79].

A general catalytic two-step sequence afforded a variety of complex indoloquinolizines. The Pictet–Spengler products **34** underwent Au(I) regioselective *6-endo-dig* intramolecular HA to give the indoloquinolizines **35**. The preference of the *6-endo-dig* over a *5-exo-dig* cyclization was attributed to the high ring strain in the spiro-products derived from the latter process (Scheme 15) [80].

An operationally simpler approach involving [Au(III])-catalyzed regioselective *6-endo-dig* ring closure led to the high diastereoselective synthesis of thiazolo fused naphthyridines **37** by the domino reaction of suitable *o*-alkynylaldehydes with L-cysteine methyl ester hydrochloride (Scheme 16) [81].

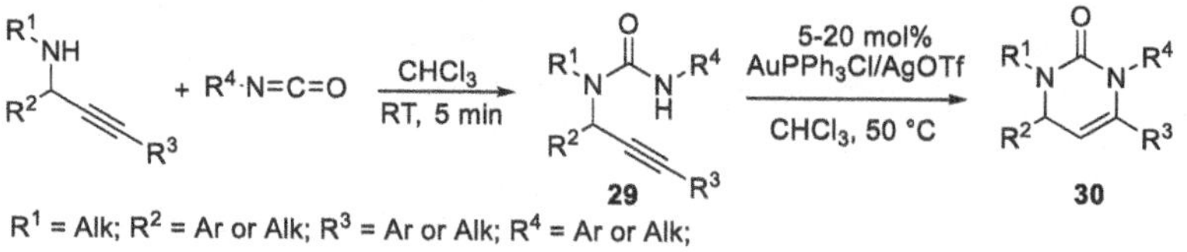

Scheme 13 Cationic gold-catalyzed protocol for the *6-endo-dig* *N*-cycloisomerization of propargyl ureas

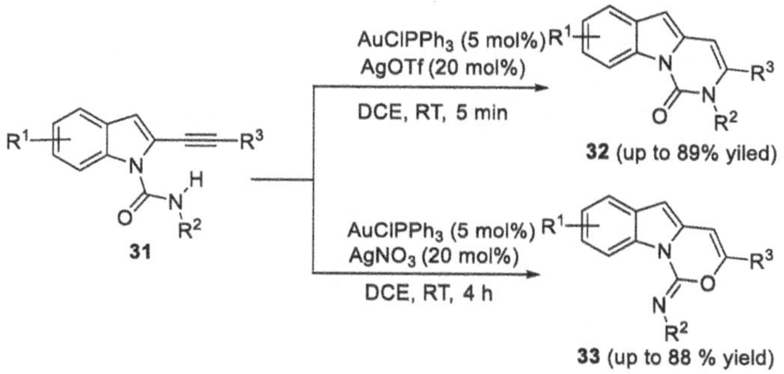

Scheme 14 Chemoselective *N*- vs. *O*-cyclization of enyne–urea systems

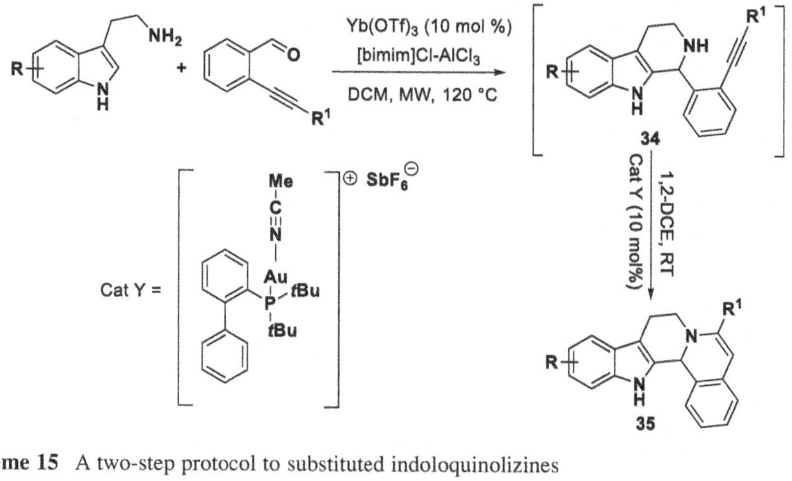

Scheme 15 A two-step protocol to substituted indoloquinolizines

A mild atom economy alternative to the classic quinolone synthesis was achieved by means of $Ph_3PAuNTf_2$-catalyzed *6-endo-dig* cyclization of alkyl- or aryl-substituted 1-(*o*-aminophenyl)-2-propyn-1-ones **38**. DFT studies suggest that the reaction outcome involve the π-activation of the carbon–carbon triple bond rather than a conjugate addition-type reaction (Scheme 17) [82].

Highly reusable gold nanoparticles supported on Al-SBA15 in combination with microwave irradiation provided a green methodology for synthesizing

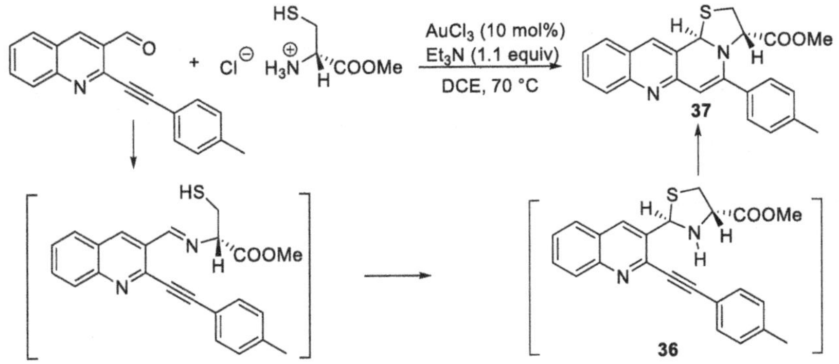

Scheme 16 Gold-catalyzed stereoselective synthesis of thiazolo fused naphthyridines

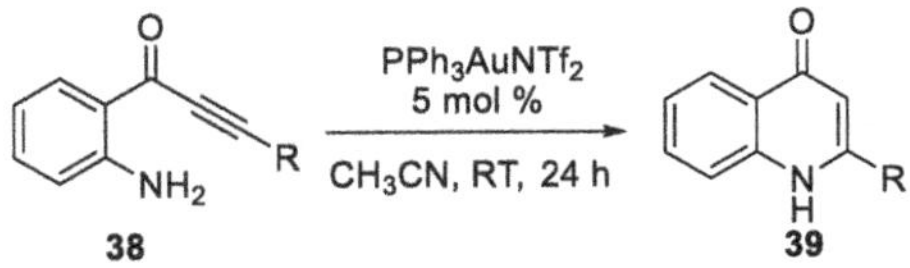

Scheme 17 Gold-catalyzed synthesis of 2-substituted-4-quinolones

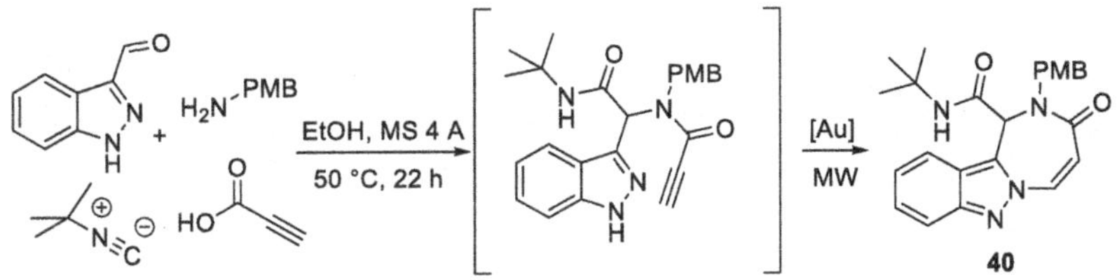

Scheme 18 Supported gold nanoparticles as efficient and reusable heterogeneous catalysts in one-pot sequential reactions

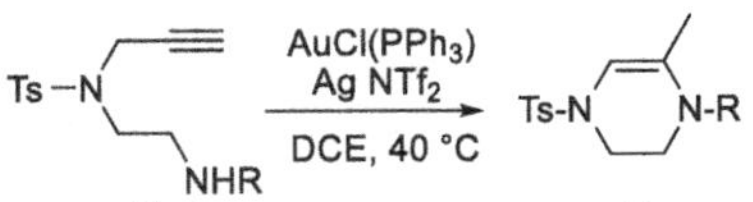

Scheme 19 Gold-catalyzed *6-exo-dig* HA of alkynylamines

unprecedented seven-membered heterocycles **40** through an intramolecular *7-endo-dig* cyclization process as the key step. The one-pot cascade reaction commenced with the Ugi reaction (Scheme 18) [83].

An efficient synthetic route to piperazine derivatives **42** was attained via gold-catalyzed regioselective *6-exo* HA of suitable alkynylamines **41** (Scheme 19) [84]. Deuterium-labeling experiments gave support to the mechanism involving the cascade cyclization/isomerization process. A similar procedure was also applied

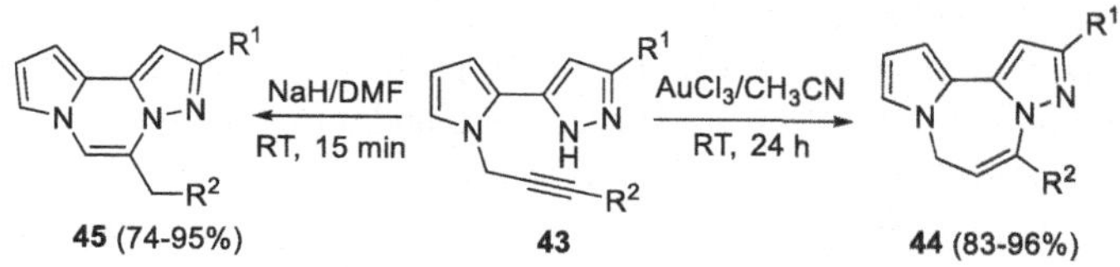

Scheme 20 Gold-catalyzed intramolecular *7-endo-dig* HA vs. *6-exo-dig* base-promoted cyclization of pyrazoles with alkyne units

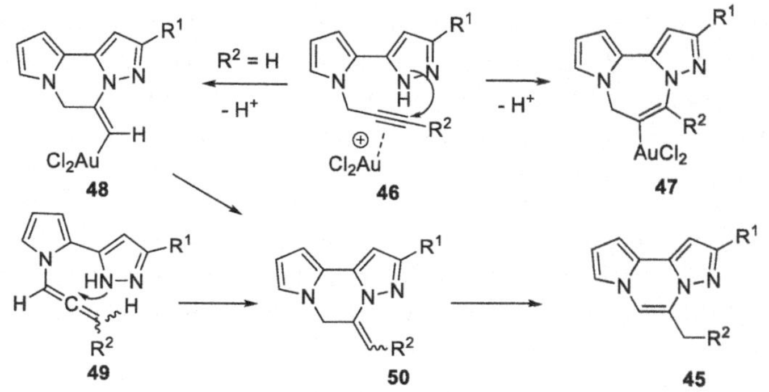

Scheme 21 Proposed reaction mechanism for the gold-catalyzed and the base-promoted cyclization reactions of pyrazoles with alkyne units

to the synthesis of a variety of lead-like piperazine scaffolds in the presence of AuCl $(PPh_3)/AgSbF_6$ in dioxane under microwave irradiation [85].

Hitherto unknown pyrazolo-pyrrolo-diazepine scaffolds **44** have been isolated in high yields at room temperature by an exclusive gold-catalyzed *7-endo-dig* cyclization reaction. The same starting substrates **43** led to the selective formation of pyrazolo-pyrrolo-pyrazine derivatives **45** under basic conditions (Scheme 20) [86].

The regioselective outcome of the reaction is subsequent to the π-activation of the alkyne unit by gold which determines the nucleophilic attack over the more positive charged C_{sp} in the π-gold complex **46**. Indeed, with the starting substrates bearing a terminal alkyne moiety, only the *6-exo-dig* heterocyclization products **45** have been isolated. These latter derivatives have been, also, isolated under the basic reaction conditions. Very likely, the base-promoted isomerization of the alkyne to give the corresponding allene **49** is responsible for the exclusive formation of the pyrazolo-pyrrolo-pyrazine derivatives **45** by the regioselective attack over the more electropositive central carbon atom in the allene moiety (Scheme 21).

3 Gold-Catalyzed Annulation Reactions of Aminoallenes

The gold-catalyzed allene HA was initially investigated by the Krause, Yamamoto, and Widenhoefer groups [87–89]. Aminoallenes represent versatile building blocks for the construction of a variety of different-sized nitrogen heterocycles

[90–92]. Their annulation reactions occur under several conditions. However, metal-catalyzed aminocyclizations achieve in most of the cases better regio- and stereoselectivity control, which is not accessible under thermal or basic conditions. A prominent review on the pivotal role to access to carbo- and heterocycles by means of gold-catalyzed cyclization of functionalized allenes has been reported [10]. Insights into the reaction mechanisms of the gold-catalyzed transformations of allenes which are based on computational studies, labeling studies, the detections of intermediates, chirality transfer, and diastereoselective product formation have been, also, summarized [93]. Chiral products were accessed in a stereoselective manner either from chiral allenes by axis-to-center chirality transfer or from achiral allenes utilizing chiral gold catalysts. Regio- and stereoselective routes to *N*-hydroxy-3-pyrrolines, 4,5-dihydroisoxazoles, and 3,6-dihydro-1,2-oxazines by gold-catalyzed cyclohydroamination of chiral allenic hydroxylamine derivatives have been reported by Krause. In all cases, *5*- or *6-endo* cyclization involved the nucleophilic attack by the nitrogen atom [94]. Gold-catalyzed dihydroamination was effective for a variety of *N*-δ- and *N*-γ-allenyl ureas affording bicyclic imidazolin-2-ones in good yield with high diastereoselectivity [95]. A diastereo-selective gold(III)-catalyzed allene cyclization allowed a formal synthesis of swainsonine [96]. The development of phosphine gold(I)-bis-*p*-nitrobenzoate complexes accomplished the enantioselective formation of vinyl-substituted pyrrolidines and piperidines [97]. The careful choice of the coordinating ligand (or counterion) was necessary to achieve high levels of enantioselectivity in the gold-catalyzed allene HA reactions focused on bis(phosphine gold) complexes derived from axial chiral scaffolds [17, 98–100]. Biaryl phosphine gold (I) complexes were also suitable catalysts for the enantioselective synthesis of pyrazolidines, isoxazolidines, and tetrahydrooxazines through intramolecular allene cyclohydroamination processes [101]. Gold(I)-catalyzed enantioselective HA of *N*-allenyl carbamates was reported by Widenhoefer [102]. However, the asymmetric HA of allenes is a continuing goal in gold catalysis. Phosphoramidates incorporating TADDOL-related diols with an acyclic backbone **52** achieved advancement as excellent ligands in the asymmetric intramolecular HA of the allene derivatives **51** (Scheme 22) [103].

Furthermore, a series of *ansa*-bridged chiral cyclophosphazane (CycloP) ligands **54** in which the chirality can be introduced at the diol bridge and/or the amido R group was prepared (Scheme 23).

The study of their activity and selectivity in the cyclohydroamination of γ-allenylsulfonamides **55** showed that the enantioselectivity was imparted by the chirality of the bridge only, but the steric demands of the CycloP ligands exerted influence on the reactivity and the selectivity of the reaction (Table 1) [104].

The cyclohydroamination of γ-allenylsulfonamides **55** was also explored to study the influence of the nuclearity of the catalyst on its catalytic performance. Only the C_3-symmetric trinuclear complex **59a** provided high enantioselectivities. It is worth noting that the monodentate complex **57a** produced racemic mixtures in the cyclohydroamination of the substrate **55a**, and the opposite enantiomer of the target pyrrolidine prevailed with substrates **55b and 55c** (Table 2) [105].

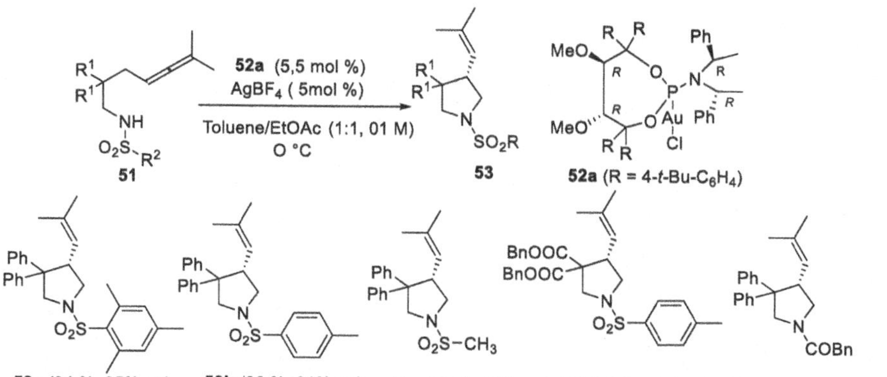

53a (94 %, 95% ee) 53b (90 %, 91% ee) 53c (86 %, 90% ee) 53d(94 %, 90% ee) 53e (94 %, 90% ee)

Scheme 22 Asymmetric gold-catalyzed pyrrolidine derivative synthesis

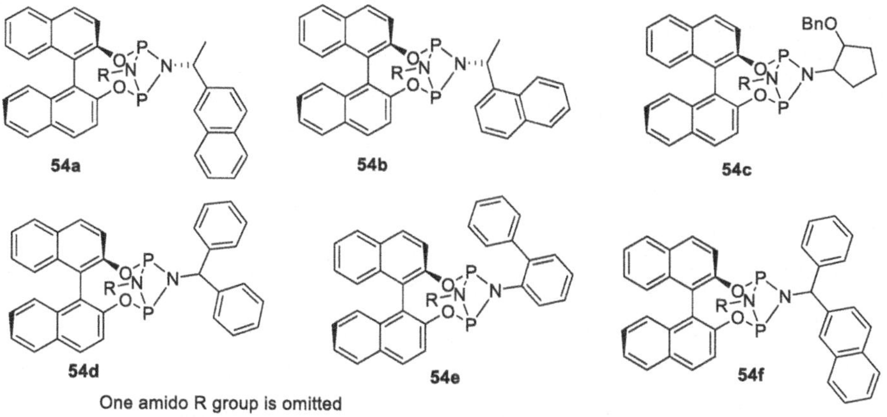

Scheme 23 List of chiral cyclophosphazane (CycloP) ligands

Activation of the cumulated double bonds by gold through coordination to either allenic double bond followed by the intramolecular nitrogen nucleophilic attack to afford regioselectively different *endo-* or *exo-*cyclization products is still receiving much attention. An interesting application is the development of a new approach to pyrrolines **61** (52–98% yield) by means of gold(I)-catalyzed tandem 3,3-rearrangement/HA reaction of γ-amino-substituted propargylic esters **60** (Scheme 24) [106].

The formation of only the *cis*-2,5-disubstituted pyrroline **61a** was observed from the ester **60a** (Scheme 25). This procedure has been successfully applied to the formal synthesis of ($\pm$) aphanorphine.

$AuCl_3$ is the most efficient catalyst in providing the 2-vinylquinazolin-4-one **63** from anthranilic allenamides **62**. The cyclization of compounds **62** occurred in lower yield with a significant amount of decomposition products under platinum-catalyzed reactions [$PtCl_4$ or $PtCl_2(CH_3CN)_2$], and palladium catalysis was much more unsuccessful. The gold-catalyzed heterocyclization took place exclusively by a *6-exo* process (Scheme 26) [107].

Table 1 Study of the activity and selectivity in the gold-catalyzed cyclohydroamination of γ-allenylsulfonamides **55**

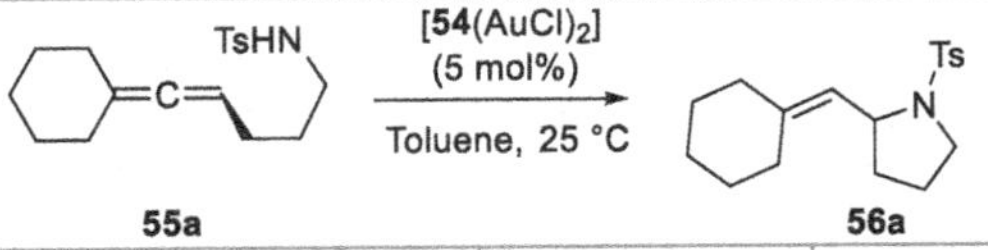

Entry	$[Ligand(AuCl)_2]$	Time (h)	Conv. (%)	e.r. (*R*/*S*)
1	(S_{binol},R_{amine})-**54a**	24	100	88:12
2	(R_{binol},S_{amine})-**54b**	37	99	15:85
3	$(R_{binol},(1R,2R_{amine})$-**54c**	16	93	19:81
4	(R_{binol},R_{amine})-**54a**	17	87	12:88
5	(*R*)-**54d**	25	87	28:72
6	(*R*)-**54e**	24	96	45:55
7	(*R*)-**54f**	45	100	23:77

Table 2 Examination of the catalytic efficiency of mono-, di-, and trinuclear gold complexes in the hydroamination of allenes **55**[a]

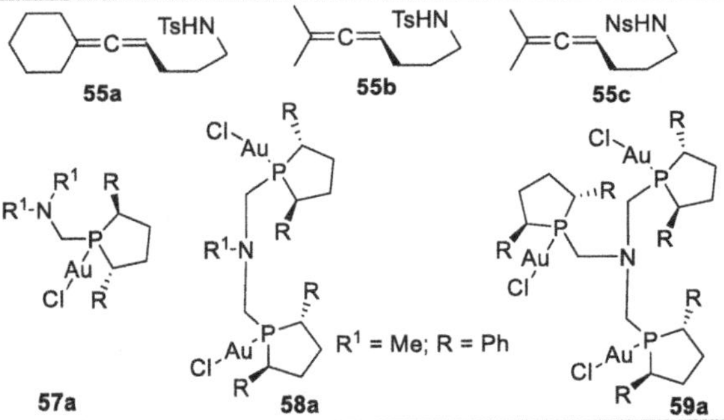

	Substrate	Silver salt	**57a** ee (%)	**58a** ee (%)[b, c]	**59a** ee (%)[b]	Time (h)
1	**55a**	AgOBz	rac	36	95	48
2	**55b**	AgOPNB	rac	36	92	20
3	**55b**	AgOBz	−21	22	94	48
4	**55b**	AgOPNB	−23	29	91	12
5	**55c**	AgOBz	−15	33	91	24
6	**55c**	AgOPNB	−11	40	87	10

[a]All reactions were carried out with 5 mmol of [Au(I)] catalyst in toluene (1 mL) at 50°C
[b]Enantioselectivities determined by chiral HPLC
[c]Reaction time: 120 h

The intramolecular HA reaction of the allenic sulfamates **64** accomplished the synthesis of six-membered cyclic sulfamidates **65** in high yields by means of the commercial available Gagosz's catalyst $Ph_3PAuNTf_2$ under mild conditions. Use of other ligands and counterions on gold gave better diastereoselectivity in some cases, but with diminished yield. It is noteworthy that the reaction was suitable for the formation of *N*-substituted quaternary centers. Very likely, the formation of the cyclic sulfamidate results from the *anti*-aminoauration of the allene moiety. The

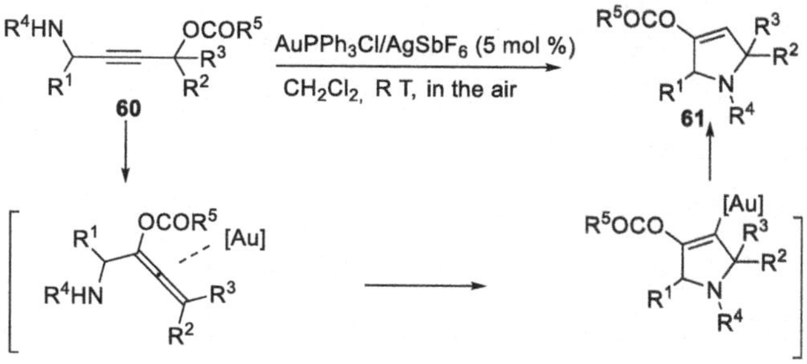

Scheme 24 Gold-catalyzed tandem 3,3-rearrangement/HA reaction of γ-amino-substituted propargylic esters to pyrrolines

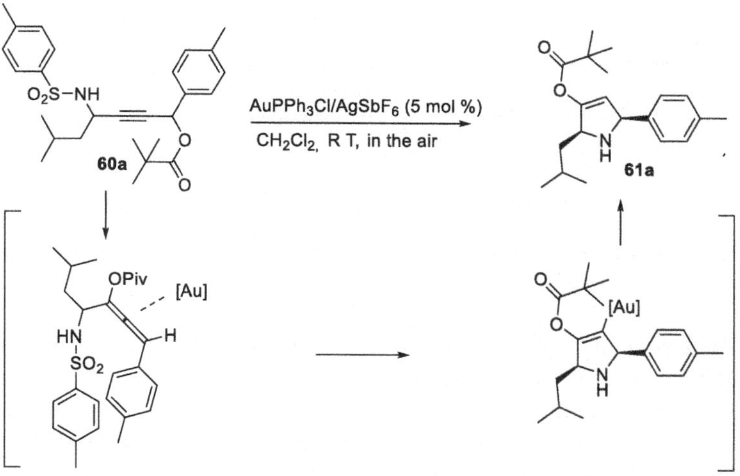

Scheme 25 Stereoselective synthesis of *cis*-2,5-disubstituted pyrrolines

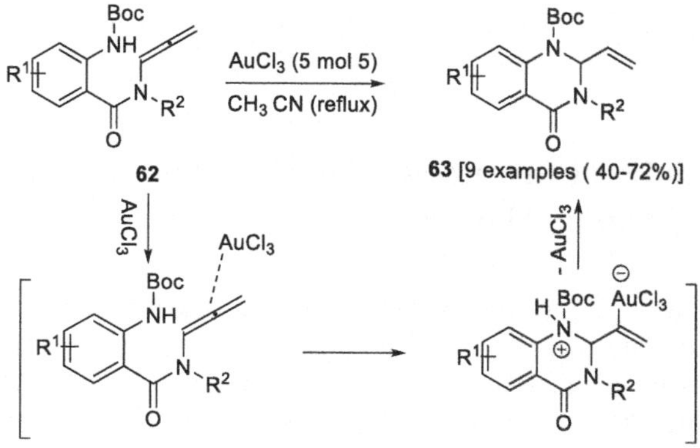

Scheme 26 Gold-catalyzed HA of anthranilic allenamides

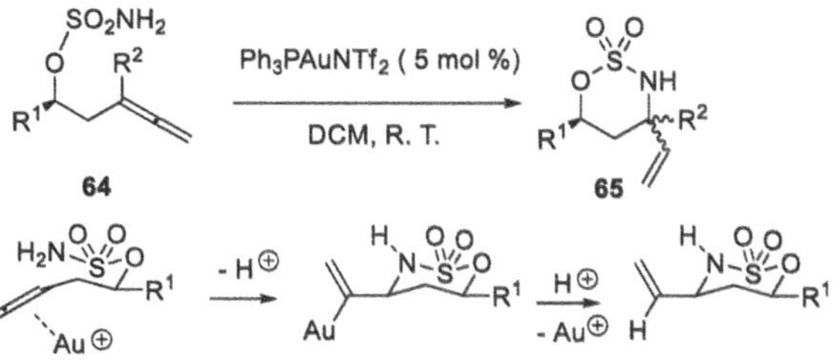

Scheme 27 Gold-catalyzed synthesis of cyclic sulfamidates by intramolecular allene HA

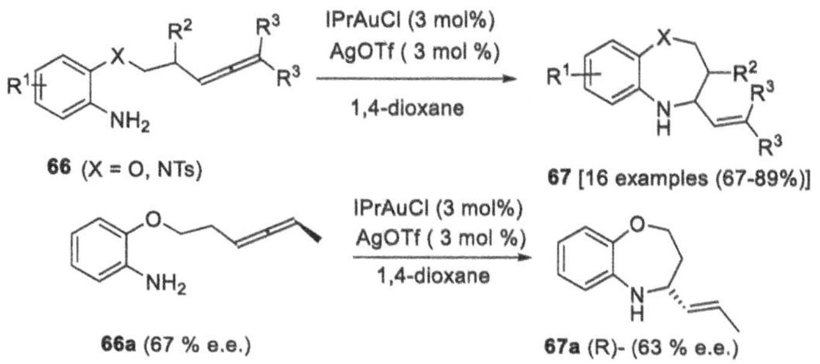

Scheme 28 Synthesis of 1,5-benzoxazepines and 1,4-benzodiazepines via gold-catalyzed *7-exo-trig* intramolecular allene HA

diastereoselective outcome was consistent with a kinetically controlled cyclization. No equilibrium existed between the diastereoisomers (Scheme 27) [108, 109].

Unprotected aniline derivatives with allene side tethers **66** afforded 1,5-benzoxazepines and 1,4-benzodiazepines **67** via a gold-catalyzed *7-exo-trig* HA reaction. The influence of different ligands, silver salts, temperature, and catalyst loading on the reaction outcome was investigated. Better results were observed with IPrAuCl in combination with AgOTf in 1,4-dioxane at 80°C. The HA chemistry was completely suppressed with protected aniline derivatives. Axial-to-central chirality transfer was shown for the reaction (Scheme 28) [110].

4 Gold-Catalyzed Intramolecular Hydroamination of Unconjugated Alkenes

In the past few years, significant exploration was devoted to extend the scope of the metal-free and late-transition metal-catalyzed HA of unactivated alkenes [111]. Despite lower reactivity with respect to alkynyl and allenyl counterparts, the gold-catalyzed HA of isolated alkenes inspired developments within organic synthesis by providing significant results in terms of selectivity/efficiency and mild reaction conditions [112]. He firstly reported on the gold(I)-catalyzed intra- and intermolecular HA of unactivated olefins under relatively mild conditions

Scheme 29 Au:PVP- catalyzed intramolecular HA of toluenesulfonamide to alkenes

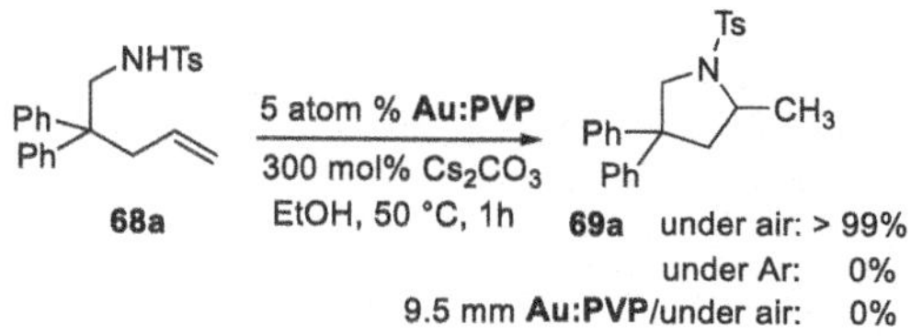

[113]. Several *N*-tosylated *γ*-amino olefins were firstly cyclized in toluene at 85°C with a catalytic amount (5 mol%) of $Ph_3PAuOTf$ (generated by mixing equal equivalents of Ph_3PAuCl and AgOTf) to afford the corresponding pyrrolidine derivatives in high yields. Carbamates also worked as nucleophiles in intramolecular gold-catalyzed addition to alkenes. The reaction was inhibited under the presence of alkylamines or aniline. These molecules may compete with olefin to bind gold(I) and thus inhibit the addition step. Moreover, the more nucleophilic/basic amines can interfere with the proton transfer pathway. Microwave applications are advantageous for gold(I)-catalyzed intramolecular HA of *ortho*-substituted benzenesulfonamides as precursors for the preparation of cyclic sulfonamides in a much shorter time than that required under conventional thermal conditions [114]. More recently, the intramolecular HA of toluenesulfonamide to alkenes was achieved through gold nanoclusters stabilized by a hydrophilic polymer, poly(*N*-vinyl-2-pyrrolidone) (**Au:PVP**), in EtOH under aerobic and basic conditions [115]. The HA product **69a** was isolated in about quantitative yield by treating **68a** with 5 atom% **Au:PVP** and an excess of Cs_2CO_3 at 50°C in EtOH under aerobic conditions (Scheme 29). The catalytic activity of **Au:PVP** was drastically influenced by the cluster size. Cyclization did not occur with the use of larger clusters of **Au:PVP** (average size, 9.5 nm) or under carefully degassed conditions. Molecular oxygen also plays an essential role in the HA process.

The formation of **68a**-D in 96% (70% D only when the reaction was carried out in CD_3CD_2OD) (**68a**-D was not detected when CH_3CH_2OD was used as the reaction medium) showed that the ethyl group was the source of hydrogen atom (Scheme 30).

Moreover, the resulting stereochemistry of the cyclized **68b** showed that the reaction proceeds via *anti*-addition of toluenesulfonamide to the alkenes. Then, the results are consistent with catalysis by **Au:PVP** via *π*-activation/*anti*-addition mechanism (Scheme 31).

N-Alkenyl carboxamides gave protected pyrrolidines, piperidines, and heterobicyclic compounds through an intramolecular *exo*-HA reaction catalyzed by a 1:1 mixture of $Au[P(t\text{-}Bu)_2(o\text{-biphenyl})]Cl$ (2) and AgOTf (5 mol%) in dioxane at 80°C [116]. Intramolecular *exo*-HA of *N*-4-pentenyl or *N*-5-hexenyl urea with a catalytic 1:1 mixture of a gold(I) *N*,*N*-diaryl imidazol-2-ylidine complex and AgOTf led to the corresponding nitrogen heterocycle in excellent yield [117]. A study of efficiency of a variety of bulky acyclic aminooxycarbene–[Au(I)] [AAOC–Au(I)] complexes **69** in the intramolecular HA of alkenyl ureas **70** showed that as the consequence of their ability to maintain coplanarity around the carbene

Scheme 30 Labeling experiments for the **Au:PVP**-catalyzed intramolecular HA of toluenesulfonamide to alkenes

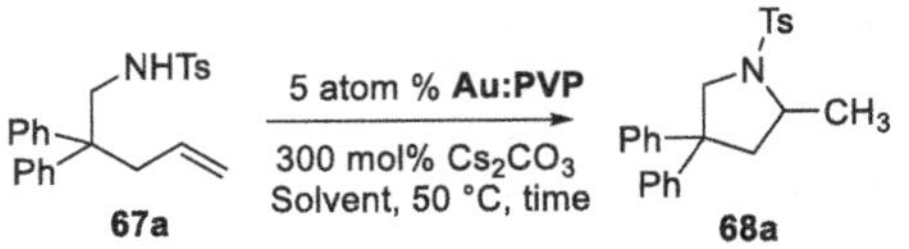

Solvent	Time (h)	Isolated yield % (%D)[a] 68a
CH_3CH_2OD	6	>99 (0)
CD_3CD_2OD	20	96 (70)

Scheme 31 Labeling experiments for the **Au:PVP**-catalyzed intramolecular HA of toluenesulfonamide to alkenes

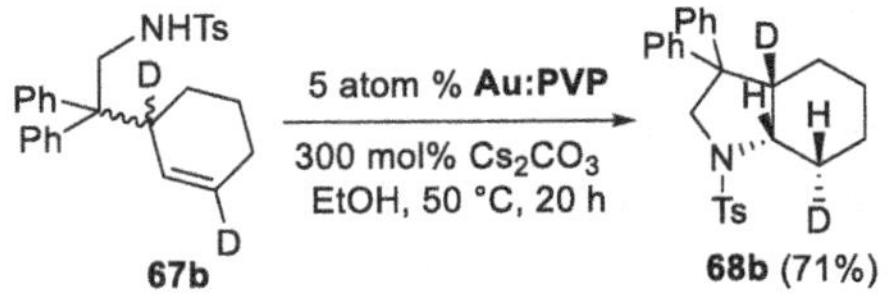

Scheme 32 HA of alkenyl ureas catalyzed by acyclic aminooxycarbene–gold (I) complexes [AAOC–Au (I)]

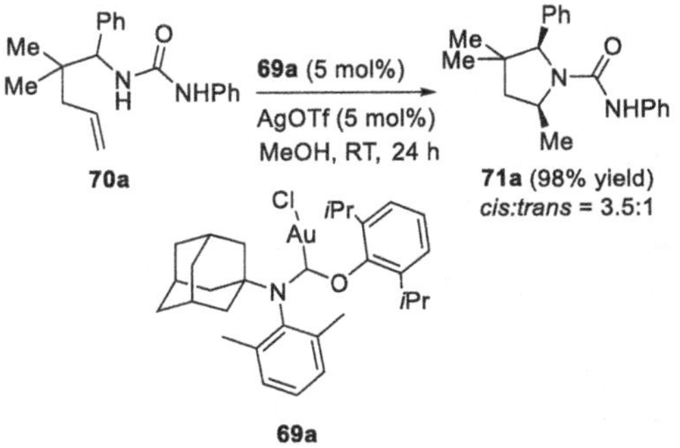

center, the reaction rate, yield, and stereoselectivity were better than those with acyclic diaminocarbene Au catalysts (Scheme 32) [118].

The Toste group reported the isolation and characterization of the alkylgold (I) complexes **73** derived from the gold-activated nucleophilic addition to an olefin of various nitrogen nucleophiles. Deuterium-labeling studies, X-ray crystal structures, and DFT analysis provided evidences for a mechanism involving *anti*-addition of the nucleophile to a gold-activated alkene. The rate of aminoauration was dramatically increased by the use of electron-poor arylphosphines, which are also shown to be favored in ligand exchange experiments. The gold was removed from alkylgold(I) complexes **73** under reducing conditions to provide the HA products **74**. Attempts at protodeauration of **73a** led only to recovery of the starting olefins **72a** (Scheme 33) [119].

The challenging extension of gold(I)-catalyzed alkene HA to include alkylamines has been firstly provided by the employment of alkylamines in their protonated form **75** (Scheme 34) [120].

Furthermore, gold nanoclusters stabilized by the poly(*N*-vinyl-2-pyrrolidone) hydrophilic polymer **[Au:PVP]** catalyzed the cycloisomerization reaction in the presence of formic acid derivatives of unactivated alkenes bearing a proximate

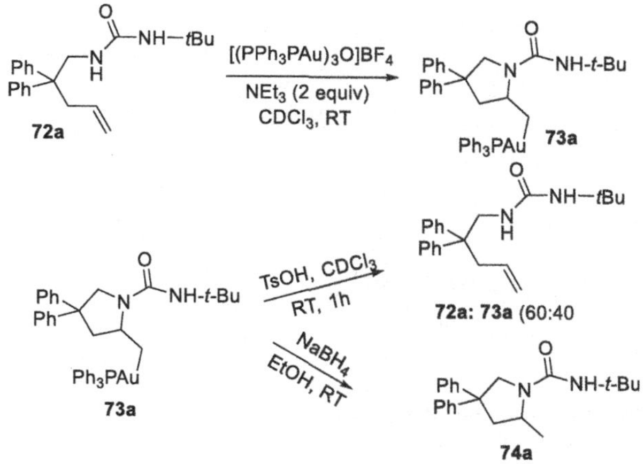

Scheme 33 Experimental evidences for the gold(I)-promoted intramolecular HA to unactivated alkenes

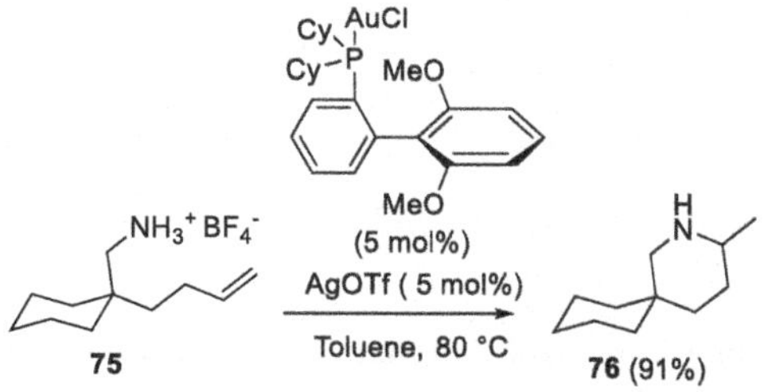

Scheme 34 Gold-catalyzed intramolecular HA of alkenes with alkylammonium salts

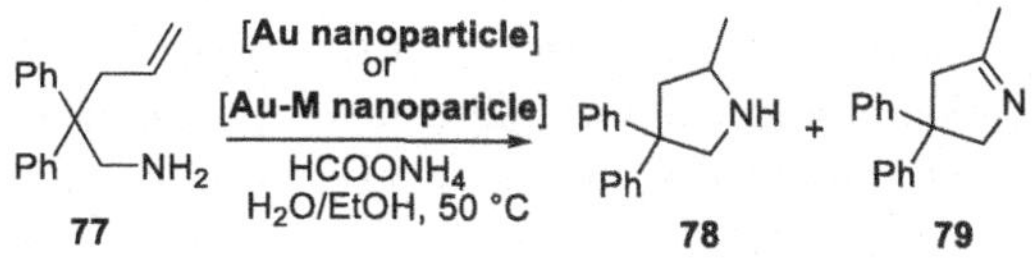

Scheme 35 Au-based bimetallic nanoparticles for the intramolecular aminoalkene HA

primary amino group [121]. Bimetallic Au-based nanoparticles [**Au-M**] exhibited also a higher activity than monometallic [**Au nanoparticles**] in the annulation reaction of the 2,2-diphenylpent-4-en-1-amine **77** to afford the five-membered Markovnikov hydroamination product **78**. Mononuclear Au nanoparticle catalytic systems (particle size, 1.3 nm) favored the formation of the hydroaminated pyrrolidine **78** as the major product (88%), whereas the imine **79** was formed in only 7% (Scheme 35) [122].

The selectivity for the formation of the hydroaminated derivatives is dependent on the alloying metal component when Au-M nanoparticles were used as the catalyst. Au nanoparticle catalysts are well known for their properties to adsorb O_2 on the Au surface to give the intermediate species **A**. After the cycloamination reaction due to the π-activation of the C–C double bond, the intermediate **C** can undergo the hydrogen transfer from $HCOONH_4$ to give the pyrrolidine **78**. Alternatively, the bimetallic nanoparticle catalysts can promote the β-hydrogen elimination of the catalyst-bound cyclic intermediate **C** to give the imine derivative **79** after isomerization. Indeed, when **77** was treated with 1,000 mol% of $DCOONH_4$ in

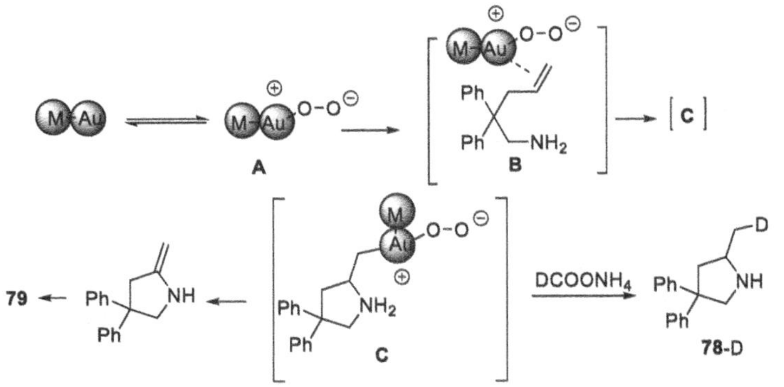

Scheme 36 Proposed mechanism of intramolecular HA of aminoalkenes under Au-based bimetallic nanoparticle catalysis

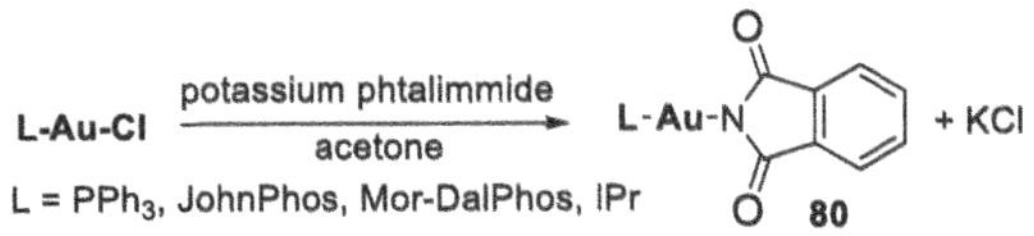

Scheme 37 Synthesis of **L-Au-**Pht

the presence of AuPt nanoparticles, effective formation of **78**-D was observed. The formation of **79** by the secondary oxidation of amine **78** was ruled out because the formation of **79**-D was not observed (Scheme 36).

Silver-mediated halogen abstraction was the most preferred method to generate cationic gold from a gold catalyst precursor (e.g., **L-Au-Cl**) because of the mild conditions needed and the relative availability of silver activators. However, the use of silver activators is not problem free. Recently, a number of gold phthalimide complexes (**L-Au-**Pht) **80** were easily synthesized from **L-Au-Cl** and potassium phthalimide (Scheme 37).

Subsequently, cationic gold was generated by using a **80**/acid combination. The reactivity of the **80**/acid system could be fine-tuned by readily available Brønsted acids/Lewis acids, each of which with a unique acid strength and counterion. Interestingly, the combination of Ph_3P-Au-Pht/$HCTf_3$ system achieved in very good yield the high demanding addition of a basic alkylamine to an alkene using a low loading (0.5%) and a simple Ph_3P ligand. Instead, the commonly used gold catalytic system such as $PPh_3PAu/AgOTf$ gave very low conversion (5%) even at high catalyst loading (Scheme 38) [123].

Moreover, the asymmetric version of the gold-catalyzed intramolecular HA of alkenes was explored. Rigid dinuclear gold complexes **81** having Au–Au interactions provided good yields and moderate enantioselectivity in the intramolecular HA of *N*-alkenyl ureas **70** via proximal and bimetallic activation of alkene and urea (Scheme 39) [124].

Olefins tethered with a NHTs functional group underwent intramolecular HA reaction with axially chiral Au(I) complexes bearing a binaphthalene scaffold with

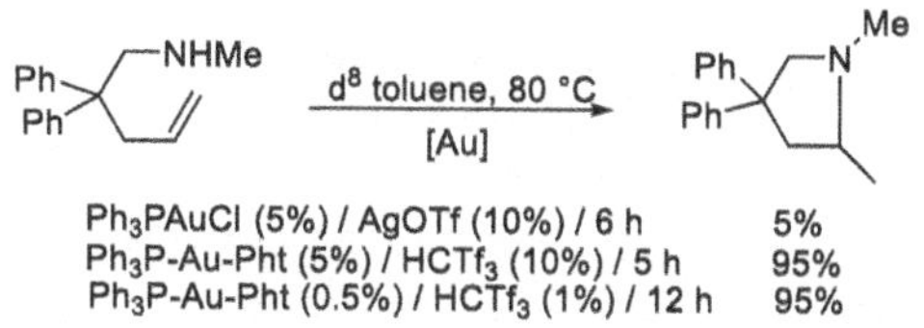

Scheme 38 Lewis acid-assisted activation of imidogold precatalysts

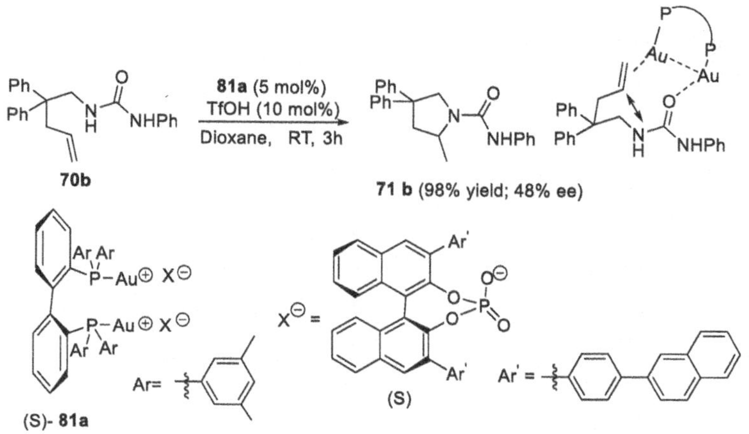

Scheme 39 Enantioselective intramolecular HA of *N*-alkenyl ureas catalyzed by *tropos BIPHEP*–gold(I) complexes

NHC or phosphine gold complexes on one side and an arene moiety on another side to give the corresponding cyclic derivative in moderate yield and up to 29% ee [125]. The intramolecular gold(I)-catalyzed asymmetric HA of alkenes by using a wide range of mononuclear gold(I) and (III) complexes in combination with silver salts was also investigated by Agbossou-Niedercorn, Michon, and co-workers [126]. Phosphoramidate-based mononuclear gold(I) catalysts are valuable alternatives to binuclear gold(I) achieving the intramolecular HA of alkenes with good conversions and average enantioselectivities at mild temperatures. Ligands based on BINOL and the Whitesell amine **82** were used. The stereoselectivity and the catalytic activity were determined by the ligand configuration. Opposite chirality between the BINOL scaffold and the amine [(*R*,*S*,*S*) or (*S*,*R*,*R*) configurations] accomplished the best results. The study of the anion influence on the conversion and enantiomeric excess showed that tetrafluoroborate and perchlorate were the most appropriate. Reaction was not running in polar solvents (Scheme 40).

Afterward, the same research groups studied the intramolecular asymmetric HA in the presence of various binuclear gold(I) catalysts based on specific SEGPHOS diphosphine ligands **83** which achieved higher yields and enantioselectivities. The addition of up to 10 equiv. of water is very relevant (Scheme 41) [127].

Surprisingly, a reversal of enantioselectivity was observed by changing the solvent from toluene to methanol. Very likely, an anion-assisted tautomerization

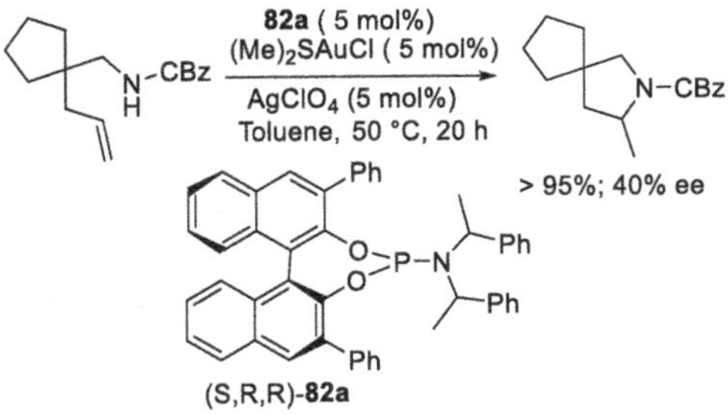

Scheme 40 Enantioselective intramolecular HA of alkenes catalyzed by phosphoramidate-based mononuclear gold(I) complexes

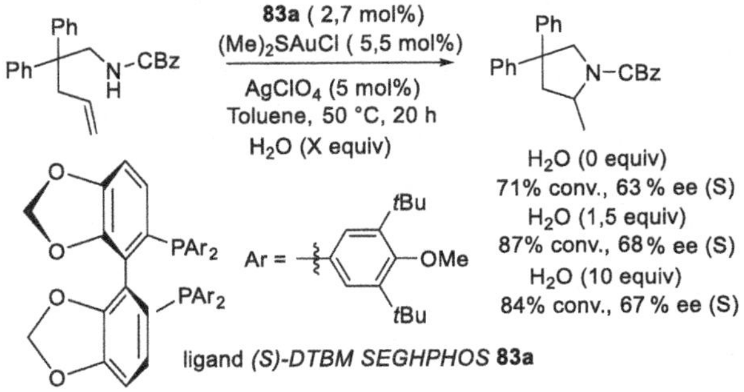

Scheme 41 Asymmetric intramolecular HA of alkenes by cationic binuclear gold(I) catalysts based on *SEGPHOS* diphosphine ligands

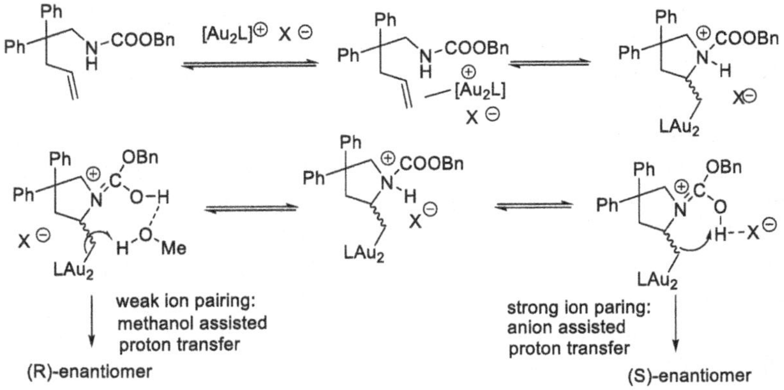

Scheme 42 Mechanistic proposal

follows the reversible cycloamination step. The polarity of the solvent determines the differentiation of the two diastereomeric intermediates in the final irreversible protodeauration step (Scheme 42).

Scheme 43 Gold-catalyzed enantioselective HA of *N*-4-pentenyl carbamates

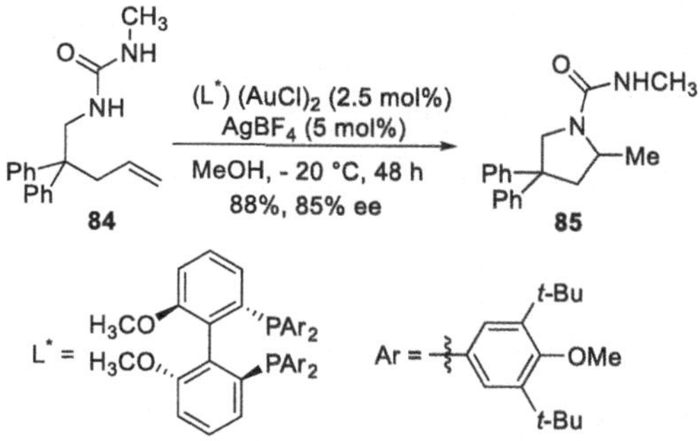

Methanol was also the most effective solvent in the intramolecular enantioselective HA of alkene with carbamates and ureas [128]. *N*-4-Pentenyl ureas and carbamates **84** gave the corresponding pyrrolidine derivatives **85** with up to 85% ee (Scheme 43). *N*-Alkyl-*N*′-4-pentenyl ureas underwent intramolecular enantioselective HA faster than did *N*-4-pentenyl carbamates. Cationic bis(gold) complex generated in situ by a 1:2 mixture of *(S)-DTBM-MeO-PIPHEP* and silver tetrafluoroborate ($AgBF_4$) achieved the highest enantioselectivities. As expected, in the polar reaction medium, enantioselectivity of intramolecular HA was insensitive to the nature of the counterion. By contrast with the intermolecular enantioselective HA of alkenes with imidazolin-2-ones, the intramolecular enantioselective HA is temperature dependent [129].

5 Synthesis of Nitrogen Heterocyclic Compounds Through Sequential Gold-Catalyzed Processes Implying a Hydroamination Step

Between the variety of ways of synthesizing carbon–nitrogen bonds, the HA reaction is very attractive for the one-pot construction of complex architectures of organic derivatives which have widespread application. Gold-catalyzed reactions are of relevant importance in promoting innovative/alternative one-flask approaches for the synthesis of elaborated structure from simple starting material [130]. However, many problems have to be faced to expand the substrate scope and the limitations associated with assembling organic nitrogen compounds. The relevance of the benzo[*b*][1, 4]diazepine nucleus as a privileged pharmacophoric motif spurred investigations to overcome the limitations of the unique gold-catalyzed synthesis of *N*-unsubstituted 1,5-benzodiazepines directly from *o*-phenylenediamine derivatives and alkynes [131]. The atom-efficient HA/cyclization process of *N*-substituted *o*-phenylenediamines **86** and terminal alkynes achieved the synthesis of 1-substituted benzo[*b*][1, 4]diazepines **87** by switching from a phosphine–Au complex to the NHC–Au(I) complex SIMesAuCl as the catalytic systems (Scheme 44) [132].

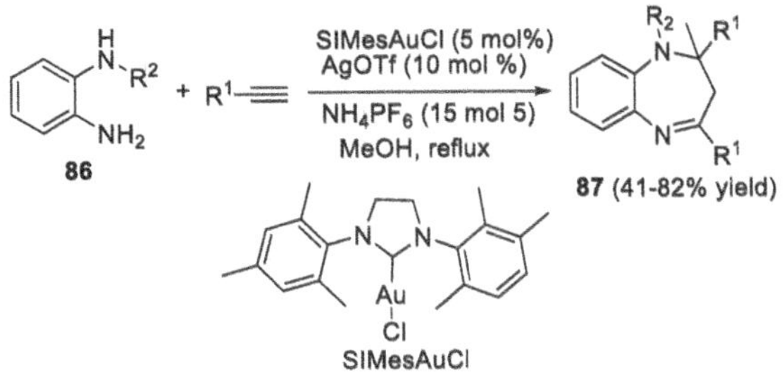

Scheme 44 Gold-catalyzed synthesis of 1-substituted benzo[*b*][1,4]diazepines

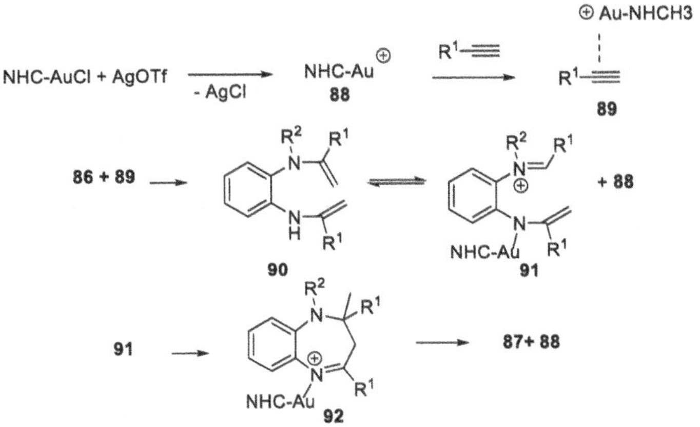

Scheme 45 Proposed mechanism for the NHC–Au(I)-catalyzed sequential HA/cyclization reaction

Other NHC-AuCl are less efficient than SIMesAuCl, and the addition of 15 mol % of NH_4PF_6 as a proton donor for the hydroamination reaction was beneficial for the reaction yield. Very likely, the activated alkyne **89** by the coordination with the cationic [Au(I)] species **88** is prone to undergo nucleophilic attack of *N*-substituted *o*-phenylenediamine **86** to give the dienamine intermediate **90**. The intramolecular addition of the enamine moiety to ketimine, one of the intermediates **91** derived from the gold-catalyzed conversion of **90**, produces the gold complex **92**. Finally, release of cationic [Au(I)] species **88** affords the 1-substituted benzo[*b*][1, 4] diazepine product **87** (Scheme 45).

The synthesis of 3-substituted 2,5-dimethylpyrazines **94** was readily performed by the gold-catalyzed sequential reactions of imine with propargylamine. Deuterium-labeling and computational studies provided understanding of the mechanism of the sequential process which was trigged by the chemo- and regioselective HA reactions of the gold-coordinated triple bond of the starting imine (Scheme 46) [133]. Attractively, the gold-catalyzed one-pot protocol can also be accomplished by the reaction of a suitable aldehyde with 2 equiv. of

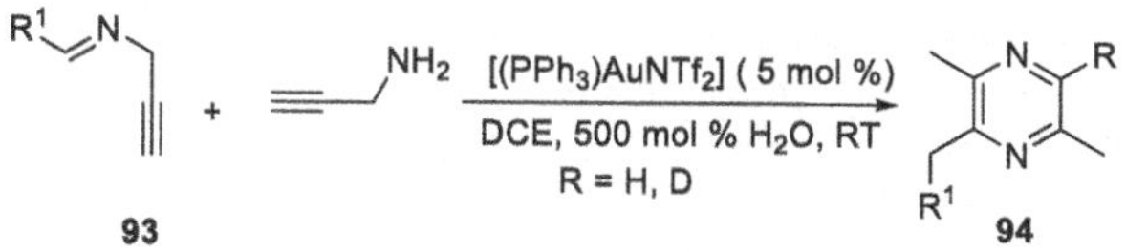

Scheme 46 Gold-catalyzed synthesis of 3-substituted 2,5-dimethylpyrazines

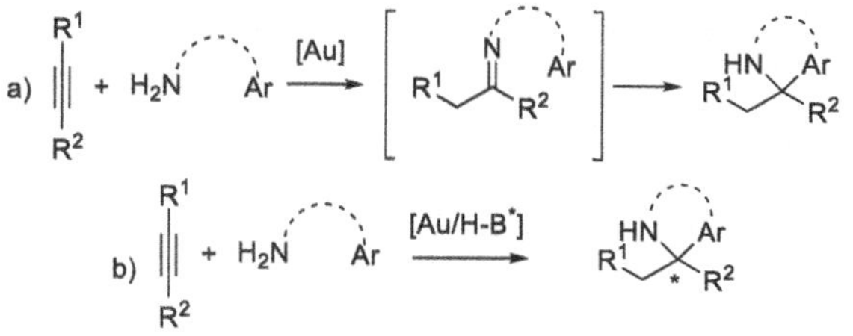

Scheme 47 Gold(I)/chiral Brønsted acid-catalyzed enantioselective HA/hydroarylation of alkynes

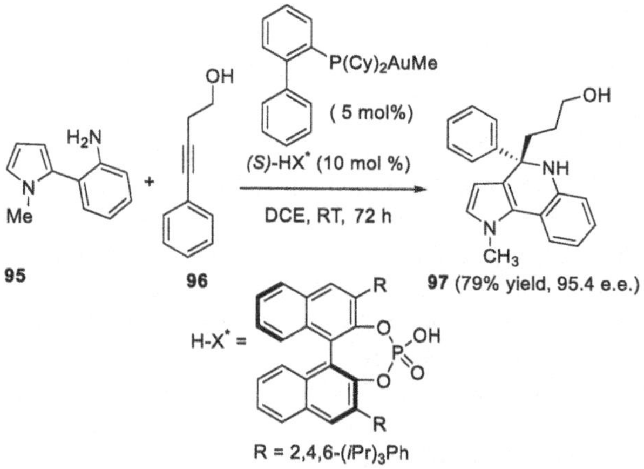

Scheme 48 Enantioselective synthesis of functionalized dihydro [3,2-*c*]quinolines

propargylamine. The efficient scale-up of the procedure to obtain gram quantities of pyrazine derivatives with a reduced catalyst loading (1 mol%) was described.

Patil and Liu groups developed a variety of sequential gold(I)-catalyzed HA/hydroarylation reactions of aromatic amines with alkynes to give multisubstituted pyrrolo[1,2-*a*]quinoxalines, indolo[3,2-*c*]quinolines, tetrahydro-4-quinazolinones, and benzo[4,5-c]quinazolines (Scheme 47a) [134, 135]. A catalytic enantioselective version of such process accomplishing the generation of the corresponding aza-heterocycles bearing a quaternary carbon center represents a very interesting advancement [136]. The success of the reaction is based on the use of $LAuX^*$ complexes, generated in situ from LAuMe and $H\text{-}X^*$ (X^* = chiral phosphate anion) (Scheme 47b).

The gram-scale asymmetric synthesis of **97** demonstrates the practicality of the method for providing such molecules in enantiomeric pure form (Scheme 48). The presence of the tethered OH group on the alkyne **96** plays a pivotal role for the enantioinduction process. DFT calculations indicate that the "*Re*-face" attack of the aryl nucleophile is kinetically preferred over the "*Si*-face" attack.

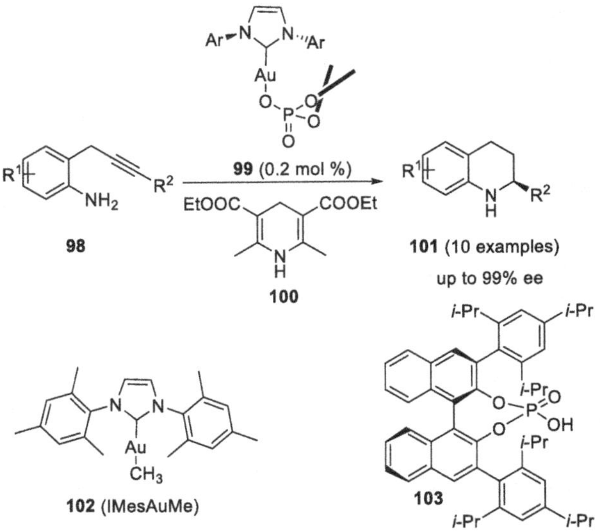

Scheme 49 Sequential HA/asymmetric transfer hydrogenation by chiral gold phosphate catalysts

The combination of gold catalysis and chiral phosphoric acid also accomplished the direct transformation of alkynylamines into chiral tetrahydroquinolines [137]. However, the request of a high chiral Brønsted acid loading hampers the practical applicability of the methodology. More advantageously, a high-yielding enantioselective cascade HA/asymmetric transfer hydrogenation reaction for directly transforming 2-(2-propynyl)aniline derivatives **98** into optically active tetrahydroquinolines **101** in one operation with moderate to good yields and excellent enantioselectivities was achieved by means of the catalysis of only 0.2 mol% of the chiral gold phosphate **99** generated in situ from IMesAuMe **102** and the chiral phosphoric acid **103**. Notably, the gold catalyst enables both the intramolecular HA as a π-Lewis acid and the asymmetric reduction as a highly effective chiral Lewis acid (Scheme 49) [138].

The coupling of gold catalysis with visible light allowed an alternative innovative approach to highly functionalize indoles **106** in moderate to good yields at room temperature by a tandem gold-catalyzed HA/radical cyclization/dehydrogenative coupling of substituted anilines **104** with alkynones **105**. It is worth noting that the reaction proceeds without any photocatalyst (Scheme 50) [139].

Autotandem gold catalysis in which one gold catalyzed two sequential processes in different modes provided functionalized pyrroles **112** in moderate to good yields in a modular fashion. The initially formed gold acetylide **109** undergoes addition to oxonium ion **110** to generate in situ the amido alkyne adduct **111** which is activated by π-coordination of the gold catalyst toward the *5-endo-dig* cyclization (Scheme 51) [140].

Gold-catalyzed cascade oxidative C–H/C–H cross-coupling, cyclization, and alkynylation of β-enamino derivatives **113** with terminal alkynes achieved an efficient approach to 3-alkynylpyrroles **116**. The procedure features complete regiocontrol, relatively wide substrate scope, and high functional group tolerance (Scheme 52) [141].

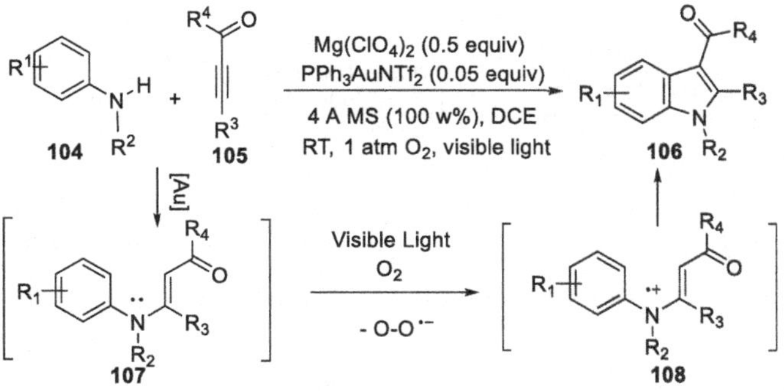

Scheme 50 Merging gold and photoredox catalysis

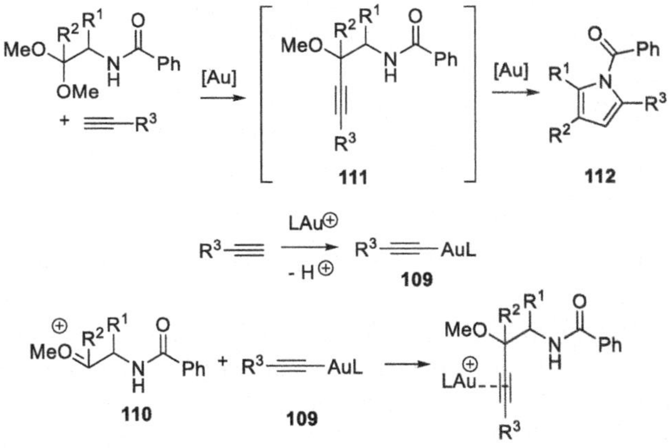

Scheme 51 Autotandem gold catalysis

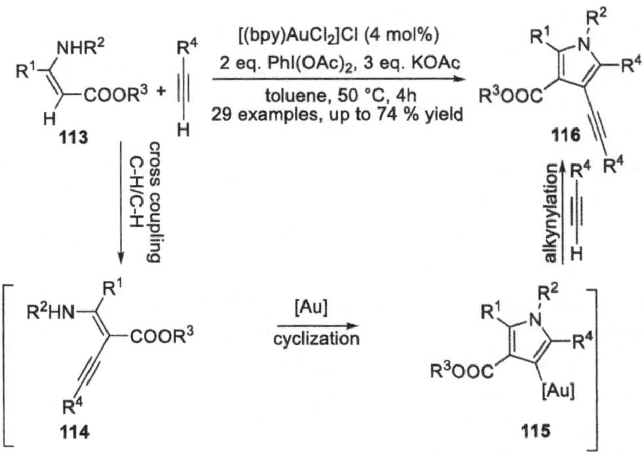

Scheme 52 Gold-catalyzed C–H/C–H cross-coupling/cyclization/alkynylation cascade reaction

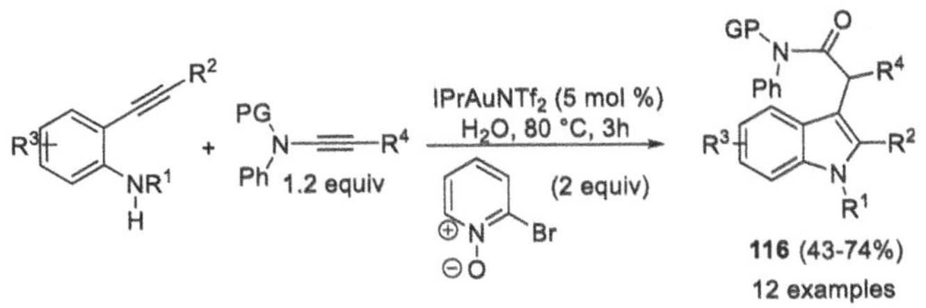

Scheme 53 Sequential gold-catalyzed cycloisomerization/intermolecular trapping of α-oxo gold carbenes

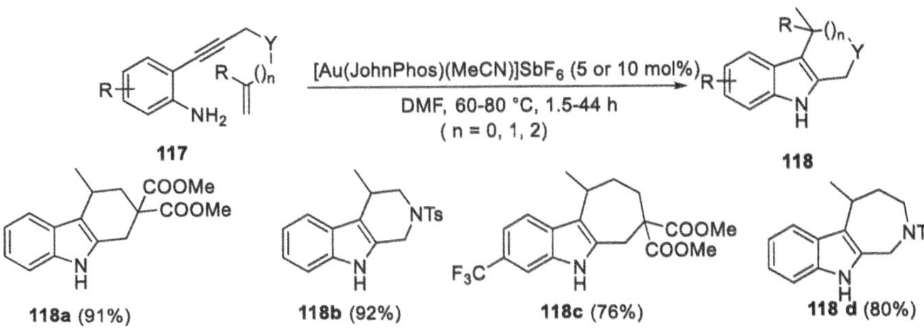

Scheme 54 Gold-catalyzed cascade reactions in the synthesis of 2,3-fused indole derivatives

The dual catalytic role of gold mediating both the cycloisomerization of 2-alkynylanilines and the intermolecular oxidation of ynamides at the same time provided a new type of concurrent tandem catalysis allowing a new approach to densely functionalize indole derivatives **116** in water (Scheme 53) [142].

Gold-catalyzed cascade reactions of 2-alkynylanilines represent a powerful tool for the direct construction of highly functionalized indoles and corresponding polycyclic derivatives ([143] and references therein) [144–146]. Gold(I)-catalyzed *5-endo* HA/X-*exo*(or *endo*)-*trig* arylation cascade reactions of 2-aminoaryl 1, X-enynes **117** led to a variety of indole-fused scaffolds **118** (Scheme 54) [147].

Aniline derivatives **119** bearing a conjugated diyne moiety underwent predominant sequential gold-catalyzed *5-endo-dig*/*7-endo-dig* bis-cyclization to give indoles fused with a seven-membered ring **120** in good yields (Scheme 55) [148]. Conversely, the formation of the indole ring was followed by *5-exo-* or *6-endo-dig* consecutive cyclization with phenylurea derivatives bearing a terminal conjugated diyne moiety at the *ortho*-position [149].

An unusual gold(I)-catalyzed bicyclization of N-propargylic sulfonylhydrazones **121** involving firstly the chemoselective cyclization of the hydrazone nitrogen instead of the favored aniline nitrogen onto the alkyne achieved an efficient method for the preparation of 5,6-dihydropyrazolo[1,5-c]quinazolines **122**. This strategy was applied to the synthesis of a potential Eg5/kinesin spindle protein inhibitor **122a** (Scheme 56) [150].

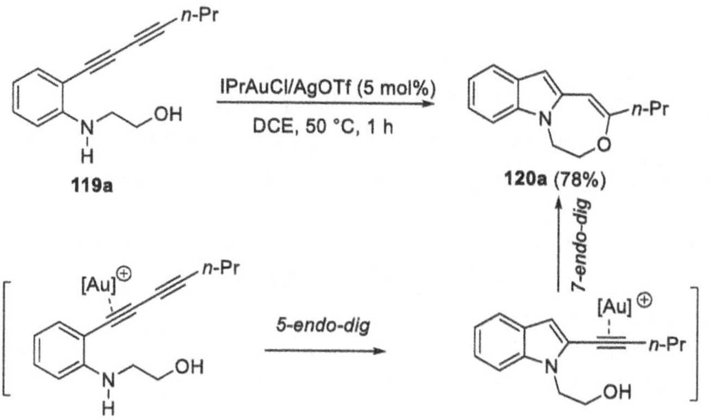

Scheme 55 Gold-catalyzed cascade cyclization of aniline derivatives bearing a conjugated diyne moiety

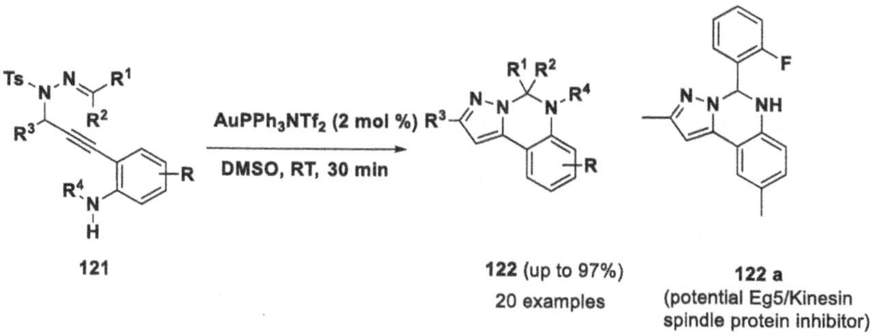

Scheme 56 Gold-catalyzed chemoselective bicyclization of *N*-propargylic sulfonylhydrazones

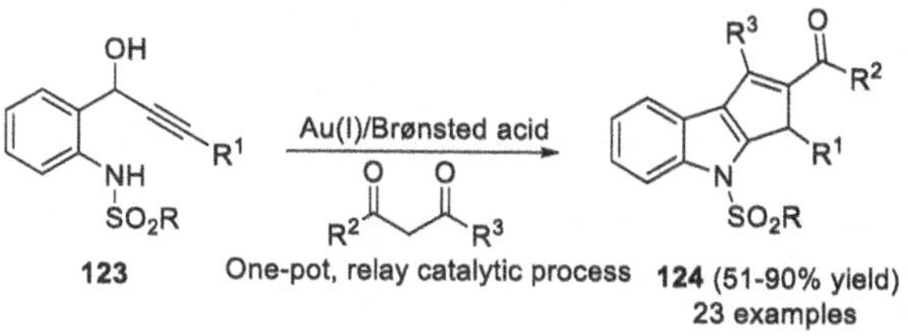

Scheme 57 Relay gold(I)/Brønsted acid catalysis

A further application of the versatile strategy which assembles directly complex chemical architectures associated with pot, step, and atom economy has been recently extended to the synthesis of cyclopentannulated indoles **124** through a relay catalytic process involving [Au(I)]/Brønsted acid in the reaction of 1-(2-aminophenyl)prop-2-ynols **123** with 1,3-dicarbonyls (Scheme 57) [151].

6 Conclusion

Gold-catalyzed addition of N-H nucleophiles to alkynes, allenes, and alkenes has emerged as a highly valuable tool to access to nitrogen heterocyclic compounds from simple precursors. As shown in this review, particular progress has been achieved by means of gold catalysis in selective cycloisomerization reactions of alkyne bearing proximate amino groups, in which the regio- and stereoselectivity of resultant products can be efficiently addressed by a suitable choice of the catalyst system. Moreover, the asymmetric gold-catalyzed HA of allenes and alkenes has undergone relevant advancement and encouraged the synthetic community to apply methodologies to the synthesis of natural products or biologically active heterocycles. Further developments are expected in the upcoming years by a better understanding of such processes.

References

1. Müller TE, Hultzsch KC, Yus M, Foubelo F, Tada M (2008) Chem Rev 108:3795
2. Müller TE, Beller M (1998) Chem Rev 98:675
3. Huang L, Arndt M, Gooβen K, Heydt H, Gooβen LJ (2015) Chem Rev 115:2596
4. Bernoud E, Lepori C, Mellah M, Schulz E, Hannedouche J (2015) Catal Sci Technol 5:2017
5. Patil NT, Kavthe RD, Shinde VS (2012) Tetrahedron 68:8079
6. Yadav JS, Antony A, Rao TS, Subba Reddy BV (2011) J Organomet Chem 696:16
7. Severin R, Doye S (2007) Chem Soc Rev 36:1407
8. Alonso F, Beletskaya IP, Yus M (2004) Chem Rev 104:3079
9. Ohno H, Chiba H, Inuki S, Oishi S, Fujii N (2014) Synlett 25:179
10. Krause N, Winter C (2011) Chem Rev 111:1994
11. Zeng X (2013) Chem Rev 113:6864
12. Huang H, Zhou Y, Liu H (2011) Beilstein J Org Chem 7:897
13. Hashmi ASK, Bührle M (2010) Aldrichimica Acta 43:27
14. Widenhoefer R, Han X (2006) Eur J Org Chem 4555
15. Toste FD, Michelet V (eds) (2014) Gold catalysis: an homogeneous approach. Imperial College Press, London
16. Ranieri B, Escofet I, Echavarren AM (2015) Org Biomol Chem 13:7103
17. Jia M, Bandini M (2015) ACS Catal 5:1638
18. Hashmi ASK, Toste FD (2012) Modern gold catalyzed synthesis. Wiley, Weinheim
19. Lo VK-R, Chan AO-Y, Che C-M (2015) Org Biomol Chem 13:6667
20. Bour C, Gandon V (2015) Synlett 26:1427
21. Fürstner L (2014) Acc Chem Res 47:925
22. Zhang D, Tang X-Y, Shi M (2014) Acc Chem Res 47:913
23. Abbiati G, Rossi E (2014) Beistein J Org Chem 10:481
24. Liu X, He L, Liu Y-M, Cao Y (2014) Acc Chem Res 47:793
25. Fensterbank L, Malacria M (2014) Acc Chem Res 47:9
26. Rudolph M, Hashmi ASK (2012) Chem Soc Rev 41:24
27. Lu B-L, Dai L, Shi M (2012) Chem Soc Rev 41:3318
28. Brenzovich WE Jr (2012) Angew Chem Int Ed 51:8933
29. Liu L-P, Hammond GB (2012) Chem Soc Rev 41:3129
30. Corma A, Leyva-Pérez A, Sabater MJ (2011) Chem Rev 111:1657

31. Hashmi ASK (2010) Angew Chem Int Ed 49:5232
32. Shapiro ND, Toste FD (2010) Synlett 675
33. Abu Sohel SM, Liu R-S (2009) Chem Soc Rev 38:2269
34. Nevado C (2010) Chimia 64:247
35. Fürstner A (2009) Chem Soc Rev 38:3208
36. Arcadi A (2008) Chem Rev 108:3266
37. Li Z, Brouwer C, He C (2008) Chem Rev 108:3239
38. Skouta R, Li C-J (2008) Tetrahedron 64:4917
39. Hashmi ASK, Rudolph M (2008) Chem Soc Rev 37:1766
40. Muzart J (2008) Tetrahedron 64:5815
41. Gorin DJ, Sherry BD, Toste FD (2008) Chem Rev 108:3351
42. Hashmi ASK (2007) Chem Rev 107:3180
43. Shen HC, Graham TM (2013) Drug Discov Today Technol 10, e3
44. Rudolf M, Hashmi ASK (2011) Chem Commun 47:6536
45. Das A, Abu Sohel SM, Liu R-S (2010) Org Biomol Chem 8:860
46. Shen HC (2008) Tetrahedron 64:7847
47. Patil NT, Yamamoto Y (2008) Chem Rev 108:3395
48. Shen HC (2008) Tetrahedron 64:3885
49. Fukuda Y, Utimoto K (1991) Synthesis 975
50. Fukuda Y, Utimoto K, Nozaki H (1987) Heterocycles 25:297
51. Patil NT, Singh V (2011) J Organomet Chem 696:419
52. Minkler SRK, Isley NA, Lippincott DJ, Krause N (2014) Org Lett 16:724
53. Ritzen B, Richelle GJJ, Brocken L, van Delft FL, Rutjes FPJT (2014) Synlett 25:270
54. Yu Y-F, Shu C, Zhou B, Li J-Q, Zhou J-M, Ye L-W (2015) Chem Commun 51:2126
55. Zeng Q, Zhang L, Yang J, Xu B, Xiao Y, Zhang J (2014) Chem Commun 50:4205
56. Yoshimanu M, Ohta K, Takahashi N (2012) Chem Eur J 18:15602
57. Barbour PM, Marholz LL, Chang L, Xu W, Wang X (2014) Chem Lett 43:572
58. Abbiati G, Marinelli F, Rossi E, Arcadi A (2013) Isr J Chem 53:856
59. Zhdanko A, Maier ME (2014) Angew Chem Int Ed 53:7760
60. Hashmi ASK, Ramamurthi TD, Rominger F (2010) Adv Synth Catal 352:971
61. Gala E, Córdoba M, Izquierdo ML, Álvarez-Builla J (2014) Arkivoc (V):319
62. Gaggioli CA, Ciancaleoni G, Biasiolo L, Bistoni G, Zuccaccia D, Belpassi D, Belazoni P, Tarantelli F (2015) Chem Commun 51:5990
63. Zeng X, Kinjo R, Donnadieu B, Bertrand G (2010) Angew Chem Int Ed 45:942
64. Morozov OS, Lunchev AV, Bush AA, Tukov AA, Asachenko AF, Khrustalev VN, Zalesskiy SS, Ananikov VP, Nechaev MS (2014) Chem Eur J 20:6162
65. Hashmi ASK (2014) Acc Chem Res 47:864
66. Obradors C, Echavarren AM (2014) Chem Commun 50:16
67. Weber D, Gagne MR (2013) Chem Sci 4:335
68. Braun I, Asiri AM, Hashmi ASK (2013) ACS Catal 3:1902
69. Cheong PH-Y, Morganelli P, Luzung MR, Houk KN, Toste FD (2008) J Am Chem Soc 130:4517
70. Gimeno A, Cuenca AB, Suárez-Pantiga S, de Arellano RC, Medio-Simón M, Asensio G (2014) Chem Eur J 20:683
71. Ciano L, Fey N, Halliday CJV, Lyman JM, Milner LM, Mistry N, Pridmore NE, Townsend NS, Whitwood AC (2015) Chem Commun 51:9702
72. Jiang Y, Tang X-Y, Shi M (2015) Chem Eur J 21:7675
73. Belger K, Krause N (2015) Org Biomol Chem 13:8556
74. Li S, Li Z, Yuan Y, Peng D, Li Y, Zhang L, Wu Y (2012) Org Lett 14:1130
75. Bouvet S, Moreau X, Coeffard V, Greck C (2012) J Org Chem 78:427
76. Pereshivko OP, Peshkov AV, Peshkov AA, Jacobs J, Van Meervelt L, Van der Eychen EV (2014) Org Biomol Chem 12:1741

77. Pereshivko OP, Peshkov AV, Jacobs J, Van Meervelt L, Van der Eychen EV (2013) Adv Synth Catal 355:781
78. Peshkov AV, Pereshivko OP, Sharma S, Meganathan T, Parmar VS, Ermolat'ev DS, Van der Eychen EV (2011) J Org Chem 76:5867
79. Gupta S, Koley D, Ravikumar K, Kundu B (2013) J Org Chem 78:8624
80. Danda A, Kumar K, Waldmann H (2015) Chem Commun 51:7536
81. Jha RR, Saunthwal RK, Verma AK (2014) Org Biomol Chem 12:522
82. Seppänen O, Muuronen M, Helaja J (2014) Eur J Org Chem 4044
83. Schröder F, Ojeda M, Erdmann N, Jacobs J, Luque R, Noël T, Van Meervelt L, Van der Eychen J, Van der Eychen EV (2015) Green Chem 17:3314
84. Yao L-F, Wang Y, Huang K-W (2015) Org Chem Front 2:721
85. James T, Simpson I, Grant JA, Sridhran V, Nelson A (2013) Org Lett 15:6094
86. Basceken S, Balci M (2015) J Org Chem 80:3806
87. Morita N, Krause N (2004) Org Lett 6:4121
88. Nishina N, Yamamoto Y (2006) Angew Chem Int Ed 45:3314
89. Zhang Z, Liu C, Kinder RE, Han X, Qian H, Widenhoefer RA (2006) J Am Chem Soc 128:9066
90. Yu S, Ma S (2012) Angew Chem Int Ed 51:3074
91. Alcaide B, Almendros P (2011) Adv Synth Catal 353:2561
92. Bates RW, Satcharoen V (2002) Chem Soc Rev 31:12
93. Yang W, Hashmi ASK (2014) Chem Soc Rev 43:2941
94. Winter C, Krause N (2009) Angew Chem Int Ed 48:6339
95. Li H, Widenhoefer RA (2009) Org Lett 11:2671
96. Bates RW, Dewey MR (2009) Org Lett 11:3706
97. LaLonde RL, Sherry BD, Kang EJ, Toste FD (2007) J Am Chem Soc 129:2452
98. Wang Y-M, Lackner AD, Toste AF (2014) Acc Chem Res 47:889
99. Nguyen BN, Adrio LA, Barreiro EM, Brazier JB, Haycock P, Hii KK, Nachtegaal M, Newton MA, Szlachetko J (2012) Organometallics 31:2395
100. Aikawa K, Kojima M, Mikami K (2009) J Am Chem Soc 48:6073
101. LaLonde RL, Wang ZJ, Mba M, Lackner AD, Toste FD (2010) Angew Chem Int Ed 49:598
102. Zhang Z, Bender CF, Widenhoefer RA (2007) Org Lett 9:1935
103. Teller H, Corbet M, Mantilli L, Gopakumar G, Goddard R, Thiel W, Fürstner A (2012) J Am Chem Soc 134:15331
104. Roth T, Wadepohl H, Wright DS, Gade LH (2013) Chem Eur J 19:12823
105. Rodríguez L-I, Roth T, Fillol JL, Wadepol H, Gade LH (2012) Chem Eur J 18:3721
106. Wang Z, Zheng H, Yang J, Xie X, She X (2015) Adv Synth Catal 357:2082
107. Broggini G, Borsini E, Fasana A, Poli G, Liron F (2012) Eur J Org Chem 3617
108. Higginbotham MCM, Kennedy L, Lindsay AG, Troester A, Bebbington MWP (2015) Tetrahedron 71:727
109. Bebbington MWP, Higginbotham MCM (2012) Chem Commun 48:7565
110. Pflästerer D, Dolbundalchok P, Rafique S, Rudolph M, Rominger F, Hashmi ASK (2013) Adv Synth Catal 355:1383
111. Bermoud E, Lepori M, Mellah M, Schulz E, Hannedouce J (2015) Catal Sci Technol 5:2017
112. Chiarucci M, Bandini M (2013) Beilstein J Org Chem 2586
113. Zhang J, Yang C-G, He C (2006) J Am Chem Soc 128:1798
114. Liu X-Y, Li C-H, Che C-M (2006) Org Lett 8:2707
115. Kitahara H, Sakurai H (2011) J Organomet Chem 696:442
116. Bender CF, Widenhoefer RA (2006) Chem Commun 4143
117. Bender CF, Widenhoefer RA (2006) Org Lett 8:5303
118. Seo H, Snead DR, Abboud KA, Hong S (2011) Organometallics 7:2011
119. LaLonde RL, Brenzovich WE Jr, Benitez D, Tkatchouk E, Kelley K, Goddard WA, Toste FD (2010) Chem Sci 1:226
120. Bender CF, Widenhoefer RA (2008) Chem Commun 2741

121. Kitahara H, Sakurai H (2010) Chem Lett 39:46
122. Singh SK, Yadav M, Beherens S, Roesky PW (2013) Dalton Trans 42:10404
123. Han J, Shimizu N, Lu Z, Amii H, Hammond GB, Xu B (2014) Org Lett 16:3500
124. Kojima M, Mikani K (2012) Synlett 57
125. Sun Y-W, Xu Q, Shi M (2013) Beilstein J Org Chem 9:2224
126. Michon C, Abadie M-A, Medina F, Agbossou-Niedercorn A (2014) Catal Today 235:2
127. Abadie M-A, Trivelli X, Medina F, Capet F, Roussel P, Agbossou-Niedercorn F, Michon C (2014) ChemCatChem 6:2235
128. Lee SD, Timmerman JC, Widenhoefer RA (2014) Adv Synth Catal 356:3187
129. Zhang Z, Lee SD, Widenhoefer RA (2009) J Am Chem Soc 131:5372
130. Michelet V (2015) Top Curr Chem 357:95–132
131. Quian J, Liu J, Cui J, Xu Z (2012) J Org Chem 77:4484
132. Guo P, Zeng X, Chen S, Luo M (2014) J Organomet Chem 751:438
133. Alcaide B, Almendros P, Alonso JM, Fernández I, Gómez-Campillos G, Torres MR (2014) Chem Commun 50:4567
134. Patil NT, Lakshimi PGVV, Singh V (2010) Eur J Org Chem 4179
135. Liu G, Zhou Y, Lin D, Wang J, Zhang L, Jiang H, Liu H (2011) ACS Comb Sci 13:209
136. Shinde VS, Mane MV, Vanka K, Mallick A, Patil NT (2015) Chem Eur J 21:975
137. Han Z-Y, Xiao H, Chen X-H, Gong L-Z (2009) J Am Chem Soc 13:9182
138. Du Y-L, Hu Y, Zhu Y-F, Tu X-F, Han Z-Y, Gng L-Z (2015) J Org Chem 80:4754
139. Cai S, Yang K, Wang DZ (2014) Org Lett 16:2606
140. Ueda H, Yamagushi M, Kameya J, Sugimoto K, Tokuyama H (2014) Org Lett 16:4948
141. Zhang S, Ma Y, Lan J, Song F, You J (2015) Org Biomol Chem 13:5867
142. Shen C-H, Li L, Zhang W, Liu S, Shu C, Xie Y-E, Fu Y-F (2014) J Org Chem 79:9313
143. Tokimizu Y, Oishi S, Fujii N, Ohno H (2015) Angew Chem Int Ed 54:7862
144. Wang T, Shi S, Pflästerer D, Rettenmeier E, Rudolph M, Rominger F, Hashmi ASK (2014) Chem Eur J 20:292
145. Chiarucci M, Matteucci E, Cera G, Fabrizi G, Bandini M (2013) Chem Asian J 8:1776
146. Cera G, Piscitelli S, Chiarucci M, Fabrizi G, Goggiamani A, Ramón RS, Nolan SP, Bandini M (2012) Angew Chem Int Ed 51:9891
147. Gimeno A, Rodriguez-Gimeno A, Cuenca AB, de Arellano CM, Medio-Simón M, Asensio G (2015) Chem Commun 51:12384
148. Naoe S, Saito T, Uchiyama M, Oishi S, Fujii N, Ohno H (2015) Org Lett 17:1774
149. Sharp PP, Banwell MG, Renner J, Lohmann K, Willis AC (2013) Org Lett 15:2616
150. Tang H-T, Xiong K, Li R-H, Ding Z-C, Ping Z-P (2015) Org Lett 17:326
151. Dhiman S, Ramasastry SSV (2015) Chem Commun 51:557

Top Heterocycl Chem (2016) 46: 87–116
DOI: 10.1007/7081_2015_5002

Published online: 3 February 2016

Synthesis of Oxygenated and Nitrogen-Containing Heterocycles by Gold-Catalyzed Alkyne Oxidation

Longwu Ye and Liming Zhang

Abstract The recent explosive development in homogeneous gold catalysis has revealed its vast potential in the synthesis of oxygenated and nitrogen-containing heterocycles. In this chapter, gold-catalyzed oxidations of alkynes are discussed in the context of constructing synthetically versatile cyclic structures of these types. In cases of oxygenated heterocycles, external nucleophilic oxidants and in particular pyridine/quinoline N-oxides are employed. In cases of N-heterocycles, both external N-heteroarene oxides and tethered tertiary amine/aniline oxides are suitable oxidants for enabling efficient ring construction. Despite exceptions and alternative reaction pathways, a frequently employed theme of these oxidative reactions involves the generation of reactive α-oxo gold carbene intermediates and their subsequent trappings. A diverse range of N-/O-heterocycles are readily accessible via this oxidative gold catalysis.

Keywords Heterocycles • Gold • Catalysis • Oxidation • Diazo compounds • Alkyne • Nucleophile • Oxetan-3-one • Azetidin-3-one • Chroman-3-one • Dihydrofuranone • 3-Coumaranone • Furan • Nitrobenzene • Nitrone • Pyridine • N-Oxide • Carbene • Tertiary amine • Azepanone • Piperidine-4-one • Aza-retro-ene • Oxazole

L. Ye
Xiamen University, Jiaxi Building 842, Xiamen 361005, China
e-mail: longwuye@xmu.edu.cn

L. Zhang (✉)
Department of Chemistry & Biochemistry, University of California, UC Santa Barbara, CA 93106, USA
e-mail: zhang@chem.ucsb.edu

Contents

Abbreviations

Ar	Aryl
Ar^F	3,5-Bis(trifluoromethyl)phenyl
Bn	Benzyl
Boc	*t*-Butoxycarbonyl
Bs	4-Bromobenzenesulfonyl
Bz	Benzoyl
Bu	Butyl
Cy	Cyclohexyl
DCE	1,1-Dichloroethane
DCM	Dichloromethane
ee	Enantiomeric excess
EWG	Electron-withdrawing group
h	Hours
IPr	1,3-Bis(2,6-diisopropylphenyl)imidazolidene
m-CPBA	*meta*-Chloroperbenzoic acid
Me	Methyl
MOM	Methoxymethyl
Ms	Methanesulfonyl
MS	Molecular sieves
NIS	*N*-Iodosuccinimide
NMR	Nuclear magnetic resonance
Ns	2-Nitrobenzenesulfonyl
ph	Phenyl
Phth	Phthaloyl
PMP	4-Methoxyphenyl
Pr	Propyl
Py	Pyridine
rt	Room temperature
TBS	*t*-Butyldimethylsilyl
Tf	Trifluoromethanesulfonyl
TFA	Trifluoroacetic acid

THF	Tetrahydrofuran
THP	2-Tetrahydropyranyl
TMS	Trimethylsilyl
Ts	Toluenesulfonyl

1 Introduction

Over the past dozen years or so, homogenous gold catalysis [1–4] has undergone explosive development, resulting in the development of a large array of novel and versatile synthetic methods. These new additions to the repertoire of synthetic chemists have facilitated synthesis of various natural products and complex molecules [5–8].

A salient feature of gold catalysts and in particular cationic gold(I) complexes is their potent soft Lewis acidities and hence abilities to activate π systems toward nucleophilic attacks under exceptionally mild conditions. Alkynes and allenes are two privileged substrate classes in gold catalysis, with the former more extensively studied. For the synthesis of heterocycles, a highly efficient yet straightforward strategy is gold-promoted cyclizations of heteronucleophiles onto alkynes, which in essence are intramolecular additions of NuH across the C–C triple bond (Scheme 1a) [9, 10]. An alternative approach is to use a nucleophilic oxidant to enable the generation of an alkenylgold intermediate, i.e., **1-A** in Scheme 1b. **1-A** could either undergo direct rearrangement or proceed via the intermediacy of a reactive α-oxo gold carbene **1-B** to afford heterocycles. This latter approach realizes oxidation of C–C triple bond by nucleophilic oxidants, being tethered or external, and is the focus of this chapter.

This chapter aims to cover comprehensively N-/O-heterocycle synthesis enabled by gold-catalyzed oxidations of C–C triple bonds. Sect. 2 will discuss the preparation of O-heterocycles, and the covered methods largely employ external nucleophilic oxidants and are grouped according to the types of alkyne substrates. On the other hand, the preparations of N-heterocycles, covered in Sect. 3, are realized by

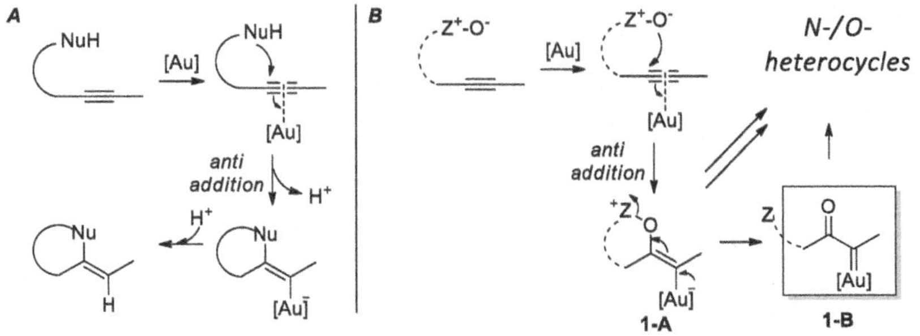

Scheme 1 Gold-catalyzed synthesis of N-/O-heterocycles: (**a**) via intramolecular additions of NuH to C–C triple bonds and (**b**) via oxidation of C–C triple bonds by nucleophilic oxidants

oxidants either tethered to or independent from the C–C triple bond. With internal oxidants, Sect. 3.1 is further divided according to the types of oxidants. With external oxidants, the reactions are grouped following the types of trapping nucleophiles.

2 Synthesis of Oxygenated Heterocycles

2.1 Reaction with Terminal Alkynes

In 2010, Zhang and co-workers reported the first generation of α-oxo gold carbene via gold-catalyzed intermolecular alkyne oxidation, thus making alkynes as surrogates of hazardous α-diazoketones [11] and enabling non-diazo approach to gold carbene chemistry [12]. In the presence of 5 mol% of $Ph_3PAuNTf_2$ and 1.2 equiv. of MsOH, the reactions of homopropargylic alcohols with 3,5-dichloropyridine N-oxide or 2-bromopyridine N-oxide allow efficient formations of dihydrofuran-3-ones in mostly good to excellent yields (Scheme 2). The use of MsOH was believed to prevent the basic pyridine by-product from deactivating the gold catalyst. In addition, a variety of functional groups including acid-sensitive ones such as OMOM and NBoc were tolerated in this catalytic system. The strategy has later been utilized effectively in the construction of bicyclic and spirocyclic scaffolds from *N*-substituted homopropargyl alcohols [13].

In a subsequent work, the same group successfully extended the scope of the reaction to simple propargylic alcohol substrates. As shown in Scheme 3, the gold-catalyzed alkyne oxidation/O–H insertion furnished in this case functionalized oxetan-3-ones in moderate to good yields [14]. Again, acids had to be added to facilitate the reaction. As such, this gold-catalyzed alkyne oxidation strategy provides an expedient and efficient access to this class of strained O-heterocycles of medical importance [15, 16], which typically demands multiple synthetic steps and/or rather functionalized substrates.

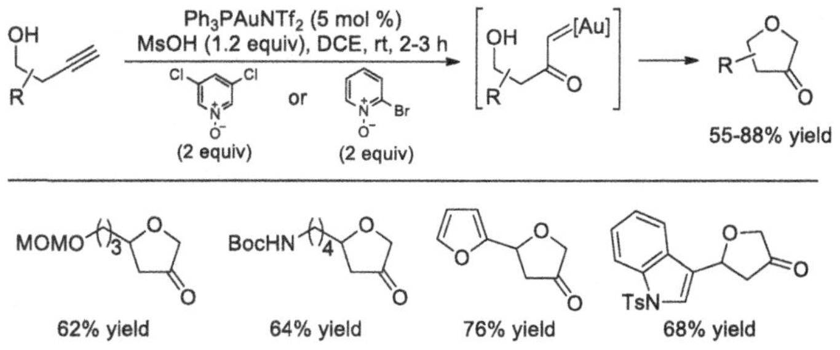

Scheme 2 Gold-catalyzed oxidative formation of dihydrofuran-3-ones

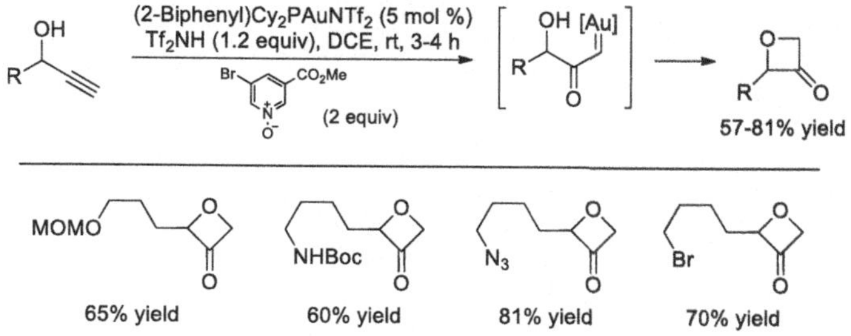

Scheme 3 Gold-catalyzed oxidative formation of oxetan-3-ones

In the case of tertiary propargylic alcohols, the alkyne terminus needs to be functionalized with an electron-withdrawing group in order to prevent ionization at the propargylic position in the presence of the acid additive (e.g., $HNTf_2$), despite the buffering by the basic N-oxide. The oxidative cyclization is, however, much slower albeit in generally higher yields (Eq. 1). Later on, Carreira and co-workers found that tertiary propargylic alcohols without EWG at the alkyne terminus could also be the suitable substrates for the preparation of azaspiro[3.3] heptanones by using $BrettPhosAuNTf_2$ (5 mol%) as catalyst and MsOH (1.5 equiv.) as the additive. This might be due to the fact that MsOH is less acidic than $HNTf_2$ and the cation at a four-membered ring is less stable than in the acyclic cases [17].

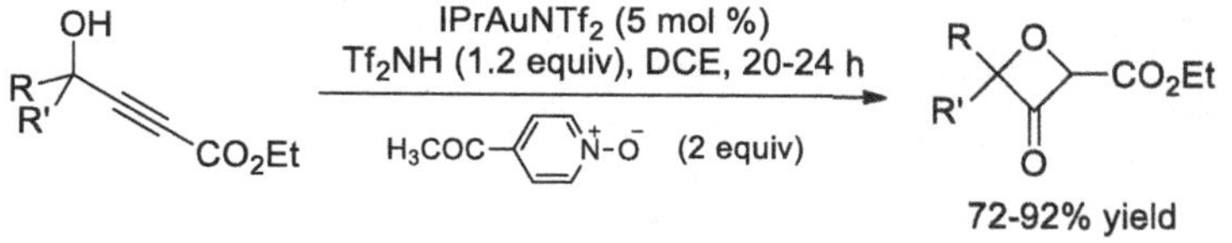

(1)

Besides the alkynyl alcohols, Ye and co-workers in 2015 reported that *O*-ethynylanisoles could also undergo facile oxidative cyclization to deliver the corresponding 3-coumaranone derivatives in moderate to good yields (Scheme 4) [18]. Of note, bulky gold catalyst and oxidant are required to achieve high efficiencies. In addition, density functional theory (DFT) studies reveal that this oxidative cyclization presumably does not proceed through gold carbene intermediates but rather an S_N2' pathway. The synthetic utility of this chemistry was also demonstrated in the facile synthesis of the flavonoid sulfuretin.

Another interesting study of the gold-catalyzed oxidative cyclization was reported by R.-S. Liu and co-workers in 2013. As outlined in Scheme 5, the treatments of 2-ethynylbenzyl ethers with 5 mol% of $[(P^tBu_2(o\text{-biphenyl}))Au]^+$ OTf^- and 1.2 equiv. of 8-methylquinoline N-oxide afford the cycloisomerized products in 48–88% yields [19]. It is worth mentioning that the competitive carboalkoxylation reaction [20] could be completely suppressed in all cases.

In addition, with enantiomerically enriched benzyl ether substrate, the reaction exhibits complete chirality transfer, as shown in Eq. (2). This result also indicates that the reaction presumably proceeds by a gold carbene pathway.

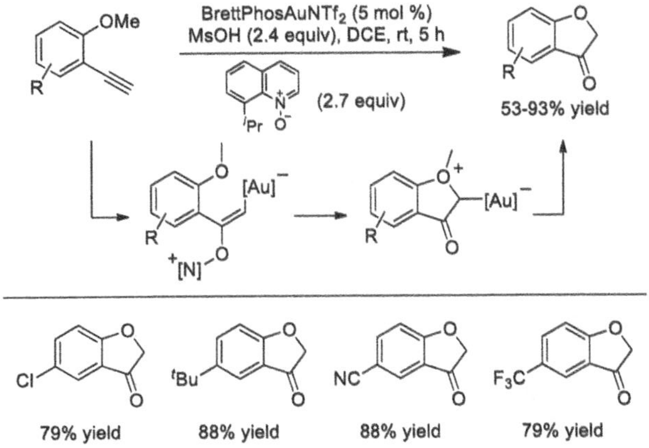

Scheme 4 Gold-catalyzed oxidative formation of 3-coumaranones

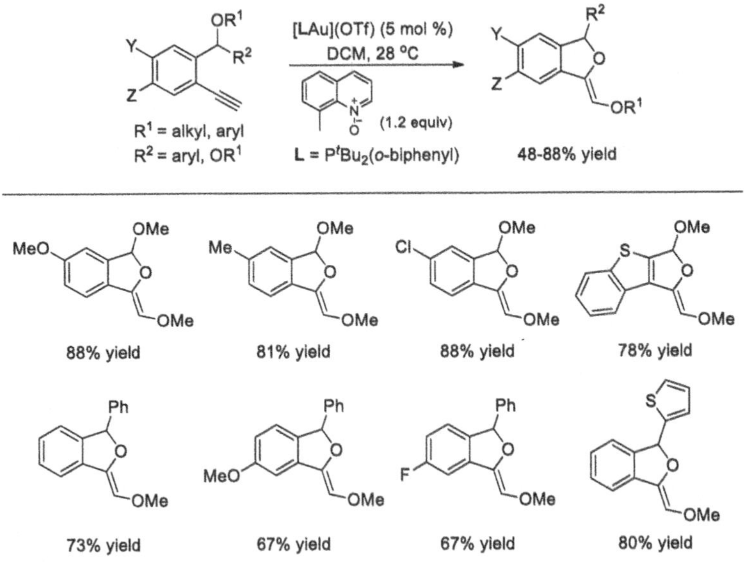

Scheme 5 Gold-catalyzed oxidative cycloisomerizations of 2-ethynylbenzyl ethers

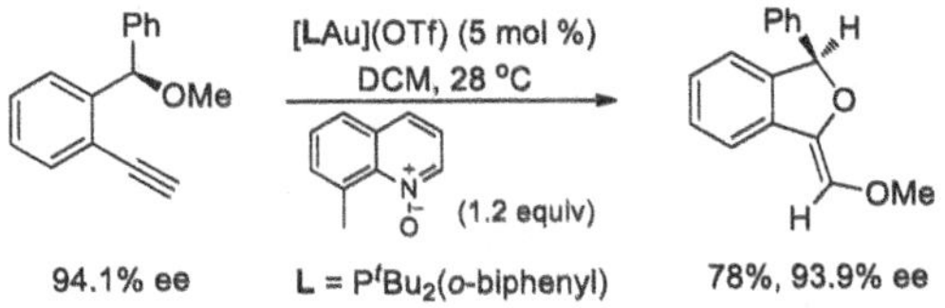

(2)

Scheme 6 outlines the proposed reaction mechanism. The α-oxo gold carbene intermediate **6-A**, generated through intermolecular alkyne oxidation, is initially trapped by the methoxy group at the carbene carbon atom to afford the oxonium species **6-B**, which is expected to form the gold enolate species **6-C** through a

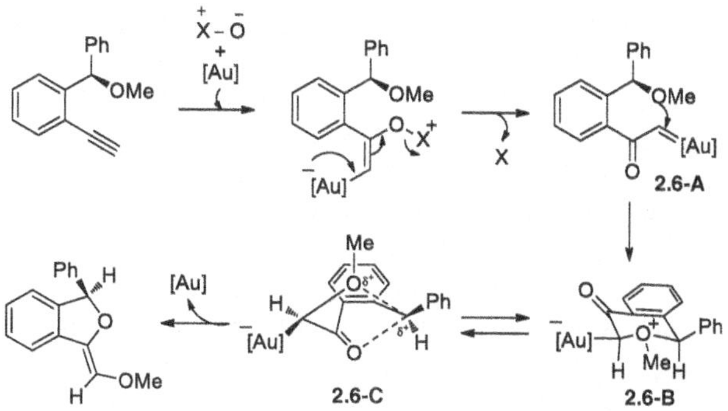

Scheme 6 Chirality transfer in the gold-catalyzed formal cycloaddition reaction

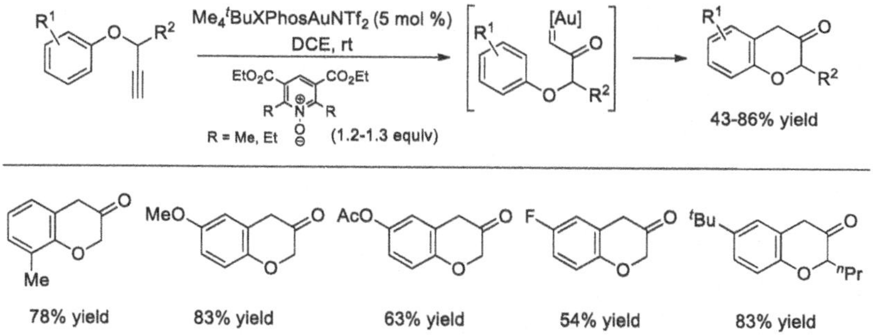

Scheme 7 Gold-catalyzed oxidative formation of chroman-3-ones

ketone/enol equilibrium. Finally, the elimination of gold species delivers the corresponding cycloadduct with *Z*-geometry because of a stable five-membered-ring transition state.

In addition to the intramolecular trapping of the α-oxo gold carbenes by oxygen, such a trapping by the aryl ring has also been explored. In 2012, Zhang and co-workers reported a gold-catalyzed tandem alkyne oxidation/Friedel-Crafts cyclization for the preparations of various chroman-3-ones from readily accessible propargyl aryl ethers [21]. As described in Scheme 7, it was found that the oxidative cyclizations of propargyl aryl ethers in the presence of 5 mol% if $Me_4{}^tBuXPhosAuNTf_2$ afford the desired chroman-3-ones in generally moderate to good yields. Notably, improved efficiency could be achieved by employing hindered and electron-deficient N-oxides as the oxidants.

2.2 *Reaction with Internal Alkynes*

In 2012, Liu and co-workers demonstrated that a [Au(I)] catalyst could be used in combination with an 8-methylquinoline N-oxide to transform 3-en-1-ynamides into substituted 2-aminofurans, thus constituting a gold-catalyzed formal [4+1] cycloaddition (Scheme 8) [22].

As depicted in Scheme 9, the formation of 2-aminofurans could be explained by an initial intermolecular alkyne oxidation to generate the alkenyl gold carbene, which in turn undergoes an oxa-Nazarov cyclization.

In 2013, an elegant method for the synthesis of dihydrofuran-3-one derivatives under gold catalysis was reported by Yang and Tang. As shown in Scheme 10, it was found that treatment of homopropargylic allylic ether substrates with 5 mol% of $IPrAuNTf_2$ as the gold catalyst, 1.1 equiv. of $Yb(OTf)_3$ as the co-catalyst, and 2.0 equiv. of 3,5-dichloropyridine N-oxide as the oxidant in DCE led to the isolation of dihydrofuran-3-ones in generally moderate to good yields [23].

The reaction presumably proceeds through a two-step alkyne oxidation–[2, 3]-sigmatropic rearrangement sequence, as shown in Scheme 11. Alternatively, the reaction might also involve the formation of α-oxo gold carbene, followed by 1,4-allyl migration and Claisen rearrangement.

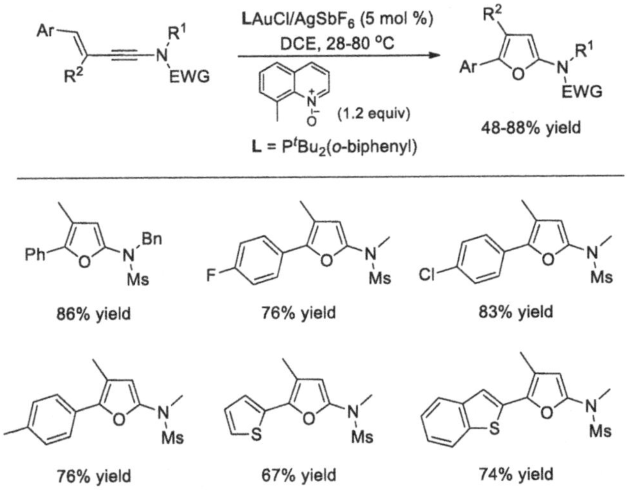

Scheme 8 Gold-catalyzed oxidative formation of 2-aminofurans

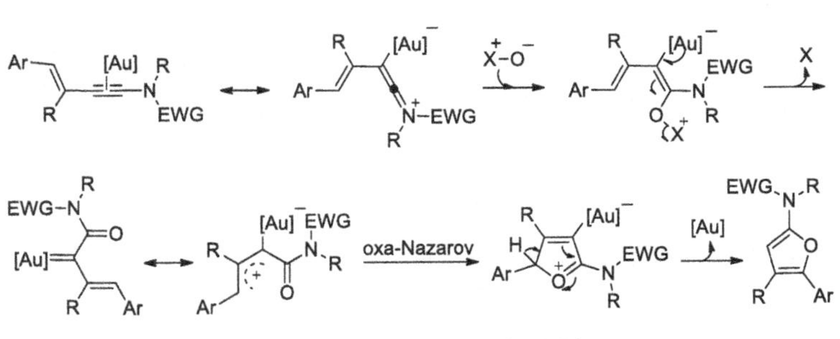

Scheme 9 Proposed mechanism for the Liu's [4+1] cycloaddition

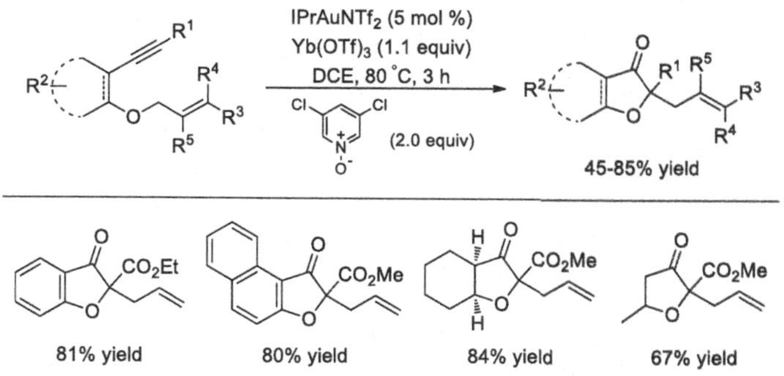

Scheme 10 Gold-catalyzed oxidative formation of dihydrofuran-3-ones by Yang and Tang

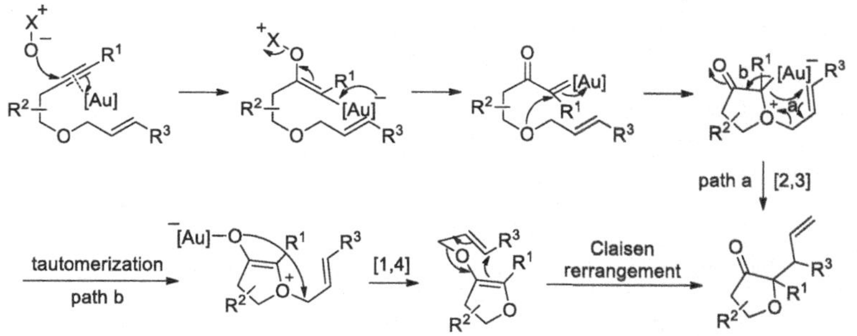

Scheme 11 Plausible mechanisms for the dihydrofuranone formation by Yang and Tang

In 2013, Metz and co-workers found that a combination of $AuCl_3$ (1 mol%) and 3,5-dichloropyridine N-oxide was a suitable system for the intramolecular oxidative domino cyclization/cycloaddition of enyne aldehydes and ketones [24]. The corresponding highly functionalized tetracyclic ketoethers were obtained in mostly good to excellent yields and with excellent diastereoselectivities (Scheme 12).

The mechanism shown in Scheme 13 was proposed to explain this oxidative tandem reaction. An initial intramolecular 6-*endo-dig* attack of the carbonyl oxygen atom on the gold-activated alkyne generates the vinyl gold species **13-A**, which could be isomerized into the carbonyl ylide intermediate **13-B**. A subsequent intramolecular 1,3-dipolar cycloaddition affords the gold carbene **13-C**, which is then oxidized by the oxidant to deliver the final tetracyclic ketoether. Of note, no products resulting from an initial 5-*exo-dig* cyclization were observed in this case.

Hashmi and co-workers demonstrated in 2014 that the oxidative cyclization of symmetric and unsymmetric 1,4-diyn-3-ols could afford the corresponding 3-formylfurans [25]. The reactions proceed at room temperature in the presence of 3 mol% of IPrAuCl/AgOTf, and generally good to excellent yields are achieved. Isotope-labeling experiments as well as DFT calculations revealed that the reaction likely undergoes an initial gold-catalyzed intermolecular oxidation to afford an α-oxo gold carbene intermediate, which is followed by sequential 1,2-alkynyl

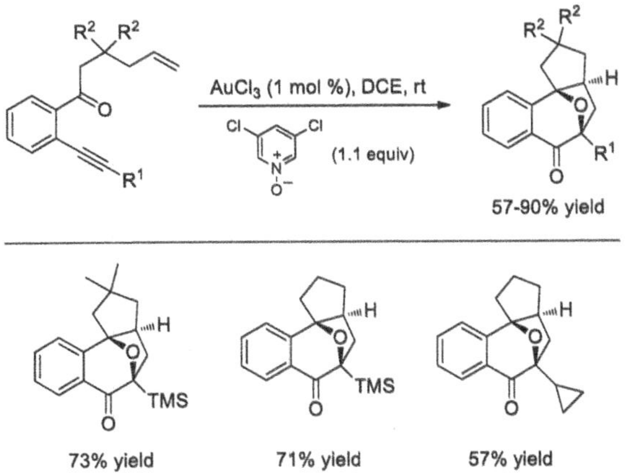

Scheme 12 Gold-catalyzed oxidative formation of tetracyclic ketoethers

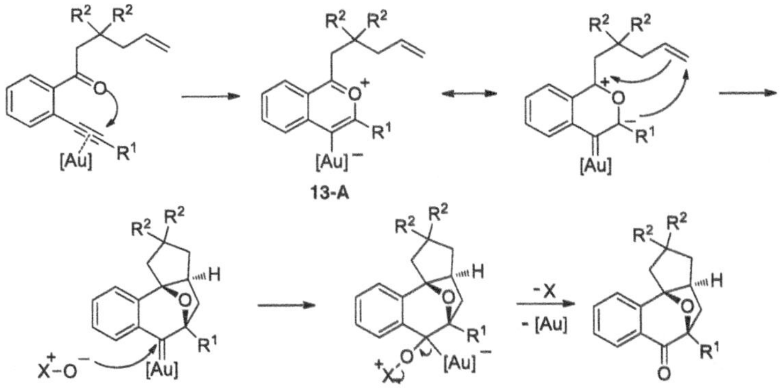

Scheme 13 Plausible mechanisms for the gold catalysis in Scheme 12

migration and cyclization (Scheme 14). This reaction achieves for the first time 1,2-alkynyl migration onto a gold carbene intermediate.

Later, the same group also extended the above tandem reaction for the synthesis of fully substituted 3-formyl-4-iodofurans, which may serve as valuable building blocks in organic synthesis (Scheme 15) [26]. Notably, mechanistic investigations revealed that the reaction might involve a direct iodination of the organo gold intermediate instead of the 3-formylfuran product obtained in the previous study (see Scheme 14).

In addition, Hashmi and co-workers found that this gold-catalyzed oxidative cyclization of dialkynes is also suitable to nitrogen-containing substrates, and the synthesis of *N*-(furan-3-ylmethylene)benzenesulfonamides can be realized (Scheme 16) [27].

Besides N-oxides, molecular oxygen is shown as oxidant in a gold-catalyzed alkyne oxidation by Liu and co-workers in 2006 in a gold-catalyzed cascade cyclization/oxidative cleavage reaction of (Z)-enynols [28]. Hence, the treatments

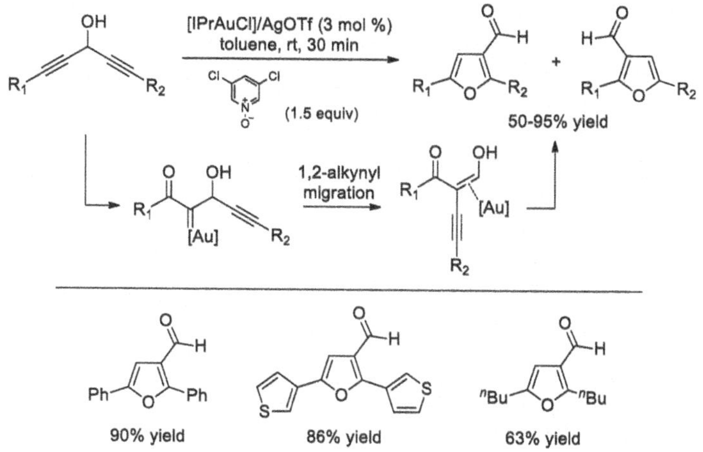

Scheme 14 Gold-catalyzed oxidative formation of 3-formylfurans

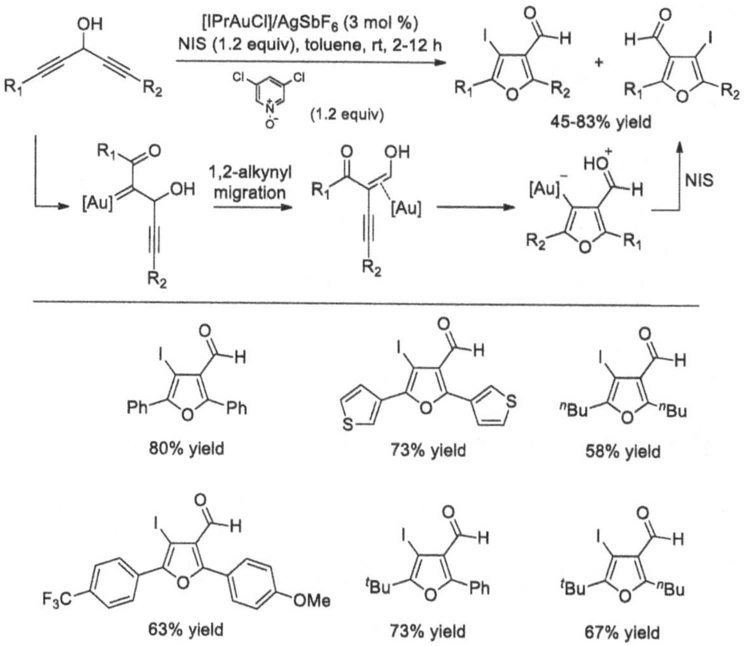

Scheme 15 Gold-catalyzed oxidative formation of 3-formyl-4-iodofurans

of(*Z*)-enynols with $Ph_3PAuCl/AgOTf$ (2 mol%) in THF at 50 °C in an atmosphere of O_2 deliver the corresponding butenolides in 41–97% yields (Scheme 17). Mechanistic studies revealed that the reaction is initiated by a gold-catalyzed 5-*exo-dig* cyclization to generate the intermediate **17-A**, which then undergoes surprising oxidative cleavage of the C–C double bond by O_2. The exact role of gold in this cleavage remains to be studied.

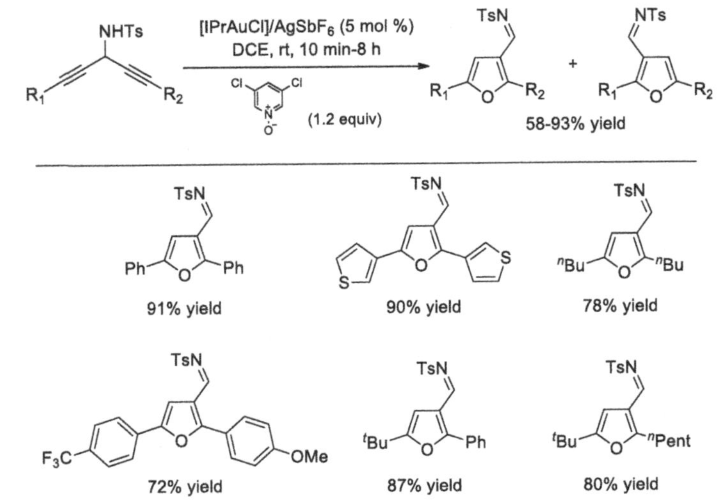

Scheme 16 Gold-catalyzed oxidative formation of *N*-(furan-3-ylmethylene)benzenesulfonamides

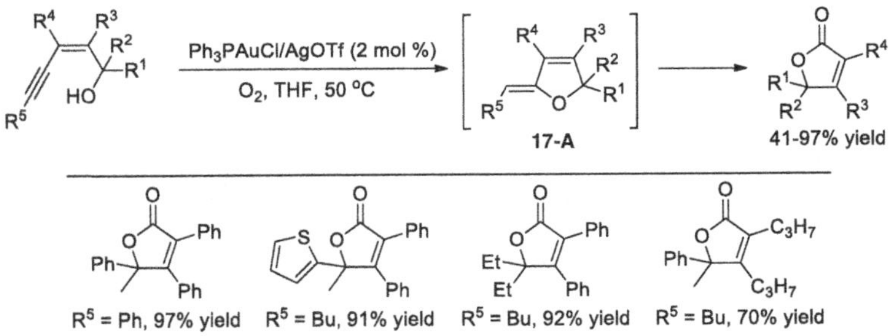

Scheme 17 Gold-catalyzed oxidative formation of butenolides

3 Synthesis of Nitrogen-Containing Heterocycles

3.1 Reactions of Alkyne Substrates with Tethered Oxidants

3.1.1 *O*-Nitro or *O*-Nitrone as Oxidants in the Reactions of Arylalkynes

Despite nitro and nitrone groups are susceptible to reduction by reductants, they are at best mild oxidants. However, they can oxidize optimally positioned C–C triple bonds in the presence of gold catalysts.

For example, *O*-(alkynyl)nitrobenzenes have been subjected to gold catalysis, in which the nitro group takes the role of an internal oxidant, while its reduced nitroso form traps the oxidized alkyne part. In 2003, Yamamoto and co-workers reported such a transformation using $AuBr_3$ as the catalyst, leading to selective formation of either isatogens or anthranils depending on the nature of the alkyne substituent (Eq. 3) [29].

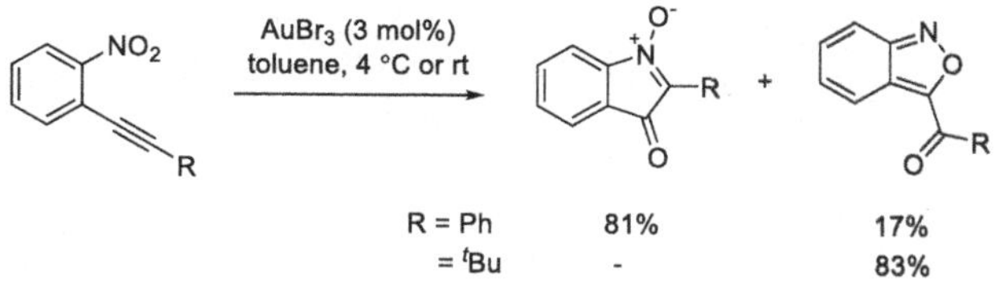

(3)

With *O*-(ethynyl)nitrobenzene substrates, the R.-S. Liu group in 2011 realized gold (I)-catalyzed reactions of these compounds with alkenyl ethers [30]. As shown in Scheme 18, the reaction offers rapid access to azacyclic products with excellent stereoselectivities. Mechanistic studies are consistent with the outlined mechanism, where an initial gold-promoted oxidation of the terminal alkyne by the *ortho*-nitro group furnishes a nitroso gold carbene species, i.e., **18-A**. **B**. This reactive intermediate then undergoes cyclization and tautomerization to render the metallated 1,3-dipole **18-B**. Its subsequent cycloaddition with a vinyl ether is highly regioselective.

With the nitrone moiety as a tethered oxidant, Shin and co-workers reported in 2008 a $AuCl_3$-catalyzed access to the azomethine ylide **19-A** via oxidation of arylalkynes (Scheme 19) [31]. Mechanistically, this redox chemistry generates an α-oxo gold carbene moiety, which is followed by its trapping by the imine generated upon nitrone deoxygenation. These intermediates subsequently react with tethered alkenes in 1,3-dipolar cycloadditions to provide rapid access to complex azacycles. Among the selected cases shown in the scheme, a tethered C–C triple bond is suitable as the dipolarophile, the benzene ring can be removed, and even a saturated two-carbon linker is allowed, albeit with a moderate yield. The last case suggests that conjugation between the redox partners is not necessary in the initial gold catalysis. With R^1 as chiral (*p*-nitropheneth)-1-yl, this gold-catalyzed redox

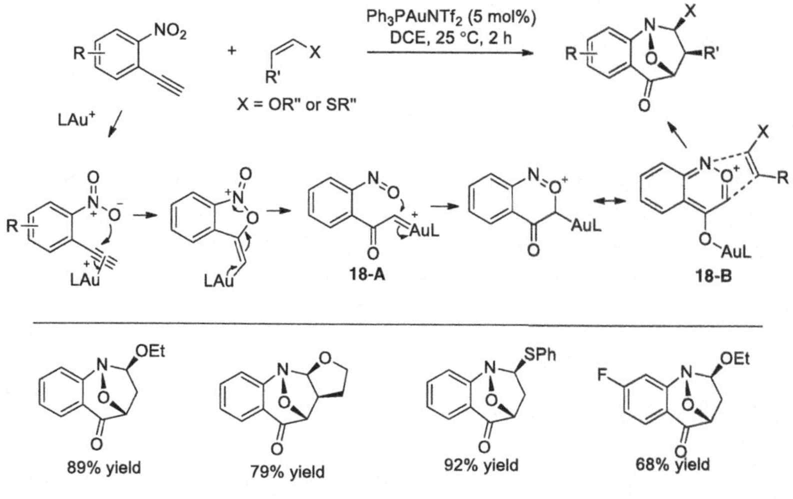

Scheme 18 Gold-catalyzed tandem redox cycloisomerizations of *o*-nitrophenylalkynes and 1,3-dipolar cycloadditions

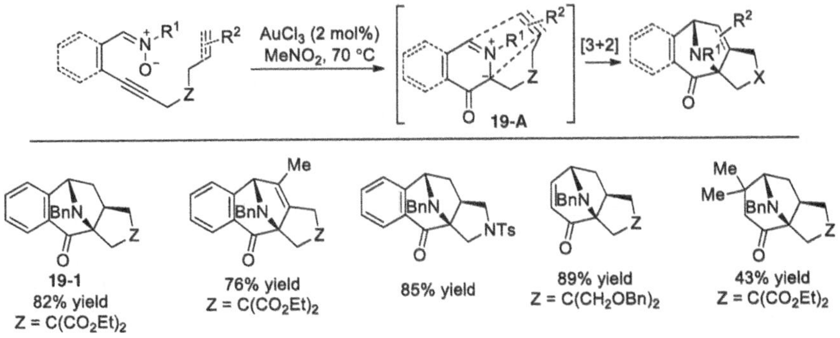

Scheme 19 Tandem gold-catalyzed intramolecular oxidations and 1,3-dipolar cycloadditions

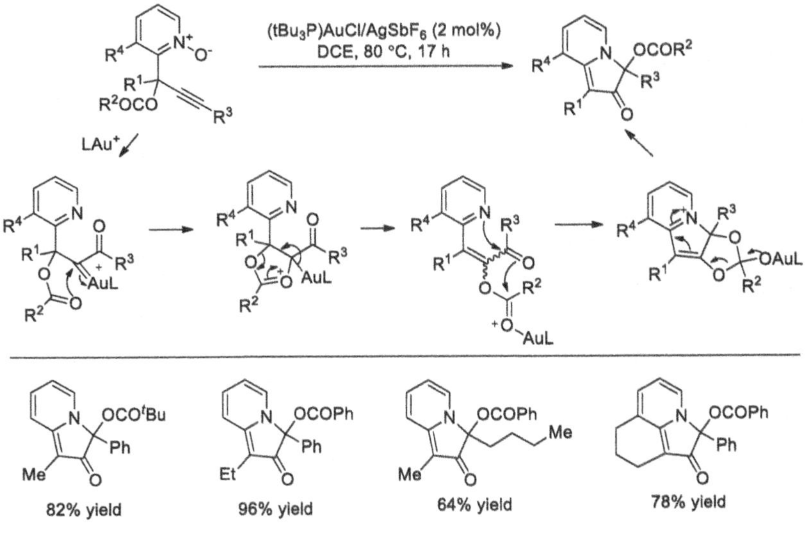

Scheme 20 Gold-catalyzed oxidative formation of indolizinones

cascade becomes highly diastereoselective (d.r. >20:1). Consequently, upon hydrogenolysis, the debenzylated **19-1** was isolated with 98.5% e.e. [32].

In 2010, the Shin group reported another application of this gold-catalyzed redox chemistry of alkynyl nitrones, where an interesting redox–pinacol–Mannich–Michael cascade enables rapid access to complex azacycles [33].

3.1.2 Pyridine N-Oxides as Oxidants

With a more nucleophilic pyridine N-oxide as oxidant, the intramolecular redox chemistry occurs regioselectively with an initial 6-*endo-trig* cyclization of an internal alkyne substrate, affording a reactive α-oxo gold carbene intermediate (Scheme 20) [34]. Its subsequent transformation, as reported by Ohe, likely involves sequential 1,2-migrations by the carboxyl group to the gold carbene center and in a thermo-promoted cyclative rearrangement. The reaction provides access to indolizinones but does not work with terminal alkyne substrates.

3.1.3 Tertiary Amine N-Oxides as Oxidants

With even more nucleophilic tertiary amine N-oxide as internal nucleophilic oxidants, several methods have been developed by Zhang and co-workers to provide rapid access to simple yet versatile N-heterocycles [35–37]. However, experimental observations as well as DFT calculations [38] suggest that these reactions do not involve α-oxo gold carbene intermediates.

In 2009, a one-pot synthesis of tetrahydrobenz[*b*]azepin-4-ones from tertiary *N*-(but-3-ynyl)anilines was realized [35]. As shown in Scheme 21, the aniline substrate was first oxidized into the corresponding N-oxide, which could decompose via the Cope elimination and was used directly for the subsequent gold catalysis. Upon the gold-promoted cyclization, the isoxazolidinium intermediate **21-A** thus formed most likely undergoes a facile [3,3]-sigmatropic rearrangement to deliver the product skeleton, analogous to the related sulfoxide case [39], instead of the gold carbene intermediate **21-B**. This mechanistic scenario is also supported by the reactions with substrates possessing electron-withdrawing groups at the alkyne terminus. In these cases, the anticipated products **21-1** were formed upon the generation of the N-oxides, without the need of any gold catalyst. The reaction can accommodate substituents of various natures on the benzene ring, albeit displaying mostly low regioselectivities with *meta*-substituents.

Expansions of the scope of the tertiary amines from anilines to aliphatic amines lead to synthetically versatile synthesis of piperidine-4-ones in a three-step, two-pot

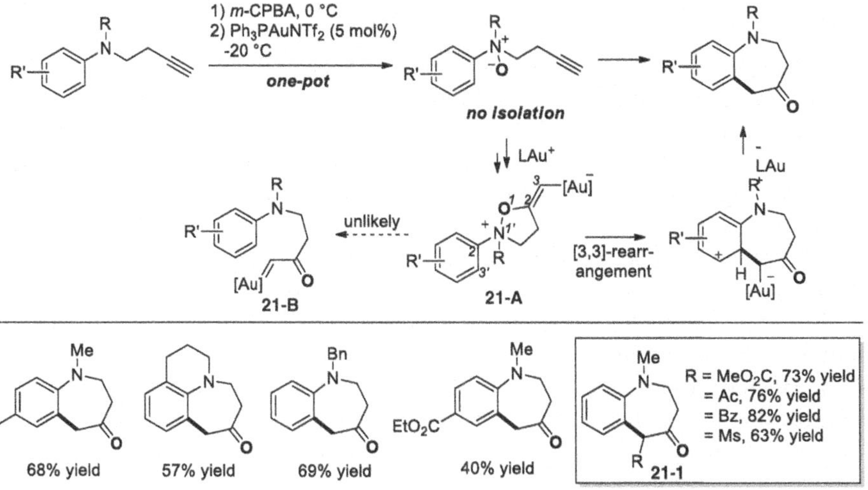

Scheme 21 Gold-catalyzed synthesis of tetrahydrobenzazepinones

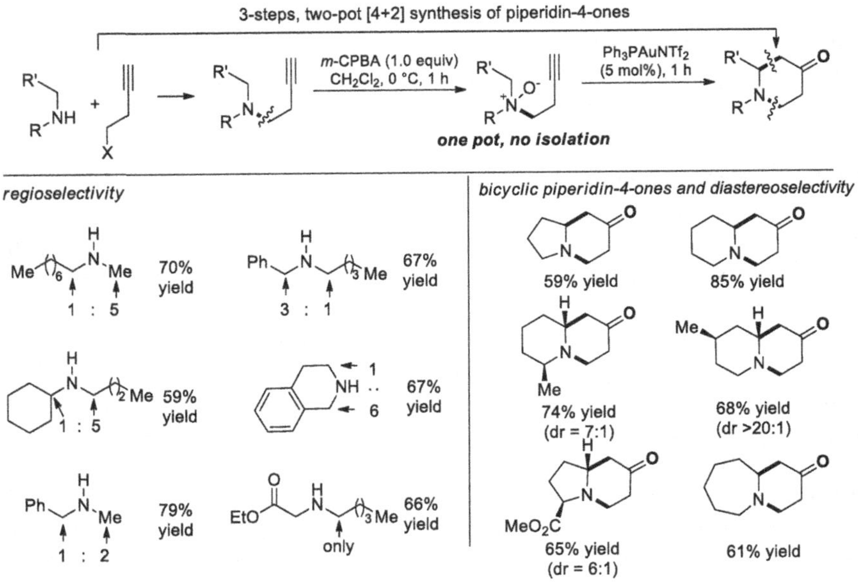

Scheme 22 Two-pot, [4+2] synthesis of piperidine-4-ones

[4+2] fashion [36]. As shown in Scheme 22, the amine substrate can be prepared in a simple S_N2 process that brings together a 4C unit and a C1 + N one. The one-pot oxidation and gold catalysis then realize the C–H insertive cyclization to afford the synthetically useful N-heterocycle. With differentiated α-C–H bonds on nitrogen, the gold catalysis displays rather surprising regioselectivities, with methyl favored over methylene, methylene over methine, and methyl over benzylic, which are inconsistent with those anticipated by metal carbene C–H insertions. The reaction displays synthetically useful diastereoselectivities in the formation of bicyclic piperidine-4-ones and has been applied as key steps in the synthesis of (+)-lentiginosine [40] and lasubine II and decinine [41].

The Zhang group later extended the chemistry to a three-step, two-pot [5+2] approach to azepan-4-ones [37]. Shown in Scheme 23 are some of the notable cases of this ready access to seven-membered N-heterocycles. Of note is the much improved regioselectivities over those observed in the above [4+2] chemistry. For example, methyl is preferred over methylene in a ratio of >20:1 as **23-1** is the only isolated product, so is methyl over benzyl in the case of **23-2**. This chemistry also displays in general excellent stereoselectivities with dr typically >20:1.

The mechanisms of these gold-catalyzed redox transformations of alkynyl tertiary aliphatic amine N-oxides are investigated by DFT calculations [38]. Scheme 24 shows the calculated key C–H fragmenting and regio-determining step for the [5 + 2] chemistry, which can be viewed as an aza-retro-ene reaction. Instead of a pathway involving an α-oxo gold carbene intermediate, this novel mechanistic pathway can readily explain the observed regioselectivities by considering selective positioning of smaller groups to the reacting axial position of the

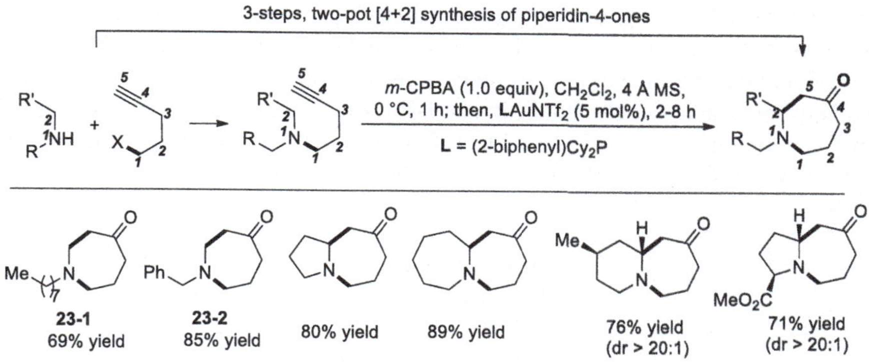

Scheme 23 Two-pot, [5+2] synthesis of azepan-4-ones

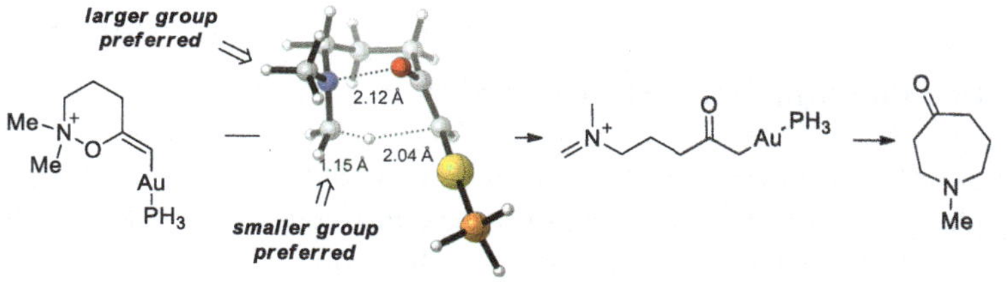

Scheme 24 Reaction mechanism involving an aza-retro-ene reaction as revealed by DFT calculations

cleaving six-membered ring of the shown transition state. The much better regioselectivities observed with [5 + 2] can also be explained qualitatively by considering that in a five-membered ring the preference for pseudoequatorial position by a larger group is to a lesser degree, which is also quantitatively supported by the calculations.

N-Sulfonylhydroxylamines can also act as internal oxidant to generate α-oxo gold carbenes. With substrates containing terminal alkynes, 3-pyrrolidinones are formed in fair to good yields (Eq. 4) [42].

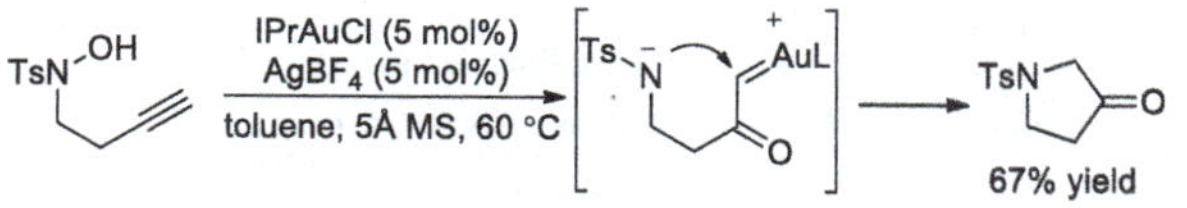

(4)

3.2 *Reactions of Alkyne Substrates with External Oxidants*

Oxidation of alkynes using external oxidants, unlike its internal counterpart, offers exceptional synthetic flexibility in the construction of N-heterocycles. Gasparrini and co-workers reported in 1993 that terminal alkynes can be oxidized by HNO_3 in the presence of a [Au(III)] catalyst, affording 3,5-disubstituted isoxazoles in moderate yields (Eq. 5) [43]. Despite this really early work, little progress was made in this type of oxidative gold catalysis, let alone its application in the synthesis of N-heterocycles until the advent of the α-oxo gold carbene approach by Zhang and co-workers (see Scheme 2) [11]. With pyridine/quinoline N-oxides as the preferred oxidants, the reactive α-oxo gold carbene **1-B** formed (see Scheme 1B) or the initial gold-containing oxidant-alkyne adduct **1-A** is capable of reacting with various types of nucleophiles to afford N-heterocycles.

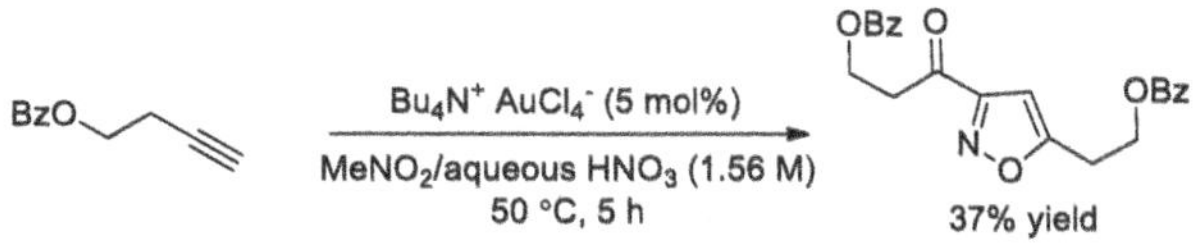

(5)

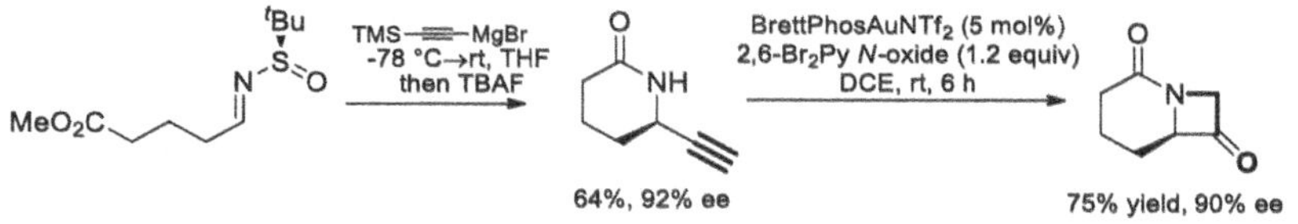

(6)

3.2.1 Reaction with Tethered Nucleophiles

With the facile generation of highly reactive α-oxo gold carbene intermediates from C–C triple bonds, a common strategy for the construction of N-heterocycles is to trap them with *N*-nucleophiles, in analogy to that using *O*-nucleophiles.

By employing this strategy, Zhang and co-workers in 2010 reported an efficient construction of azetidin-3-ones from *N*-propargylic amides [44]. As shown in Scheme 25, these amides of high e.e., readily accessible via the *tert*-butyl sulfinamide chemistry [45], cyclize smoothly to deliver the strained N-heterocycle in mostly good yields. Carboxamides are not suitable due to the interference of its non-oxidative cyclization [46]; however, with the carbonyl group tied back, the reaction in Eq. (6) proceeded smoothly, affording the bicyclic lactam in a decent 75% yield. Of note is that the combination of bulky BrettPhos ligand and sterically demanding and more reactive 2,6-dibromopyridine N-oxide is optimal and, moreover, allows the avoidance of acid additives [11].

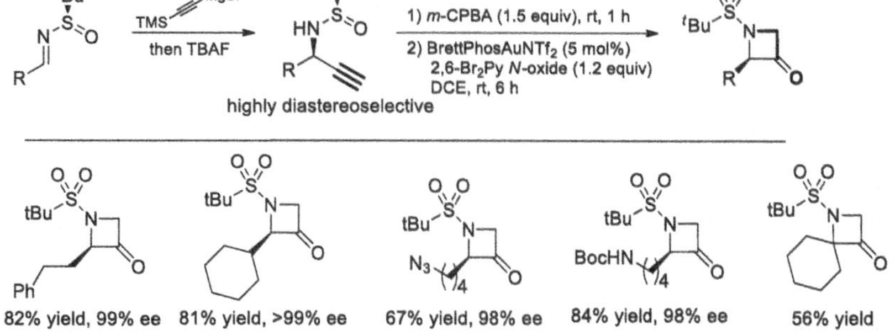

Scheme 25 Oxidative cyclization of *N*-propargyl sulfonamides

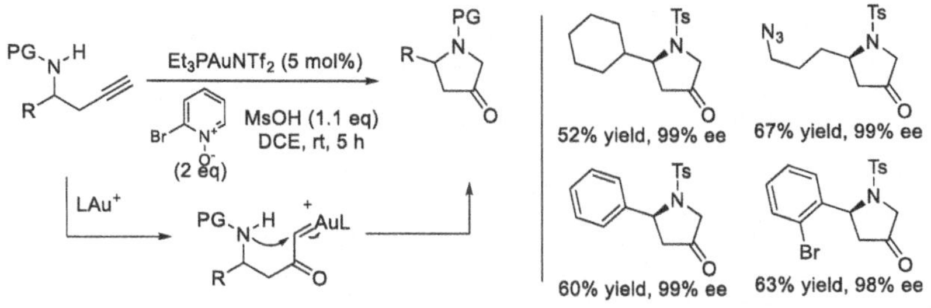

Scheme 26 Synthesis of pyrrolidin-3-ones via oxidative gold catalysis

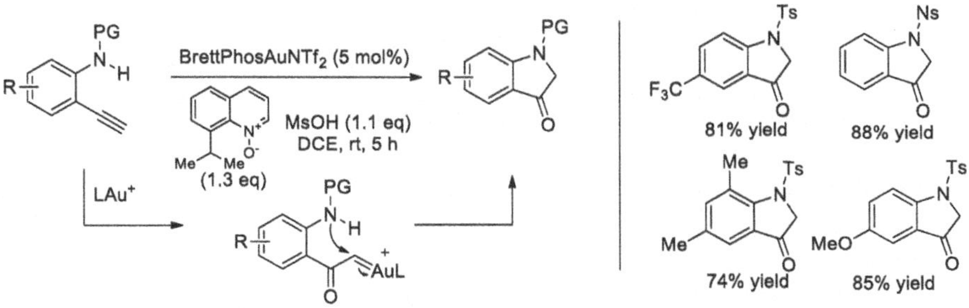

Scheme 27 Synthesis of oxindoles via oxidative gold catalysis

Ye and co-workers have applied a similar strategy in the facile construction of pyrrolidin-3-ones (Scheme 26) [47] and 3-oxindoles (Scheme 27) [48]. In the former case, chiral products were obtained from readily available chiral *N*-homopropargylic amides, and one of them was converted into (−)-irniine.

In the latter case, the typically observed non-oxidative 5-*endo*-dig cyclization en route to indole is outcompeted by the oxidative catalysis, highlighting the facile nature of the latter process; however, with internal alkyne substrates, the indole formation was exclusively observed.

Instead of *N*-based nucleophiles, the α-oxo gold carbene intermediate can be trapped by optimally tethered *C*-nucleophiles. In these cases, the nitrogen atom is part of the tether. For example, J. Zhang and co-workers reported that *N*-allyl propiolamides undergo facile gold-catalyzed oxidative cyclopropanations, affording azabicyclo[n.1.0]hexanes in fair to good yields (Scheme 28) [49]. It is notable that the reactions of all-carbon 1,6-enynones also work well and have recently been coerced into highly enantioselective processes by chiral gold phosphoramidite complexes [50]. Mechanistic investigations suggest that the initial gold-catalyzed adduct of alkyne and N-oxide might bypass the gold carbene intermediate and directly form the observed product. *N*-allylynamides can also undergo oxidative cyclopropanation to afford related azabicyclo[3.1.0] hexanes [51].

A synthetically valuable application of this oxidative gold catalysis was reported by Y. Liu and co-workers, where 2-alkynyl-1,2-dihydropyridines or quinolines

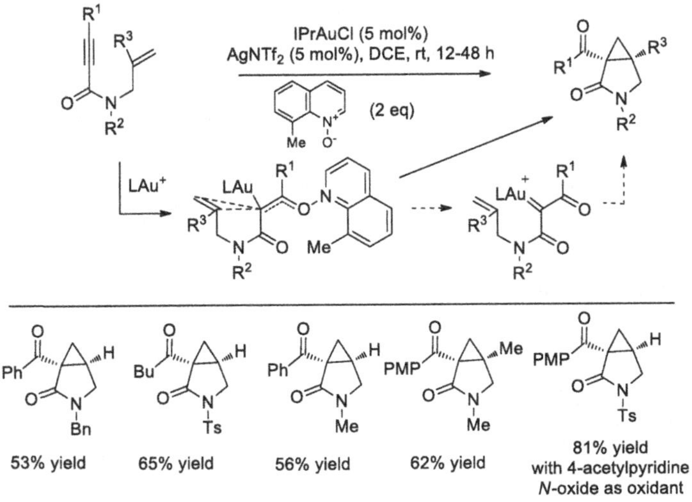

Scheme 28 Gold(I)-catalyzed intramolecular oxidative cyclopropanation

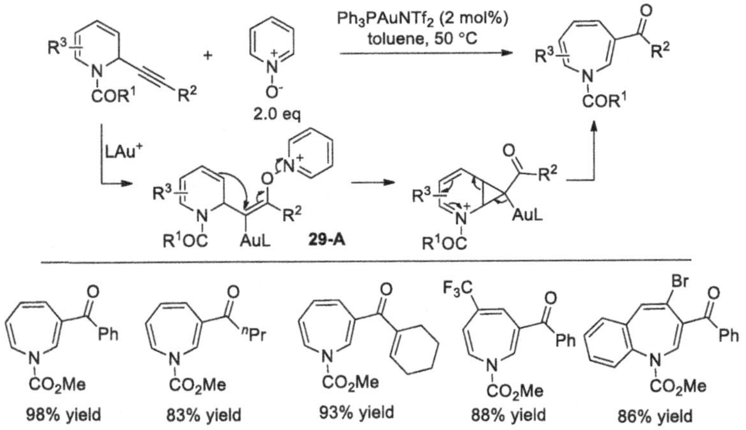

Scheme 29 Gold-catalyzed oxidative construction of functionalized azepine or benzazepine scaffolds

undergo ring expansion to afford functionalized azepines or benzazepines (Scheme 29) [52]. DFT calculations support a mechanism where no gold carbene is formed and the initial adduct **29-A** is directly attacked by the vicinal dihydropyridine π electrons.

With tethered arenes as nucleophile, 3-acyloxindoles (Scheme 30) are prepared from *N*-arylpropiolamides with a broad reaction scope and mostly excellent yields [53]. The related oxidations of *N*-arylynamides, however, afford mostly low yields

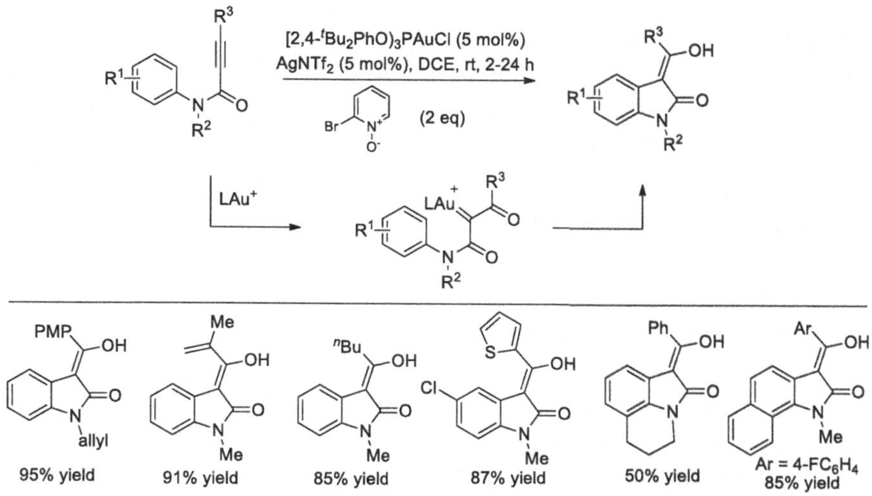

Scheme 30 Gold-catalyzed oxidative synthesis of 3-acyloxindoles

of oxindoles [54]. The high efficiencies in the former case might be due to the higher nucleophilicities of the bisacyl-substituted gold carbene moiety, albeit the intermediacy of this species might be questionable in light of the above DFT-revealed mechanism.

3.2.2 Reaction with External Nucleophiles

The α-oxo gold carbene intermediate, if generated, appears to be highly electrophilic. As a result, trapping it with external nucleophiles proves to be rather challenging. In fact, the study by Xiang and Zhang revealed that it can abstract chloride from chlorinated solvents such as DCE and DCM [55].

Instead of optimizing reaction conditions to avoid side reactions with solvents, Zhang and co-workers used nitrile nucleophiles as solvent. For example, by running the oxidation in acetonitrile, the 2,5-disubstituted oxazole **31-1** was formed in nearly quantitative yields with various gold catalysts (Scheme 31) [56]. The optimal oxidant is 8-methylquinoline N-oxide, which allows omitting acid additive in the reaction. The nitrile ylide **31-B**, formed upon trapping of the gold carbene **31-A** with solvent acetonitrile, is most likely the key intermediate leading to the formation of the versatile heteroaromatic ring.

This formal [2+2+1] annulation worked well with various terminal alkynes and a range of cheap nitriles. When the nitrile is too scarce and/or expensive to be used as solvent, it was found that the chemistry could still afford a serviceable yield if a moderate excess of the nitrile (3 equiv.) is used in the absence of any solvent (Eq. 7). In addition, the excess nitrile could be recovered.

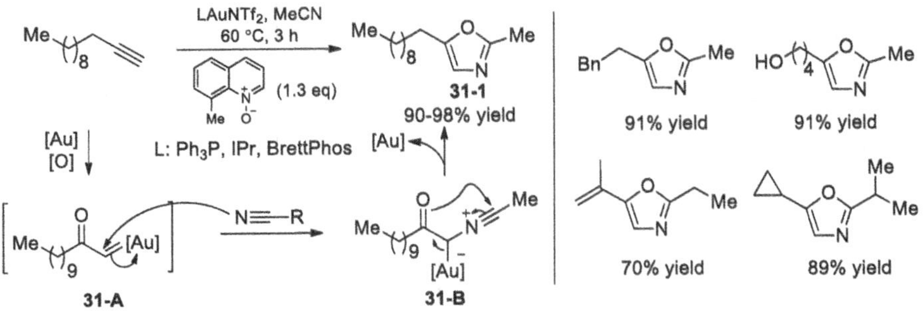

Scheme 31 Gold-catalyzed [2+2+1] annulation to 2,5-disubstituted oxazoles

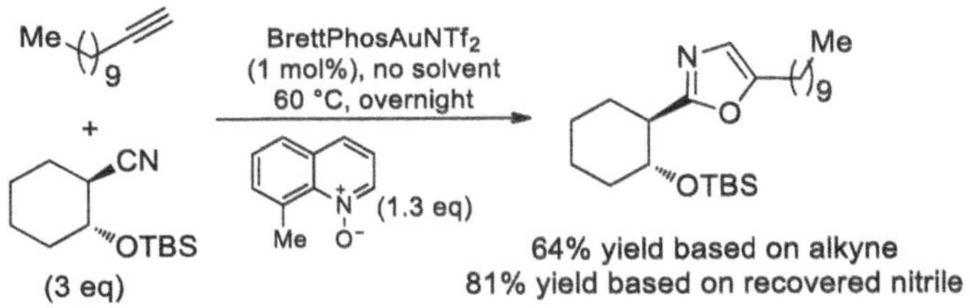

(7)

Trapping the oxidatively generated α-oxo gold carbene intermediates with stoichiometric carboxamides was realized, however with the *P*,*N*-bidentate ligand Mor-DalPhos as the metal ligand [57]. Under the optimized conditions, the reaction of 1-dodecyne and 4-methoxybenzamide affords the 2,4-disubstituted oxazole **32-1** in 87% yield (Scheme 32) [58]. The net result of this reaction is an oxidative [3+2] annulation between an amide and a terminal alkyne. The reaction scope is general to terminal alkynes but limited to conjugated carboxamides.

The unique role of Mor-DalPhos in promoting this reaction was attributed to its *P*,*N*-bidentate nature and hence its ability to use both coordinating heteroatoms to enable the formation of a tris-coordinated gold carbene.

3.2.3 Oxidants or Parts of Them Incorporated into Products

A major issue in oxidative gold catalysis is that the highly electrophilic α-oxo gold carbene generated might be attacked again by the nucleophilic N-oxide, thereby leading to further oxidation. Screening of these oxidants is often an integral part of reaction discovery/optimization in order to minimize this side reaction. Liu reported in 2014 that a facile intramolecular trapping of the gold carbene species could pave the way for a productive participation of another molecule of the oxidant in the reaction [59]. As shown in Scheme 33, from an enynyl nitrile substrate, the gold-catalyzed oxidation of its C–C triple bond affords the α-oxo gold carbene intermediate **33-A**, which is trapped by the tethered cyano group. The nitrilium intermediate **33-B** thus formed would be attacked by another molecule of the

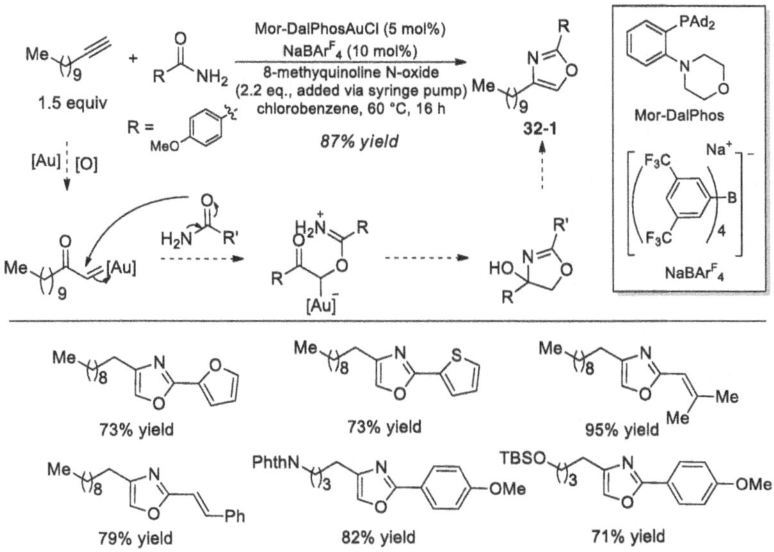

Scheme 32 One-step oxidative [3+2] annulation toward the synthesis of 2,4-disubstituted oxazoles enabled by Mor-DalPhos

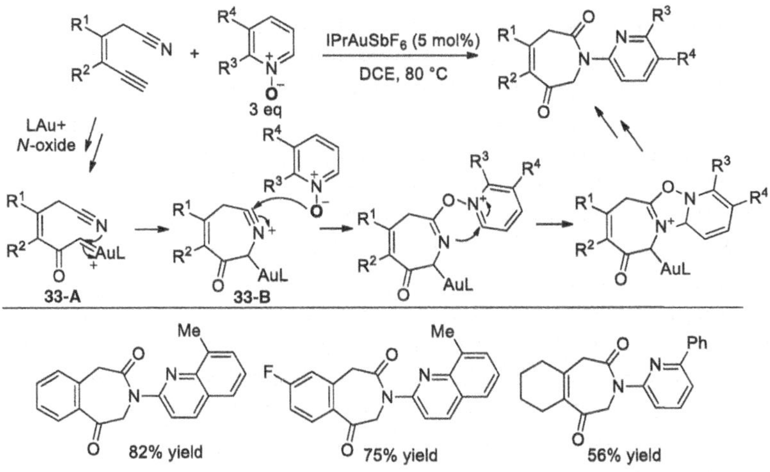

Scheme 33 Oxidative transformations of cyanoenynes

oxidant, which is followed by rearrangement to yield the observed seven-membered lactam product.

Ye reported a formal [3+2] cycloaddition between ynamides and isoxazoles to construct functionalized pyrroles (Scheme 34) [60]. Of unique feature is that an isoxazole instead of a more reactive N-oxide is employed to oxidize an activated ynamide into the α-oxo gold carbene intermediate **34-A**. With $R^4 = H$, the cyclized

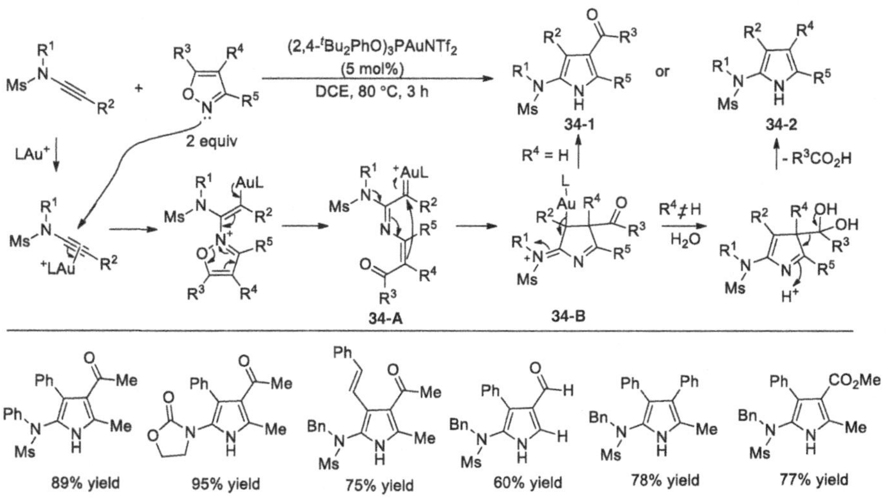

Scheme 34 Gold-catalyzed formal [3+2] cycloaddition between ynamides and isoxazoles

intermediate **34-B** undergoes deprotonative aromatization to deliver the 4-acylpyrrole **34-1**. However, when R^4 is an aryl, an alkyl, or a carboxyl group, hydrolytic decarboxylation occurs, and **34-2** was formed. The reaction displays a broad scope with regard to the isoxazoles, but the ynamides are limited to those with R^2 being nonalkyls except cyclopropyl due to the competition of 1,2-C–H insertion by the gold carbene **34-A**. Of note is that the pyrrole products are fully substituted and should be of synthetic utility.

In gold-catalyzed oxidation of alkynes by pyridine/quinoline N-oxides, the N-heteroarene is generated as by-product along with the desired reactive α-oxo gold carbene intermediates. The reaction between them, especially facilitated by close proximity in the solvent cage, is usually detrimental but can become productively en route to atom-economic preparation of azacycles.

For example, Liu reported in 2013 that the by-product 8-methylquinoline was incorporated into one of the isolated products, which is a complex polycyclic pyrrolidine (Scheme 35) [61]. The addition of external 8-methylquinoline allows improved yields, and the reaction tolerates variations at quinoline 8-position, the EWG, and the unsaturated tether. Mechanistically, the gold carbene reacts with quinoline to generate a metal-less azomethine ylide, which then undergo concerted intramolecular [3+2] cycloaddition to afford the product. Of note is that the reaction exhibits excellent stereoselectivity as only one diastereomer was observed.

Another example is the gold-catalyzed redox synthesis of imidazo[1,2-*a*]pyridines reported by Toste and co-workers [62]. As shown in Scheme 36, the oxidant 2-aminopyridine N-oxide in the presence of one equivalent of TFA and a [Au(III)] catalyst oxidizes terminal alkynes to the corresponding α-oxo gold carbene, which

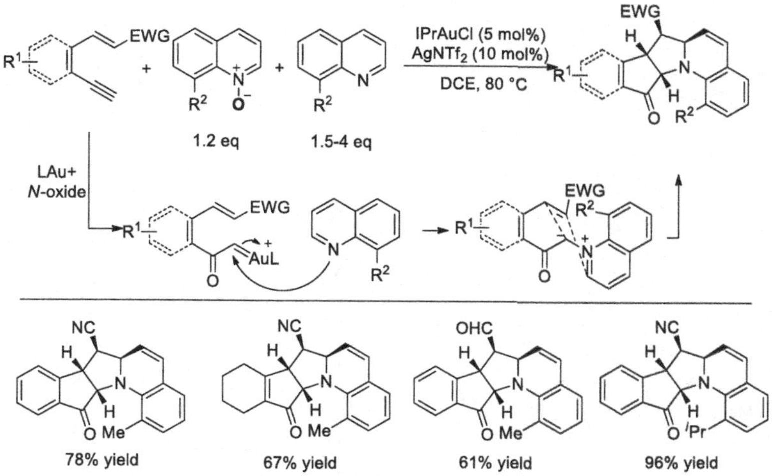

Scheme 35 Gold-catalyzed oxidative cycloaddition

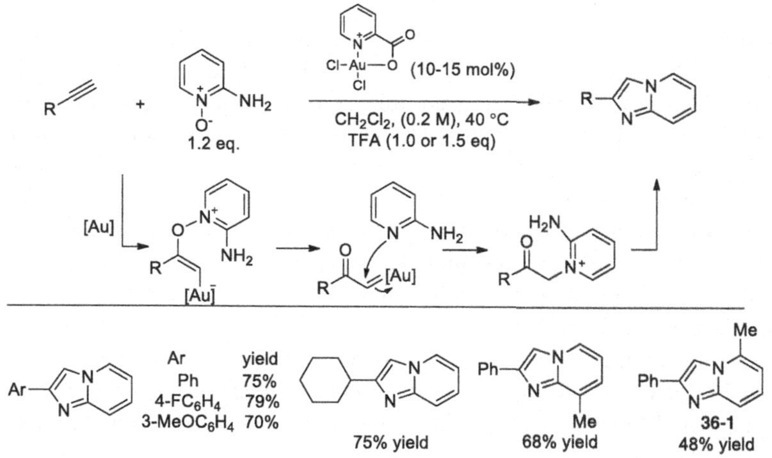

Scheme 36 Gold-catalyzed redox synthesis of imidazo[1,2-*a*]pyridines

is then effectively trapped by the pyridine by-product en route to the bicyclic N-heteroarene. The acid additive is essential for the good reaction yields. With regard to the reaction scope, internal alkynes do not undergo the reaction, terminal alkynes except those of steric hindrance are generally allowed, and various ring substituents on the pyridine 3,4 and 5 positions of the oxidant are tolerated. The reaction with 6-methyl-2-aminopyridine N-oxide, however, leads to a low yield of **36-1**.

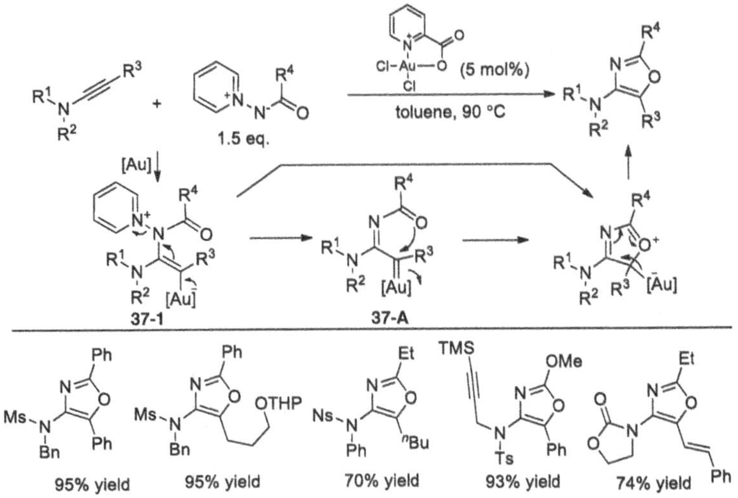

Scheme 37 Synthesis of 2,4,5-trisubstituted oxazoles via a gold-catalyzed formal [3+2] cycloaddition

Though not strictly oxidation, Davies and co-workers have reported several versatile [3+2] cycloadditions of ynamides/3-alkynylindoles and pyridinium *N*-aminides, affording oxazoles or fused imidazoles and thereby in essence realizing oxidation of these activated C–C triple bonds [63–65]. For example, the reactions of ynamides and acylaminides offer atom-economic access to fully substituted oxazole products (Scheme 37) [63]. Mechanistically, imino gold carbene intermediates **37-A** might be formed, although their subsequent trapping by the acyl group must be facile in order to outcompete the often observed 1,2-C–H insertion. Alternatively, the initial adducts **37-1** might undergo cyclative fragmentations to bypass **37-A**. Overall, the aminides behave as nitrene transfer reagents or nitrene precursors with tethered nucleophiles and react formally as 1,3-*N*,*O*-dipoles. The reaction is fairly general and can accommodate substrates with various types of substituents and functional groups including acid-labile THP and additional unsaturated C–C bonds.

The activated ynamides in the above reaction can be replaced by indol-3-ylalkynes, and the gold catalysis affords 4-(indol-3-yl)oxazoles with excellent regioselectivities in fair to good yields (Eq. 8) [64]. On the other hand, stronger reaction conditions including a more acidic $(2,4\text{-}^t Bu_2PhO)_3PAuSbF_6$ and 120 °C in *m*-xylene are required, and indol-3-ylalkynes with alkyl groups terminating the other end led to complex mixtures. Electron-rich tertiary anilines can replace indole for moderate efficient reactions.

By using pyridinium *N*-(heteroaryl)aminides as the oxidants and formally 1,3-*N*,*N*-dipoles, the above strategy can be applied to a direct synthesis of imidazole–diazines and imidazole–pyridines (Eq. 9) [65]. The reaction has a broad reaction scope and tolerates a range of functional groups.

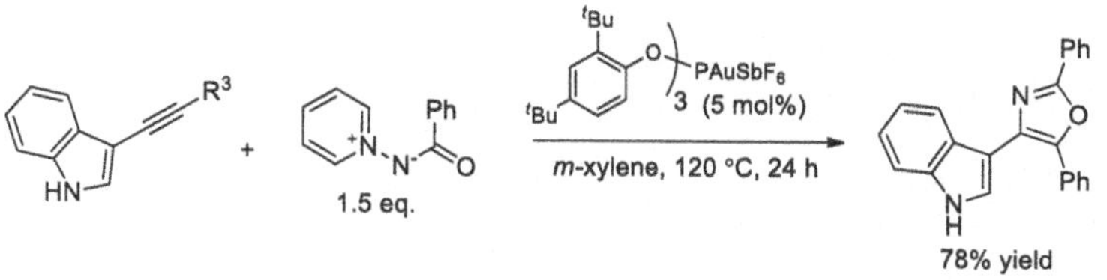

(8)

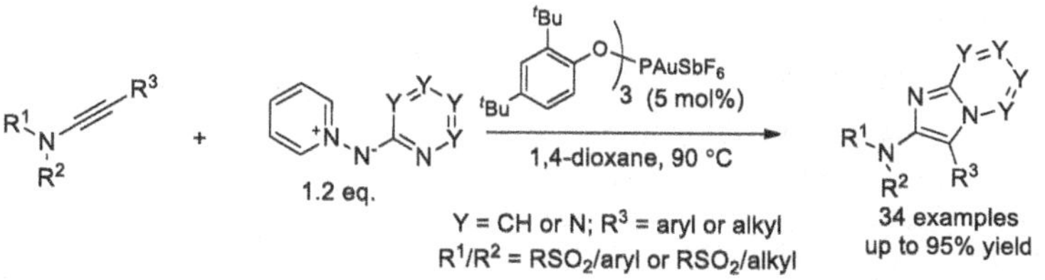

(9)

4 Conclusions

Homogeneous gold catalysis has become a versatile platform for reactivity discovery and method development. By using tethered or external nucleophilic oxidants, alkyne substrates can be transformed into various N-/O-heterocycles of high synthetic value via oxidative gold chemistry. In cases of oxygenated heterocycles, external nucleophilic oxidants and in particular pyridine/quinoline N-oxides are frequently employed; on the other hand, both external N-heteroarene oxides and tethered tertiary amine/aniline oxides are suitable oxidants for enabling efficient constructions of N-heterocycles. Despite exceptions, the reaction mechanisms frequently invoke an initial gold-promoted/catalyzed formation of alkyne-oxidant adduct, which could undergo facile redox fragmentation to generate reactive α-oxo gold carbene intermediates, the trappings of which by various nucleophiles offer versatile access to a diverse range of N-/O-heterocycles. Alternatively, the initial adduct could undergo rearrangements without the carbene intermediacy to deliver heterocycles. Further advances in heterocycle synthesis are expected by employing the oxidative gold catalysis.

References

1. Hashmi ASK (2007) Chem Rev 107:3180
2. Fürstner A, Davies PW (2007) Angew Chem Int Ed 46:3410
3. Dorel R, Echavarren AM (2015) Chem Rev 115:9028
4. Arcadi A (2008) Chem Rev 108:3266
5. Hashmi ASK, Rudolph M (2008) Chem Soc Rev 37:1766
6. Rudolph M, Hashmi ASK (2012) Chem Soc Rev 41:2448
7. Zhang Y, Luo T, Yang Z (2014) Nat Prod Rep 31:489
8. Alcaide B, Almendros P, Alonso JM (2011) Molecules 16:7815
9. Patil NT, Yamamoto Y (2008) Chem Rev 108:3395
10. Shen HC (2008) Tetrahedron 64:7847

11. Ye L, Cui L, Zhang G, Zhang L (2010) J Am Chem Soc 132:3258
12. Zhang L (2014) Acc Chem Res 47:877
13. Painter TO, Bunn JR, Schoenen FJ, Douglas JT, Day VW, Santini C (2013) J Org Chem 78:3720
14. Ye L, He W, Zhang L (2010) J Am Chem Soc 132:8550
15. Wuitschik G, Rogers-Evans M, Müller K, Fischer H, Wagner B, Schuler F, Polonchuk L, Carreira EM (2006) Angew Chem Int Ed 45:7736
16. Burkhard JA, Wuitschik G, Rogers-Evans M, Müller K, Carreira EM (2010) Angew Chem Int Ed 49:9052
17. Burkhard JA, Guerot C, Knust H, Carreira EM (2012) Org Lett 14:66
18. Shu C, Liu R, Liu S, Li J-Q, Yu Y-F, He Q, Lu X, Ye L-W (2015) Chem Asian J 10:91
19. Pawar SK, Wang C-D, Bhunia S, Jadhav AM, Liu R-S (2013) Angew Chem Int Ed 52:7559
20. Dube P, Toste FD (2006) J Am Chem Soc 128:12062
21. Wang Y, Ji K, Lan S, Zhang L (2012) Angew Chem Int Ed 51:1915
22. Dateer RB, Pati K, Liu R-S (2012) Chem Commun 48:7200
23. Fu J, Shang H, Wang Z, Chang L, Shao W, Yang Z, Tang Y (2013) Angew Chem Int Ed 52:4198
24. Gross T, Metz P (2013) Chem Eur J 19:14787
25. Wang T, Shi S, Hansmann MM, Rettenmeier E, Rudolph M, Hashmi ASK (2014) Angew Chem Int Ed 53:3715
26. Wang T, Shi S, Rudolph M, Hashmi ASK (2014) Adv Synth Catal 356:2337
27. Wang T, Huang L, Shi S, Rudolph M, Hashmi ASK (2014) Chem Eur J 20:14868
28. Liu Y, Song F, Guo S (2006) J Am Chem Soc 128:11332
29. Asao N, Sato K, Yamamoto Y (2003) Tetrahedron Lett 44:5675
30. Jadhav AM, Bhunia S, Liao H-Y, Liu R-S (2011) J Am Chem Soc 133:1769
31. Yeom HS, Lee JE, Shin S (2008) Angew Chem Int Ed 47:7040
32. Jeong J, Yeom H-S, Kwon O, Shin S (2011) Chem Asian J 6:1977
33. Yeom H-S, Lee Y, Jeong J, So E, Hwang S, Lee J-E, Lee SS, Shin S (2010) Angew Chem Int Ed 49:1611
34. Murai M, Kitabata S, Okamoto K, Ohe K (2012) Chem Commun 48:7622
35. Cui L, Zhang G, Peng Y, Zhang L (2009) Org Lett 11:1225
36. Cui L, Peng Y, Zhang L (2009) J Am Chem Soc 131:8394
37. Cui L, Zhang L (2010) Chem Commun 46:3351
38. Noey EL, Luo Y, Zhang L, Houk KN (2012) J Am Chem Soc 134:1078
39. Lu B, Li Y, Wang Y, Aue DH, Luo Y, Zhang L (2013) J Am Chem Soc 135:8512
40. Cui L, Zhang L (2010) Sci China B 53:113
41. Shan Z-H, Liu J, Xu L-M, Tang Y-F, Chen J-H, Yang Z (2012) Org Lett 14:3712
42. Yeom H-S, So E, Shin S (2011) Chem Eur J 17:1764
43. Gasparrini F, Giovannoli M, Misiti D, Natile G, Palmieri G, Maresca L (1993) J Am Chem Soc 115:4401
44. Ye L, He W, Zhang L (2011) Angew Chem Int Ed 50:3236
45. Robak MT, Herbage MA, Ellman JA (2010) Chem Rev 110:3600
46. Hashmi ASK, Weyrauch JP, Frey W, Bats JW (2004) Org Lett 6:4391
47. Shu C, Li L, Yu Y-F, Jiang S, Ye L-W (2014) Chem Commun 50:2522
48. Shu C, Li L, Xiao X-Y, Yu Y-F, Ping Y-F, Zhou J-M, Ye L-W (2014) Chem Commun 50:8689
49. Qian D, Zhang J (2011) Chem Commun 47:11152
50. Qian D, Hu H, Liu F, Tang B, Ye W, Wang Y, Zhang J (2014) Angew Chem Int Ed 53:13751
51. Wang K-B, Ran R-Q, Xiu S-D, Li C-Y (2013) Org Lett 15:2374
52. Chen M, Chen Y, Sun N, Zhao J, Liu Y, Li Y (2015) Angew Chem Int Ed 54:1200
53. Qian D, Zhang J (2012) Chem Commun 48:7082
54. Yang L-Q, Wang K-B, Li C-Y (2013) Eur J Org Chem 2013:2775
55. He W, Xie L, Xu Y, Xiang J, Zhang L (2012) Org Biomol Chem 10:3168
56. He W, Li C, Zhang L (2011) J Am Chem Soc 133:8482

57. Lundgren RJ, Peters BD, Alsabeh PG, Stradiotto M (2010) Angew Chem Int Ed 49:4071
58. Luo Y, Ji K, Li Y, Zhang L (2012) J Am Chem Soc 134:17412
59. Karad SN, Liu R-S (2014) Angew Chem Int Ed 53:5444
60. Zhou A-H, He Q, Shu C, Yu Y-F, Liu S, Zhao T, Zhang W, Lu X, Ye L-W (2015) Chem Sci 6:1265
61. Huple DB, Ghorpade S, Liu R-S (2013) Chem Eur J 19:12965
62. Talbot EPA, Richardson M, McKenna JM, Toste FD (2014) Adv Synth Catal 356:687
63. Davies PW, Cremonesi A, Dumitrescu L (2011) Angew Chem Int Ed 50:8931
64. Chatzopoulou E, Davies PW (2013) Chem Commun 49:8617
65. Garzon M, Davies PW (2014) Org Lett 16:4850

Top Heterocycl Chem (2016) 46: 117–174
DOI: 10.1007/7081_2015_5004

Published online: 12 February 2015

Synthesis of Heterocyclic Compounds via Gold-Catalysed Enyne Rearrangements

María Teresa Quirós and María Paz Muñoz

Abstract Syntheses of heterocycles using different gold-catalysed rearrangements of enynes are discussed in this chapter. The term skeletal rearrangement has been used in a broad sense to include reactions involving cyclopropyl gold carbene intermediates formed by initial enyne cyclisation, which can undergo many different transformations to give a wide range of heterocyclic structures. Other transformations involving rearrangement of propargylic esters and [3,3]-rearrangement (concerted or stepwise comprising metallic intermediates), as well as special cases, have also been covered. References to earlier work in this area and to recent reviews have been included, but the focus of the chapter is to present recent developments, interesting cases and an overview on how subtle differences in the enyne starting materials, the catalyst used or the reaction conditions can alter the reaction pathway increasing the structural diversity towards complex heterocyclic structures of high value.

Keywords Enynes • Gold • Heterocycles • Mechanisms • Rearrangements

Contents

M.T. Quirós and M.P. Muñoz (✉)
School of Chemistry, University of East Anglia, Norwich NR4 7TJ, UK
e-mail: m.munoz-herranz@uea.ac.uk

Abbreviations

Ac	Acetyl
Ar	Aromatic
Bz	Benzoyl
Cbz	Carboxybenzyl
dba	Dibenzylideneacetone
DCE	1,2-Dichloroethane
dr	Diastereomeric ratio
ee	Enantiomeric excess
equiv. = eq.	Equivalent
er	Enantiomeric ratio
Et	Ethyl
et al.	et alia = and others
*i*Pr	Isopropyl
IPr	1,3-Bis(2,6-diisopropylphenyl)imidazol-2-ylidine
JohnPhos	(1,1′-biphenyl-2-yl)-di-*tert*-butylphosphine
LUMO	Lowest unoccupied molecular orbital
Me	Methyl
Men	Menthyl
Mes	Mesitylene 1,3,5-trimethylbenzene
MS	Molecular sieves
*n*Pr	Normal propyl group
Nu	Nucleophile
Ph	Phenyl
Piv	Pivaloyl
PMP	*para*-methoxyphenyl
Qn	Quinoline
RT	Room temperature
SPhos	2-Dicyclohexylphosphino-2′,6′-dimethoxybiphenyl
*t*Bu	*tert*-butyl
TES	Triethylsilyl
Tf	Triflate trifluoromethanesulfonyl
TIPS	Triisopropylsilyl
TMS	Trimethylsilyl
Ts	Tosyl toluenesulfonyl
vs	Versus

1 Introduction

In the field of metal catalysis, gold has had a late development, since it was considered to be mainly inactive as a catalyst due to the stability of the metal. However, gold has demonstrated by far its utility in the field of catalysis, and it has become one of the most used metals for the activation of multiple carbon–carbon bonds [1].

A good example of the late discovery of the catalytic properties of gold is the rearrangement of enynes, in which many other metals were explored before gold aroused as the best choice for this transformation.

Enynes are interesting compounds in which an alkene and an alkyne appear together in the same molecule. Under metal catalysis, one of these functionalities, usually the alkyne, can coordinate to the metal, changing its electronic properties and being susceptible to the nucleophilic attack of the other unsaturated partner, which is the case of cycloaddition reactions. But enynes are attractive compounds not only for this reason. The reactivity of alkene and alkyne in the presence of other (internal or external) functional groups can be exploited separately to generate reaction intermediates that will then react with the other unsaturated partner. Those are only some examples of how extensive the reactivity of enynes can be.

The subject of this chapter is the gold-catalysed rearrangements of enynes utilised for the synthesis of heterocyclic structures. New methodologies for the preparation of heterocyclic derivatives will always be an essential research topic, since the vast majority of the drugs commercialised nowadays include at least one heterocycle in their structure [2]. Besides, with the constant demand from pharmaceutical industry for new potent and resistant drug-type structures to be tested in old and new diseases, atom economic methodologies offering the orthogonal synthesis of complex structures are of crucial importance. In this context, gold-catalysed rearrangement reactions of enynes have proven to be a great tool for the synthesis of a wide pool of carbo- but also heterocyclic structures with potential for biological activity, from simple furan-type to more complex heteropolycyclic structures, core of natural products, as will be illustrated with the synthesis of Englerin A in Sect. 3.2.4 of this chapter (Scheme 45).

Within this chapter, we will explore different transformations involving gold-catalysed rearrangements of enynes for the preparation of a variety of heterocycles, starting from the basis of how gold interacts with enynes. The chapter is organised according to the type of reaction occurring with special emphasis in the proposed mechanisms. Thus, initially we will review what we have called skeletal rearrangement of enynes. In this section, the term rearrangement has been used in a broad sense, so reactions covered as skeletal rearrangement comprise the transformations occurring through formation of cyclopropyl gold intermediates. In the next section, the special case of enynes in which the alkyne group forms part of a propargylic ester will be treated, since these reactions occur through an initial acyloxy migration, and the intermediate formed will be the one that reacts with the alkene moiety. Subsequently, reactions of enynes implicating a [3,3]-

rearrangement will be discussed. The term [3,3]-rearrangement will be used to cover concerted (true [3,3]-sigmatropic rearrangement) and non-concerted mechanisms involving metal-containing intermediates. Finally, we will comment some peculiar gold-catalysed transformations involving enynes.

References to earlier work in this area and to recent reviews have been included, but the focus of the chapter is to present recent developments, interesting cases and an overview on how subtle differences in the enyne starting materials, the catalyst used or the reaction conditions can alter the reaction pathway increasing the structural diversity towards complex heteropolycyclic structures of high value.

2 Gold–Enyne Coordination

Gold complexes, specially cationic gold(I) complexes, have been used as catalyst in many processes [3, 4]. Some of the features that make cationic gold(I) complexes interesting from a catalytic point of view are their high electronegativity, Lewis acid character, potential to stabilise cationic intermediates and aurophilicity. The explanation of these characteristics lies on relativistic effects [5].

Amongst the most studied processes using gold catalysis are the reactions involving unsaturated compounds. These reactions often occur through coordination of the metal to the π system of the unsaturation [6–8]. In the case of the enynes, focus of this chapter, there are two unsaturated groups suitable for coordination with the metal, the alkene and the alkyne. Examples of both types of complexes using gold (gold–alkene and gold–alkyne complexes) have been reported in the literature [9–17].

The reaction of enynes with a metal catalyst could imply the coordination of the metal to both unsaturated moieties at the same time, followed by oxidative cyclometalation forming a metallacycle. This has been reported in the case of platinum or rhodium amongst other metals [18, 19]. However, in the case of the gold(I)-catalysed reaction, the simultaneous coordination of alkyne and alkene is not possible, due to the linear bicoordinated geometry of the gold complex, so the reaction has to occur through a different manifold.

In the reaction of enynes, gold seems to activate uniquely the alkyne group forming a gold–alkyne complex that undergoes the corresponding reaction. Although alkenes usually bind more strongly with gold than alkynes, the reactivity of enynes is only explained by the gold activation of the alkyne group. This is not surprising, since many examples probing the alkynophilicity of gold have been reported [20]. Echavarren et al. explained this behaviour with experimental and computational evidences [21]. Regardless of the higher affinity of the alkenes in the gold coordination, the nucleophilic attack to the gold–alkyne complexes is thermodynamically more favoured, due to the lower LUMO energies in the coordinated alkyne compared to the alkene, which explains the preference of gold for activating the alkyne moiety.

The coordination of the gold to the alkyne can be explained through the Chatt–Dewar–Duncanson bonding model [22, 23]. The bond formed between these two partners has two components, the σ-donating and the π-accepting interactions. The first one is the result of the donation from one of the π bonding orbitals of the alkyne to an unoccupied d orbital of the metal. The second interaction appears due to the overlap of an occupied d orbital in the metal with the alkyne π* orbitals, which is also known as backdonation from the metal to the alkyne (Fig. 1). The nature of the alkyne and the metal will define which of these interactions is predominant, and therefore, it will establish the reactivity of a given complex [24]. Usually, in the case of gold–alkyne complexes, the σ-donor character of the ligand is predominant. As a result, the alkyne will donate part of its electronic density to the gold, being the electrophilicity of the coordinated alkyne enhanced.

As mentioned before, examples of gold–alkyne complexes have been isolated, such as the ones shown in Fig. 2. In these complexes, **1–3**, the η^2-coordination (or π-coordination) of the alkyne to the gold changes slightly the geometry of the triple bond and the substituents in both sides of the alkyne bend away from the linear arrangement.

It is worth noting that, in the case of asymmetrical substitution of the triple bond, the η^2-coordinated gold atom is not always positioned in the exact centre of the alkyne. In some cases, a displacement of the gold towards one of the carbons is observed. This phenomenon is called "slippage", and it is emphasised in the event of a nucleophilic attack, when the movement of the metal away from its symmetrical position leads to a η^2-η^1 deformation (Scheme 1) [17, 25, 26]. This effect is also observed in gold–alkene complexes [27–29].

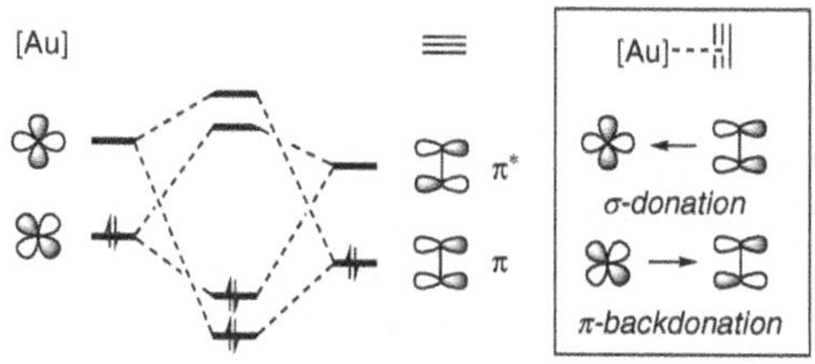

Fig. 1 Bonding model of the gold–alkyne coordination [22, 23]

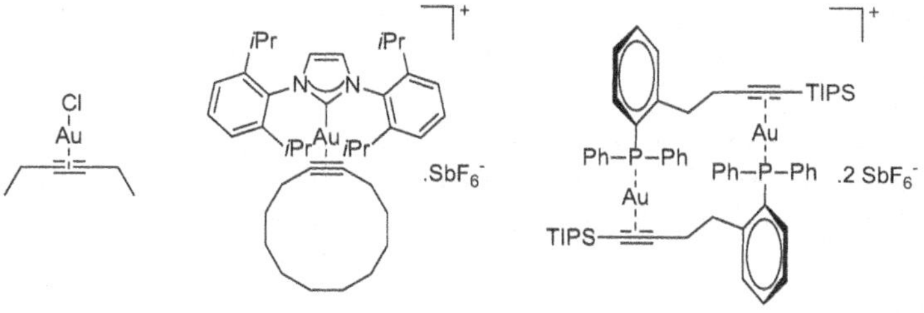

Fig. 2 Examples of reported isolated gold–alkyne complexes [13, 23, 24]

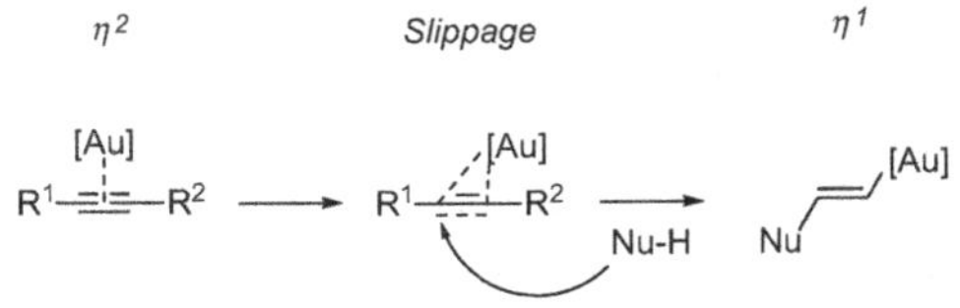

Scheme 1 The slippage phenomenon as the source of η^2-η^1 deformation

Before concluding this brief overview of the gold–alkyne coordination, it is important to mention that the η^2 is not the only way in which gold can coordinate to an alkyne. In the case of terminal alkynes, σ-gold–alkyne complexes [30–32] and the coordination of more than one gold centre to the alkyne have been reported [33–37]. Formation of σ,π-di-gold–alkyne complexes in which one of the gold centres is coordinated in an η^2 (π) way to the alkyne whereas the other gold centre is σ (η^1)-bonded to the terminal position of the alkyne has been achieved, and these complexes have proved to be important in different catalytic processes. Furthermore, Russell's group has recently demonstrated that the gold–alkyne coordination can go beyond that, being a terminal triple bond able to accommodate up to three gold centres (or four in the case of the acetylene) [38].

3 Gold-Catalysed Enyne Cycloisomerisation by Skeletal Rearrangement

3.1 Mechanistic Insights

The first reports on the metal catalysed cycloisomerisation of enynes involved palladium catalysis and were described by Trost [39]. Since then, this reaction has attracted considerable attention as a powerful methodology that allows the synthesis of highly complex molecules in an easy, selective and atom economic manner. Furthermore, a wide range of different scaffolds can be obtained from relatively simple starting materials changing the reaction conditions. Different aspects of the reactivity of enynes have already been extensively studied and reviewed [40–44].

Numerous metals have been studied in this transformation, but since the pioneer work of the groups of Echavarren, Toste and Fürstner, gold, and in particular gold (I), complexes have arisen as the preferred choice, giving the best activity and selectivity [45–47].

In the cycloisomerisation of enynes, several skeletal rearrangements can take place, without incorporation of any external molecule, leading to the formation of a variety of cyclic structures, many of them containing heterocycles, through different reaction pathways. The intriguing mechanism of these rearrangements has been the starting point of an increasing number of studies [48–53]. It has been proposed that these transformations involve cyclopropyl gold carbene intermediates **7**, which formation could be explained schematically as shown in Scheme 2. As it has been mentioned before, this process starts with the coordination of the gold catalyst to the

Scheme 2 Formation of cyclopropyl gold carbenes

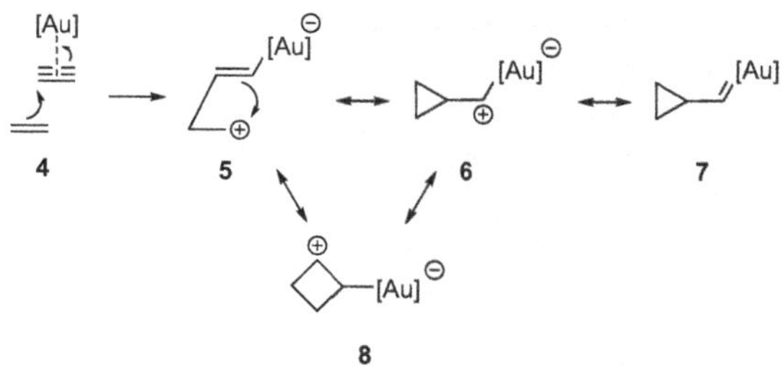

alkyne moiety of the enyne, followed by the nucleophilic attack of the alkene group to the coordinated alkyne. The formed cationic intermediate can be described as different mesomeric structures **5–8**.

These cyclopropyl gold carbenes **7** are often used to describe the reactivity observed in enynes [54]. However, they are highly distorted structures. Fürstner et al. calculated that the C-[Au(I)] distance is more in accordance with a single bond (2.02–2.04 Å). These structures are therefore probably better described as gold-stabilised carbocations **6** rather than gold carbenes, which have been supported by NMR experiments [55–57]. On the other hand, the existence of gold carbenic species non-stabilised by heteroatoms has been confirmed, and some of these structures have been synthesised and characterised by different methodologies [58–60].

Probably, amongst the cycloisomerisations of enynes, the most explored process is the one involving 1,6-enynes, being this reaction synthetically useful from the point of view of the preparation of heterocycles. The mechanism of these transformations is not straightforward, and many ramifications could occur giving rise to different products (Scheme 3) [50, 61–63].

After coordination of the gold to the alkyne, the nucleophilic addition of the alkene could occur, affording two key cyclopropyl gold carbene intermediates depending on the type of cyclisation. *Anti*-5-*exo-dig* cyclisation would form a fused five-membered cycle **10**, whereas 6-*endo-dig* cyclisation would form intermediate **15**. These two key gold carbenes could then evolve through different pathways.

In the case of gold carbene **10**, two different routes have been proposed to explain the experimental results. The first route is the single cleavage rearrangement and implies the formal cleavage of the double bond, where the terminal carbon of the alkene migrates to the terminal carbon of the alkyne, leading to 1,3-dienes of type **12**. The second route is the double cleavage rearrangement and involves the cleavage of both alkene and alkyne to obtain 1,3-dienes of type **14** [64]. The substitution of the initial enyne, the tether group or the ligands in gold affects the selectivity, driving the process towards one or the other product. As a general trend, unsubstituted alkynes or alkynes with electron-donating substituents react through the single cleavage pathway, while 1,6-enynes with electron-withdrawing substituents in the alkyne favour the double cleavage rearrangement [52].

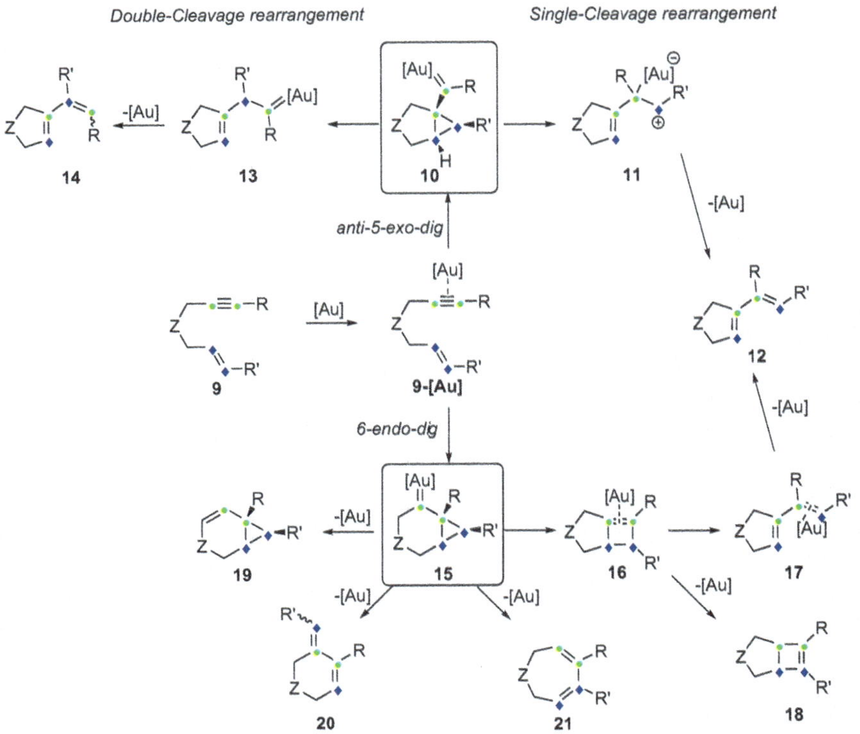

Scheme 3 Possible mechanistic scenarios for the cyclisation of 1,6-enynes

Labelling experiments have reinforced the existence of these two routes. Chatani's and Echavarren's groups carried out deuteration experiments with enynes containing the deuterium in the terminal positions of the alkene and also ^{13}C labelling experiments with the ^{13}C in the terminal position of the alkene. These experiments support the proposed mechanisms of single and double cleavage (Scheme 4) [65, 66]. Furthermore, the formation of the gold carbenes **10** and **13** has been confirmed by trapping experiments with nucleophiles [67] and alkenes [68, 69] or in the presence of oxidants (Scheme 4) [70].

These two routes described in Scheme 3 can be also rationalised through a common intermediate **22**, which would form after the cleavage of the C–C bond from the initial alkene in the intermediate **10** (Scheme 5). This intermediate would evolve through a second cleavage, and depending on the bond that breaks next, cycles **12** or **14** could be formed.

On the other hand, intermediate **15** (Scheme 3) can suffer different rearrangements to form a range of cyclic products. The cyclopropyl ring expansion would form gold–cyclobutene **16**, which could undergo ring opening to form diene **12** or isomerisation and demetalation to form bicyclo[3.2.0]heptenes **18** [65]. Alternatively, proton elimination from cyclopropyl gold carbene **15** would give rise to bicyclo[4.1.0]heptenes **19**.

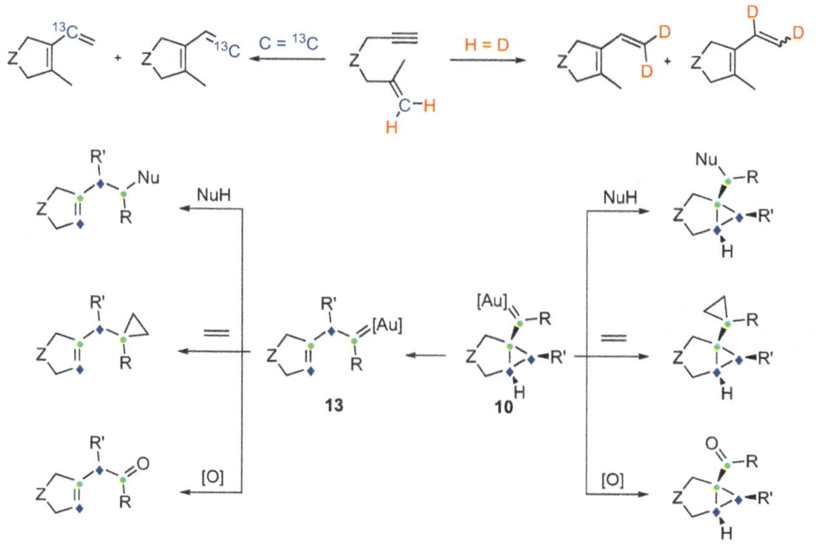

Scheme 4 Labelling experiments and experiments to trap the intermediate carbenes **10** and **13**

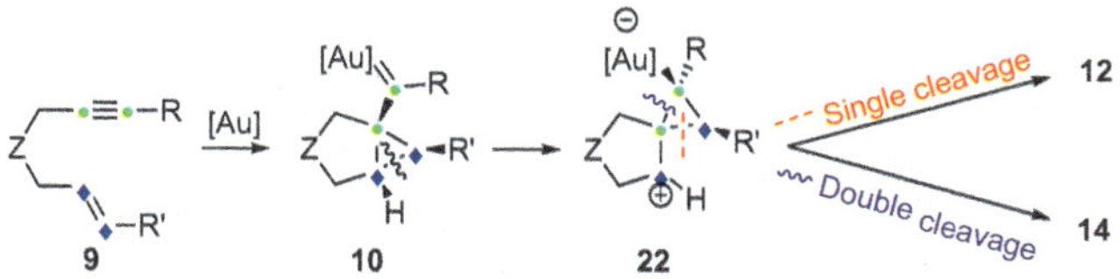

Scheme 5 Single vs double cleavage rearrangements

Furthermore, dienes **20** could be obtained through rearrangement of intermediate **15** to a second cyclopropyl intermediate [41]. However, a second explanation for the formation of this product **20** has been postulated. This alternative pathway, supported by DFT calculations, would occur through a rearrangement of intermediate **10** (Scheme 6) [71].

Finally, ring expansion of the bicyclic cyclopropene in intermediate **15** would form cycloheptadiene **21** (Scheme 3) [72].

Besides the pathways mentioned so far, there is still another possibility based on the conformation of the formed carbene **10**, which is the *syn*-5-*exo-dig* cyclisation (Scheme 7). Although this could be a competing pathway for the formation of cyclobutenes **16** and single cleavage products, DFT calculations have shown that the *syn*-5-*exo* cyclisation does not compete with the *anti*-5-*exo* or the 6-*endo* cyclisation pathways in any of the cases studied [65].

As mentioned in the previous section, when the triple bond of the enyne is a terminal alkyne, the possibility of forming di-gold–alkyne complexes exists. These complexes have proven to be involved in the mechanism of the gold-catalysed cycloisomerisation of 1,5-allenynes [73]. Encouraged by these observations, Fensterbank et al. carried out a study to figure out the implication of acetylide-

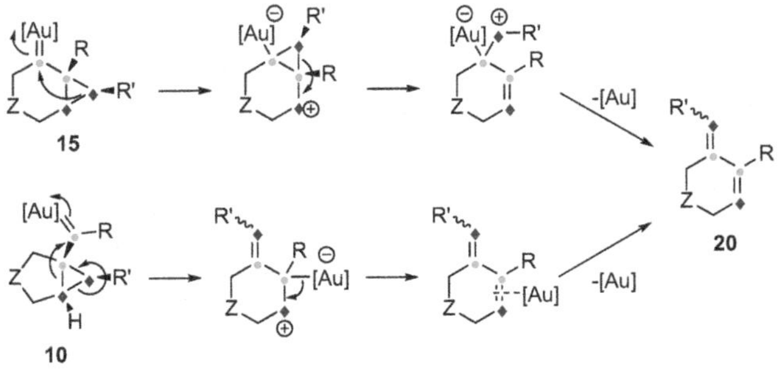

Scheme 6 Formation of diene **20**

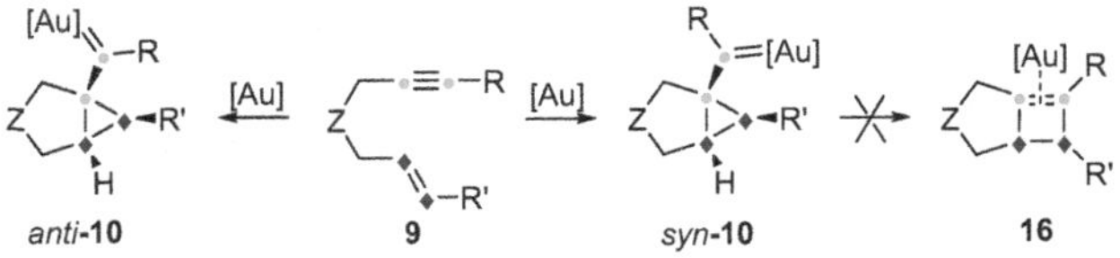

Scheme 7 *Anti-5-exo-dig* vs *syn-5-exo-dig* cyclisation

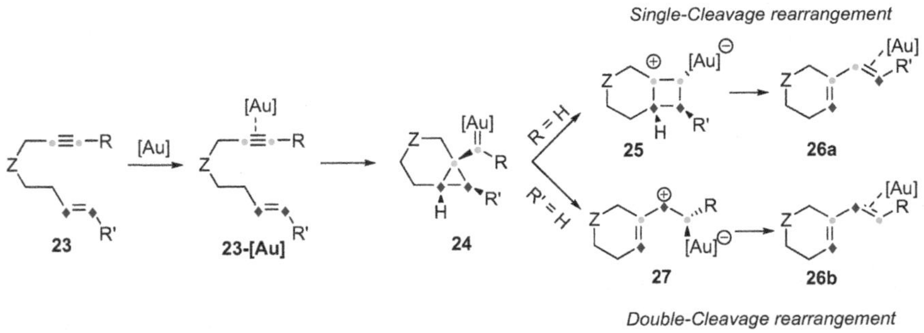

Scheme 8 Single vs double cleavage in 1,7-enynes

gold or di-gold species in the skeletal rearrangement of 1,6-enynes [74]. However, according to their results, those species are unlikely to appear as active intermediates in this process.

Although less common, the cycloisomerisation of 1,7-enynes **23** is also an important method for the synthesis of heterocyclic structures. In this case, the possibility of single versus double cleavage also exists, although it is possible to control the chemical pathway by changing the substitution pattern in the initial enyne (Scheme 8). In all cases, six-membered cycles **26a** or **26b** are formed [65].

3.2 *Formation of Heterocycles by Skeletal Rearrangement of Enynes*

3.2.1 Skeletal Rearrangement of 1,6-Enynes

Skeletal rearrangement of enynes allows the formation of a wide variety of heterocyclic scaffolds by changing the nature of the tether between alkene and alkyne, its length or the reaction conditions, as it is demonstrated by the diversity of pathways in the reaction manifold. As the general mechanism of the gold-catalysed cyclisation of enynes has already been described in Sect. 3.1, no comments will be made on the mechanism of every example unless it differs from what has already been explained. To keep simplicity in the explanation, the different heterocycles that could be obtained will be presented in the same order in which they appeared in Sect. 3.1.

1,6-Enynes have been the preferred substrates in the studies of reactivity between alkene and alkyne. As mentioned before, the cyclisation of these substrates could proceed through two different pathways: 5-*exo-dig* or 6-*endo-dig* cyclisations. On one hand, from the 5-*exo-dig* cyclisation, there are two possible ramifications: the single and the double cleavage, both of them giving rise to the formation of five-membered cycles of the type **12** or **14** (Scheme 3, Sect. 3.1). This reaction has been extensively studied using gold as a catalyst when the tether between alkene and alkyne is of the type $Z{=}CE_2$, being E a functional group such as CO_2R or SO_2R [47, 52]. However, in the case of Z being a heteroatom, typically nitrogen as the group NTs, examples are scarce, with the single cleavage pathway observed in all the cases. In those cases, 2,5-dihydropyrroles, like **29** and **32** (Scheme 9), are always formed in the presence of other heterocycles that come from the 6-*endo-dig* pathway (**30** and **33**, Scheme 9) [72, 75, 76]. Recently, Bastin, César and coworkers synthesised a new gold catalyst able to direct the reaction towards the formation of dihydropyrrole **32** as the main product without the use of any additives and in a non-polar solvent (b, Scheme 9) [75].

Gold-catalysed single cleavage rearrangements are, in most of the cases, stereospecific, being the configuration of the alkene retained in the process [54].

On the other hand, the 6-*endo-dig* cyclisation allows the formation of several different heterocycles. For example, even though this kind of cyclobutenes had been postulated to be highly strained, the bicyclo[3.2.0]heptane core **18** (Scheme 3, Sect. 3.1) can be formed under gold catalysis from 1,6-enynes [65, 77]. Chung et al. obtained a related structure **36** as by-product in the gold-catalysed reaction of 1,6-enyne **34** in a 1:1 mixture together with the tricycle **35** (a, Scheme 10) [78].

Chung et al. also reported a synthesis of bicyclo[3.2.0]hept-6-en-2-ones from a gold(I)-catalysed cycloisomerisation of amide- or ester-tethered 1,6-enynes **37** (b, Scheme 10) [79]. The introduction of a carbonyl group in the position between Z and the alkyne alters the reaction pathway allowing the formation of the corresponding aza- or oxabicyclo[3.2.0]hept-6-en-2-ones **38** as the only products.

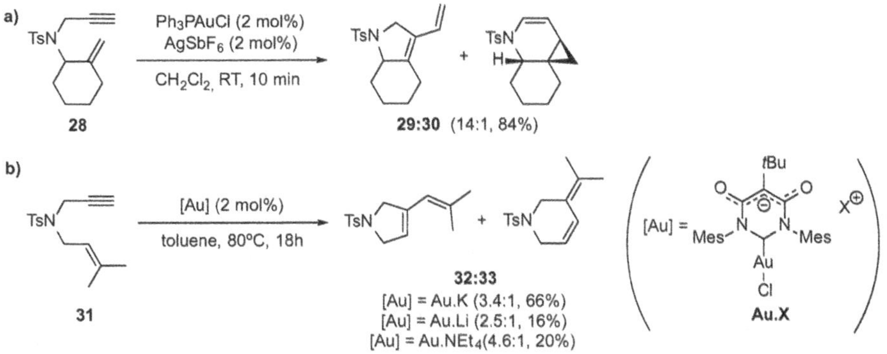

Scheme 9 Formation of dihydropyrroles from 1,6-enynes

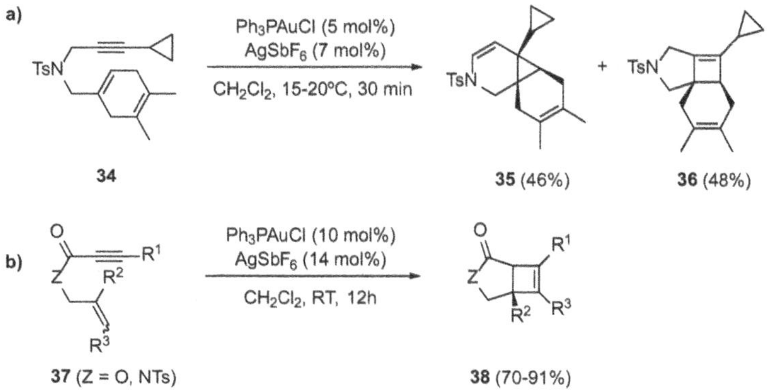

Scheme 10 Formation of bicyclo[3.2.0]heptanes from 1,6-enynes

Bicyclo[4.1.0]heptenes (core **19** in Scheme 3, Sect. 3.1) are possibly the most common heterocyclic scaffold obtained from 1,6-enynes, with the tether between alkene and alkyne being an oxygen or a nitrogenated group. This reaction has mainly been explored using platinum as a catalyst [80–83], but there are also examples using gold catalysis [47, 72, 84]. For instance, Chung et al. studied the preparation of this scaffold using a gold(I)-catalysed cyclisation of enynes containing an olefinic cycle [78]. They discovered that the introduction of an olefinic ring led to an increase in the yields of the reaction, obtaining the corresponding tricyclic products **40** in excellent yields (a, Scheme 11). As just mentioned, using the same substrates bicyclo[3.2.0]heptane **36** could also be formed (Scheme 10), but this bicycle was only obtained when the R substituent was a cyclopropyl group. Furthermore, Shi's group reported the cyclisation of alkylidenecyclopropanes **41** to form the corresponding 3-oxa- or 3-azabicyclo [4.1.0]heptenes **42** (b, Scheme 11) [85]. Surprisingly, similarly substituted enynes with carbon tethers led to completely different products [86, 87]. A related interesting example was developed by Guo et al. In this case, the starting 1,6-enynes are

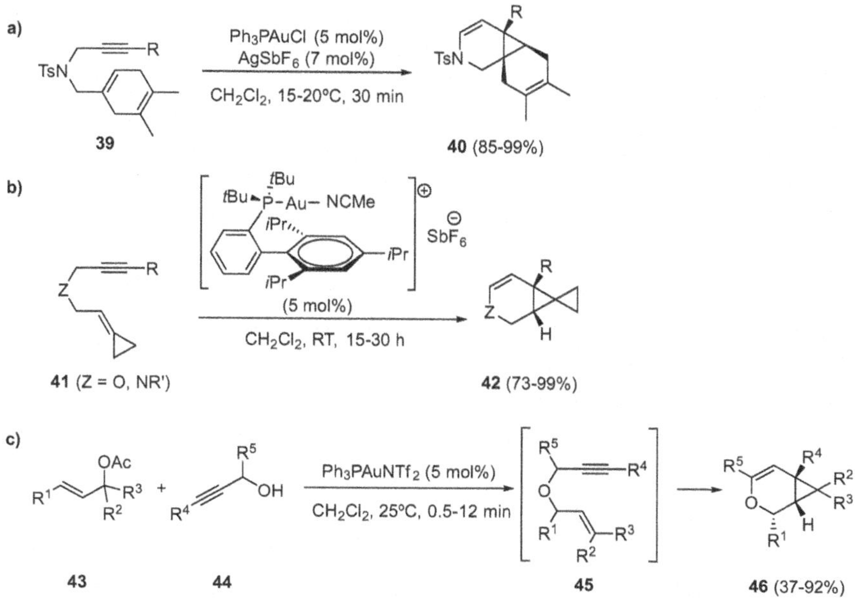

Scheme 11 Formation of bicyclo[4.1.0]heptenes from 1,6-enynes

formed in situ from allylic acetates and propargylic esters and then undergo cycloisomerisation forming heterocycles **46** as the only products (c, Scheme 11) [88].

Although the development of effective asymmetric versions of this methodology has turned out to be challenging, different enantioselective variants have been reported in the literature. Examples of the ligands used for this purpose are displayed in Fig. 3, and some representative examples of the asymmetric gold-catalysed synthesis of bicyclo[4.1.0]heptenes are summarised in Table 1.

Chiral phosphorus ligands seem to be the best choice, since they provide higher enantiomeric excesses. The more commonly explored ligands are the biphosphine ligands of the type **L1** or **L2** (Fig. 3). These have been used to form 3-oxa- or 3-azabicyclo[4.1.0]heptenes with good yields and enantiomeric excesses (Table 1). More recently, other ligands such as phosphoramidite **L3** or helicenes **L4** and **L5** have emerged as appealing alternatives for this methodology. It is worth noting that when bisphosphine ligands are used (such as **L1** and **L2** in Fig. 3), di-gold complexes are formed by coordination of one gold atom per phosphorus atom in the ligand.

The relevance of this synthetic strategy has been validated with the synthesis of the antidepressive agent (−)-GSK1360707 using this enantioselective procedure (Scheme 12) [91, 95, 96]. Starting from the enyne **47**, the direct precursor of GSK1360707, bicycle **48**, was obtained with 88% yield and enantiomeric excess of 95%.

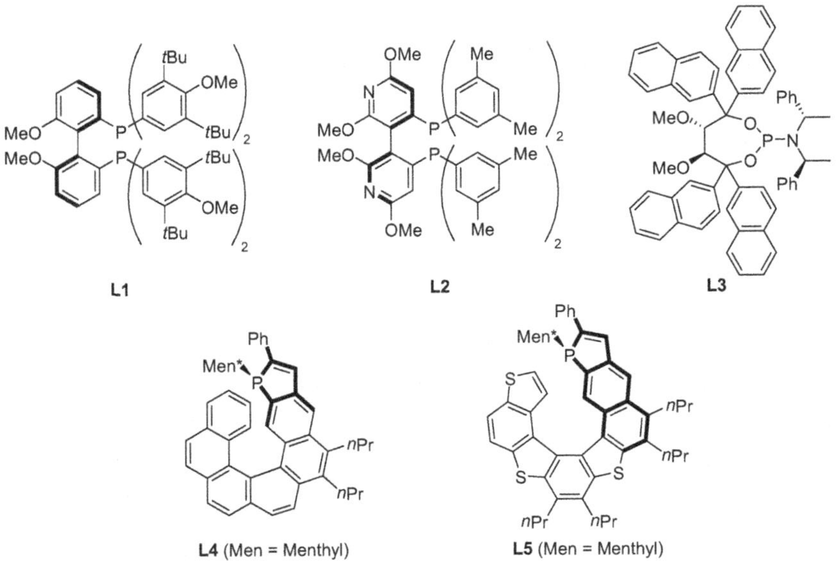

Fig. 3 Ligands employed for the asymmetric gold-catalysed synthesis of bicyclo[4.1.0]heptenes

Table 1 Enantioselective synthesis of bicyclo[4.1.0]heptenes from 1,6-enynes

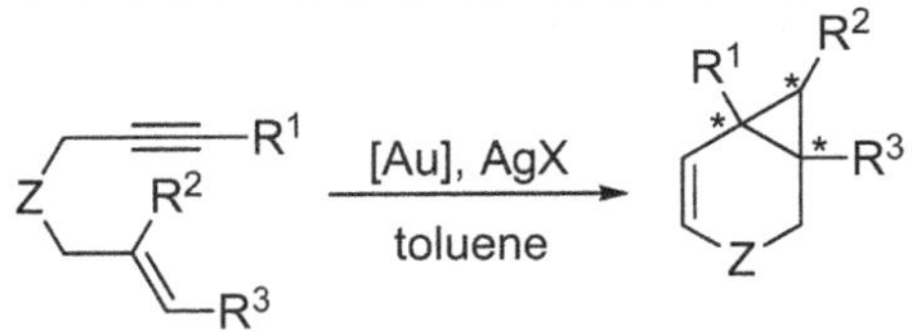

Entry	Z	Conditions	Yield (%)	ee (%)	Reference
1	O, NTs	**L1**Au_2Cl_2 (1–3 mol%), AgOTf (2–6 mol%)	24–74	90–98	[89]
2	O, NTs	**L1**Au_2Cl_2 (3 mol%), AgOTf (6 mol%)	7–64	13–99	[90]
3	O, NR	**L3**AuCl (5.5 mol%), $AgBF_4$ (5 mol%)	53–94	34–99	[91]
4	NTs	**L4**AuCl (4 mol%), $AgBF_4$ (4 mol%)	70[a]	86–88	[92]
5	NTs	**L5**AuCl (4 mol%), $AgBF_4$ or $AgNTf_2$ (4 mol %)	–[b]	78–96	[93]
6	O	**L2**Au_2Cl_2 (5 mol%), AgOTf (10 mol%)	45–93	60–78	[94]

[a]Only one yield reported for R^1 = 1-cyclohexenyl, R^2, R^3 = H
[b]Isolated yields not reported

1,6-Enynes can also be used as precursors for the synthesis of other six-membered rings, particularly 3-methylen-1,2,3,6-tetrahydropyridines (core **20** in Scheme 3, Sect. 3.1). However, there are not many examples of those and in some cases their formation is not completely regioselective. Echavarren et al. reported the first example of formation of these heterocycles. The use of a

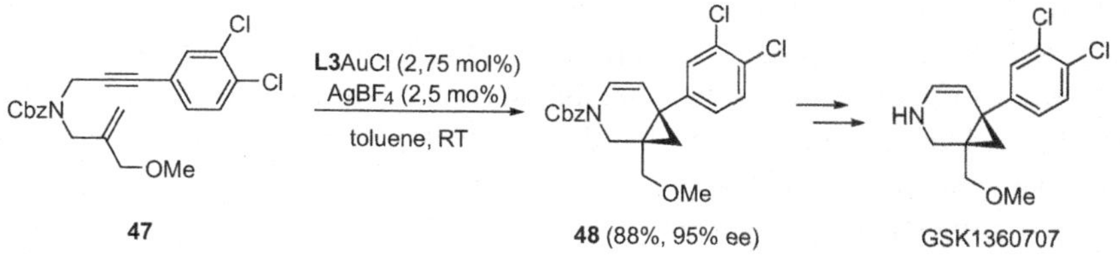

Scheme 12 Synthesis of GSK1360707. For structure of **L3**, see Fig. 3

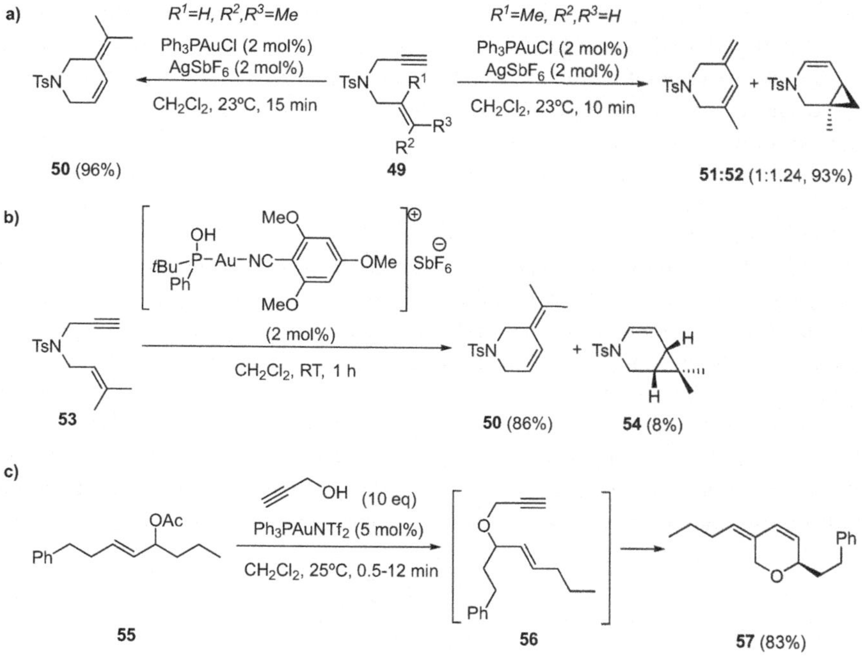

Scheme 13 Formation of 3-methylen-1,2,3,6-tetrahydropyridines from 1,6-enynes

1,6-enyne substituted only in the terminal position of the alkene afforded the corresponding tetrahydropyridine **50** regioselectively, while a 1,6-enyne bearing a methyl group in the internal position of the alkene led to the formation of the corresponding dihydropyridine **51** along with the product of β-hydrogen elimination **52** (a, Scheme 13) [47]. In a more recent example, Fensterbank et al. prepared heterocycle **50** as the major product using secondary phosphine oxide–gold (I) complexes as catalyst (b, Scheme 13) [84]. Furthermore, Chen et al. prepared dihydropyran **57** by forming the corresponding enyne in situ with the same approach employed in the synthesis of bicyclo[4.1.0]heptenes **46** (Scheme 11). In this case, the propargylic alcohols used as precursors were not substituted in the terminal position of the alkyne (c, Scheme 13) [88].

Formation of oxepines from 1,6-enynes has also been observed in some cases (core **21** in Scheme 3, Sect. 3.1). Echavarren et al. described for the first time the preparation of this heterocycle, although it was always obtained in low yield [72]. Later, Fürstner et al. discovered that the choice of solvent had a high impact in the outcome of the reaction and they were able to prepare oxepine **59** quantitatively using methanol as a solvent (Scheme 14) [91].

As it was mentioned in the previous section, there are some examples in which intermediate gold carbenes have been trapped by the use of an external oxidant, opening the possibility for the synthesis of bicyclo[3.1.0]hexanes. This synthesis has been explored in its racemic and asymmetric versions [70, 97]. In this context, Zhang et al. reported a enantioselective synthesis of azabicyclo[3.1.0]hexanone **61**, starting from enyne **60** and using a gold(I) catalyst with a chiral phosphoramidite ligand **L6** and 8-methylquinoline *N*-oxide as the external oxidant (Scheme 15) [98]. Moreover, the authors carried out some mechanistic studies, which suggest the formation of β-gold-vinyloxyquinolinium species as intermediates in the process and also the contribution of these species to the enantioselectivity of the cyclopropanation.

Gold carbene intermediates have also been trapped adding olefins to the reaction media [99]. A case worth mentioning is that in which the olefins are already included in the structure of the molecule. The cyclopropanation between the carbene and the pendant olefin will thus be intramolecular, creating a new branch in the reaction manifold for the formation of interesting heterocycles. The group of Echavarren developed some of the first examples of this reactivity. They synthesised tetracycles **63** and **65** in which the heteroatom can be either in the tether joining the alkene and alkyne that react in the first place or in the chain that connects the alkene that undergoes the cyclopropanation with the gold carbene

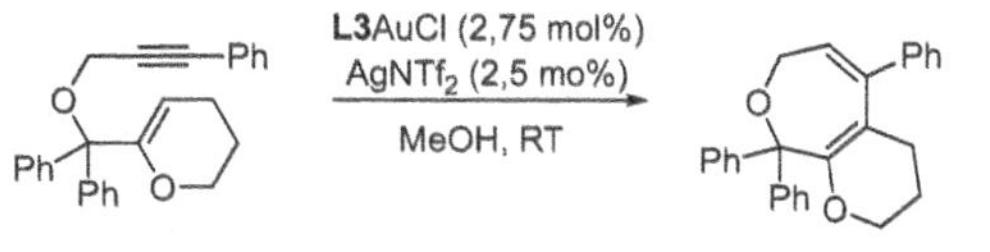

Scheme 14 Formation of oxepines from 1,6-enynes. For structure of **L3**, see Fig. 3

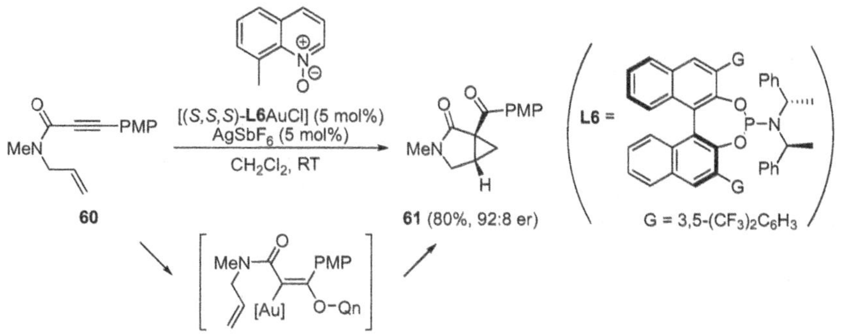

Scheme 15 Formation of azabicyclo[3.1.0]hexanone from 1,6-enynes

(a, Scheme 16) [47, 69]. In a related example, Chung et al. synthesised tetracyclo [3.3.0.0^{2,8}.0^{4,6}]octanes **68** by the use of enynes with a cyclohexadienyl moiety **67** (b, Scheme 16) [100]. It is worth noting the divergence in reactivity, from a single to a double cyclopropanation, observed with the change of the reaction conditions when compared to the reaction of enynes **39** (a, Scheme 11).

When the reaction is carried out with enynes bearing a vinyl substituent in the terminal position of the alkyne, bicyclo[4.3.0]nonanes can be obtained, resulting from a 6-*endo-dig* cycloisomerisation and a subsequent vinylcyclopropane–cyclopentene rearrangement [101]. This procedure has been explored with platinum [102], but a gold-catalysed enantioselective example has recently appeared in the literature using as the starting material enyne **69** to form octahydro-1*H*-indeno [2,1-*c*]pyridine **70** in high enantiomeric excess (Scheme 17) [93].

An unusual example was described in 2009 by Echavarren's group. When enyne **71**, with an allyloxy substituent, was submitted to gold(I) catalysis, it was

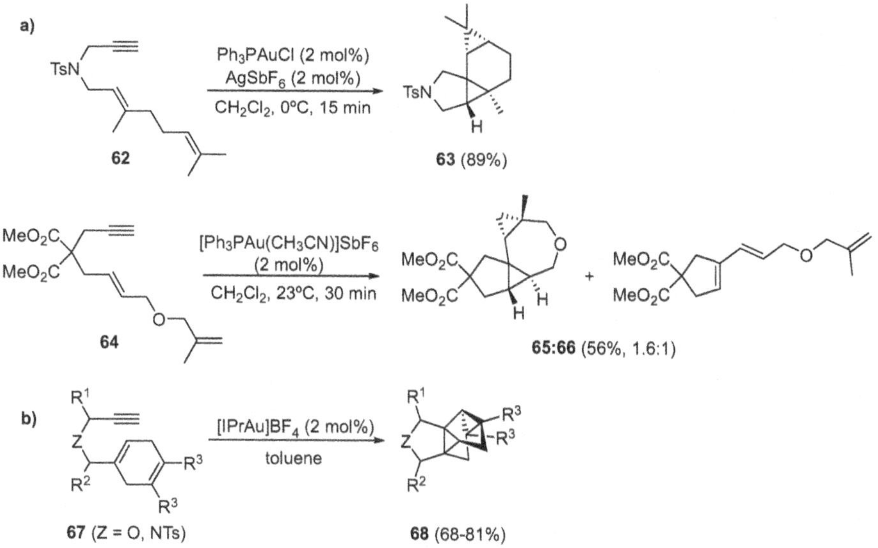

Scheme 16 Formation of tetracycles from 1,6-enynes with an extra alkene

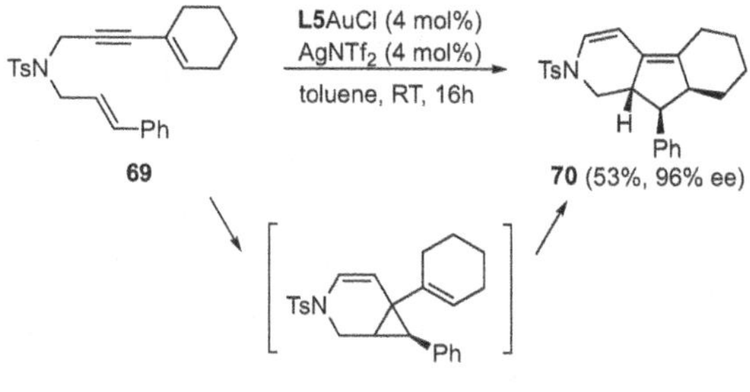

Scheme 17 Formation of bicyclo[4.3.0]nonanes from 1,6-enynes with an extra alkene. For structure of **L5**, see Fig. 3

discovered that the final product was tricycle **72** as a single isomer (Scheme 18) [103]. The formation of this heterocycle can be elucidated as a 1,5-migration of the oxygenated group from the intermediate cyclopropenyl gold carbene, followed by an intramolecular cyclopropanation.

A particular case is that in which the alkyne is substituted with a cycloheptatriene. For example, reaction of enyne **73** with a pendant cycloheptatriene gives rise to the formation of the tautomer barbaralanes **74** and **74′**, which rapidly interconvert between them at room temperature through a strain-assisted Cope rearrangement (Scheme 19) [104]. The mechanism of this transformation is fascinating since it implies the formation of a 9-barbaralyl gold carbene intermediate. This carbene comes from the reaction of the corresponding gold–alkyne complex and one of the double bonds of the cycloheptatriene. The cyclopropanation between this gold carbene and the remaining olefin would form the final products.

As it has become clear, variations in the ligands on the gold, the reaction conditions or the morphology of the starting materials can have a high impact in the process outcome, following the cyclisation of 1,6-enynes less conventional pathways and resulting in the formation of attractive and sometimes unexpected heterocyclic structures. Interestingly, Cossy et al. studied the reactivity of

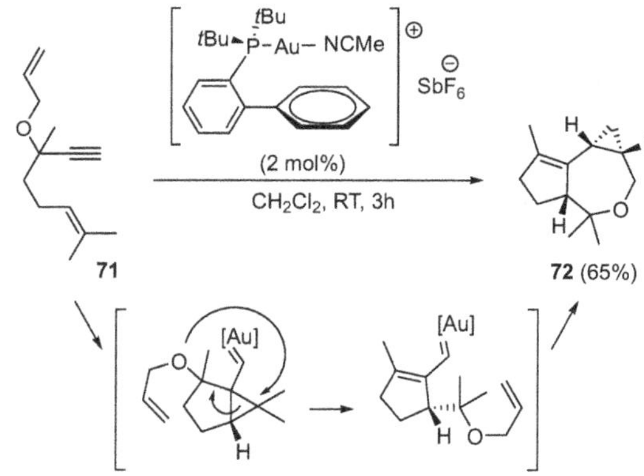

Scheme 18 Formation of hexahydro-1*H*,2*H*-3-oxa-cyclopropa[e]azulene from 1,6-enynes with an extra alkene

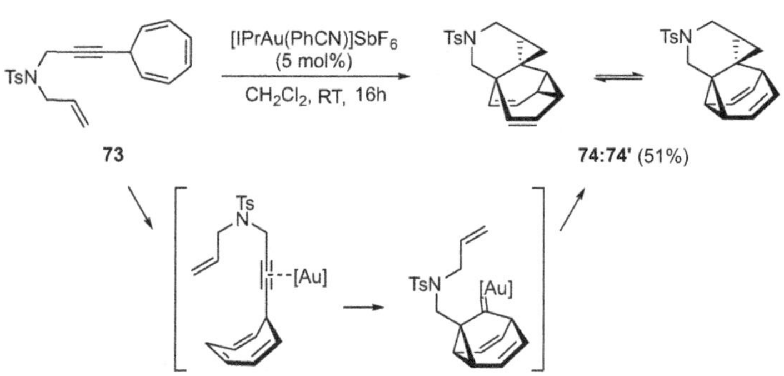

Scheme 19 Formation of barbaralanes from 1,6-enynes with a cycloheptatriene

1,6-enynes changing the position of the tether. Particularly, the reaction of ene-ynamides **75** bearing a propargyl alcohol moiety led to the formation of aldehydes or ketones **76** with a 2-azabicyclo[3.1.0]hexane framework (Scheme 20) [105, 106]. The mechanism of this transformation occurs through a gold carbene similar to **10** (Scheme 3, Sect. 3.1).

Another intriguing example is the formation of the tetracyclic acetal **78** from enyne **77** that was described by Echavarren et al. (Scheme 21). However, to the best of our knowledge, formation of this heterocycle using gold catalysis has only been reported twice [72, 91]. The mechanism postulated by these authors proceeds initially through a 6-*endo-dig* cyclisation of the gold–alkyne complex **79** to give gold carbene **80**. The ring opening of this intermediate would form oxonium cation

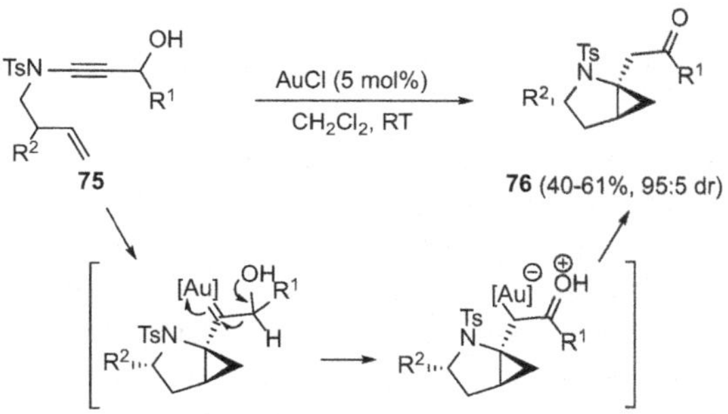

Scheme 20 Formation of 2-azabicyclo[3.1.0]hexanes from 1,6-enynes

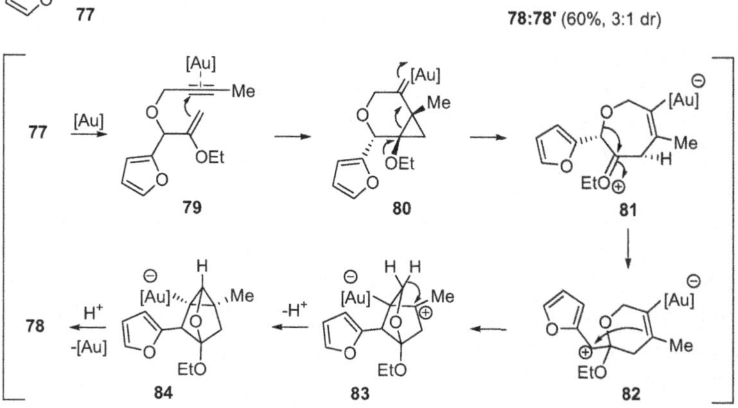

Scheme 21 Formation of 3-oxa-tricyclo[2.2.1.0^{2,6}]heptanes from 1,6-enynes

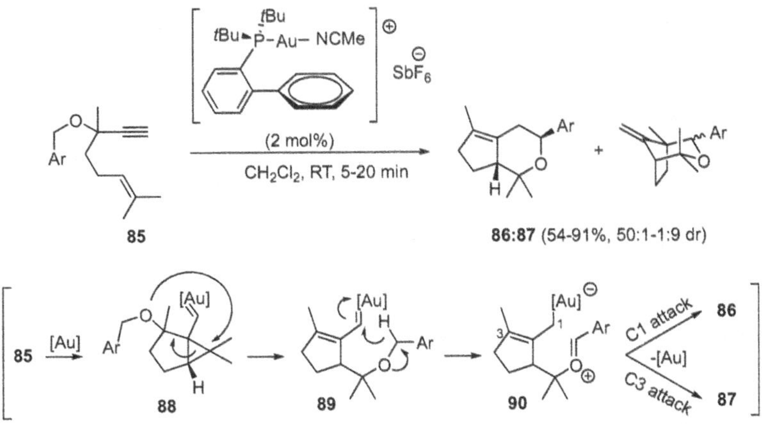

Scheme 22 Formation of oxabicyclo[4.3.0]nonane and oxabicyclo[3.2.1]octane from 1,6-enynes

81, whose ring contraction would afford carbocation **82**. The reaction of this cation with the alkenyl–gold group would furnish intermediate **83**, which after the formation of a cyclopropane ring would lead to **84**. A final protodemetalation would afford the tetracyclic acetals **78** and regenerate the gold catalyst.

When the enyne bears a benzyloxy substituent, an additional cyclisation pathway can take place. Echavarren et al. explored that situation with enynes of the type **85**, and they obtained oxabicyclo[4.3.0]nonane **86** as a single diastereoisomer (Scheme 22) [103]. Oxabicyclo[3.2.1]octane **87** was also formed as a subproduct, in different ratios depending on the aromatic substituent. The formation of both compounds **86** and **87** could be explained through a 1,5-migration of the oxygenated substituent from **88**, similar to the example with allyloxy substituents described in Scheme 18, but followed in this case by a formal C–H insertion of a proton from the $ArCH_2O$-group. The allyl gold intermediate **90** formed in that way would react with the oxonium cation either at C1 to give **86** or at C3 to afford **87**.

The addition of a cyclopropyl group adjacent to the alkene creates a new reaction pathway and allows the formation of octahydro-cyclobuta[*a*]pentalenes. Thus, the gold-catalysed reaction of enyne **91** afforded tricycles **92** and **92′** in good yield, being the diastereoisomer **92** the major product of the reaction (Scheme 23) [107]. The formation of both cycles could be explained through a common intermediate, which is the corresponding cyclopropyl gold carbene **93**. This carbene could evolve via two different pathways. Pathway *A* is the one favoured in the case of AuCl and involves the ring expansion of the cyclopropyl substituent to form an alkenyl–gold complex bearing an oxonium cation **94**. The reaction of these two groups would form the tricyclic structure **95** by a Prins reaction, which upon demetalation would lead to product **92**. Pathway *B*, preferred in the case of cationic gold complexes, will occur through the cyclopropyl-stabilised cation **96**. This intermediate will evolve via a ring expansion forming a mixture of **94** and **94′**. This mixture would follow the same route already explained to form the final products **92** and **92′**.

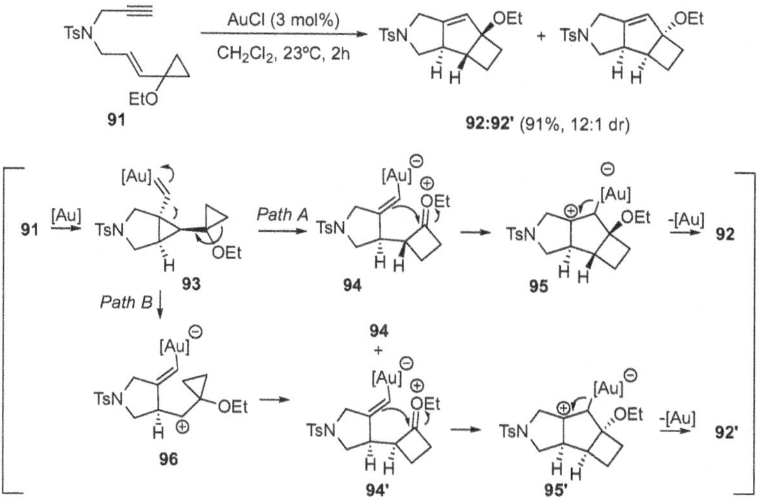

Scheme 23 Formation of octahydro-4-aza-cyclobuta[*a*]pentalenes from 1,6-enynes

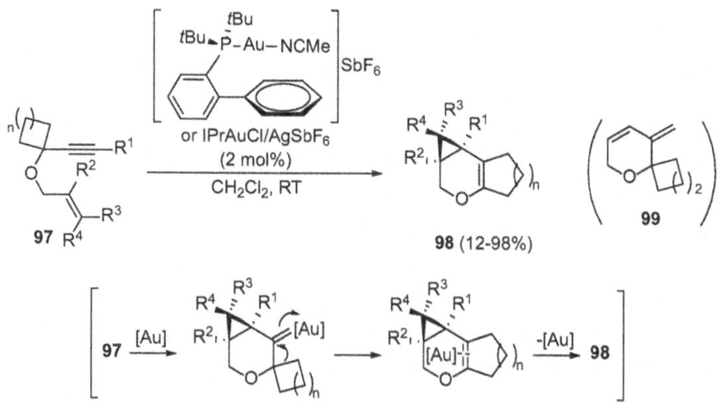

Scheme 24 Formation of fused bicyclo[4.1.0]heptenes from 1,6-enynes

Interestingly, Fensterbank et al. discovered that the reaction of oxygen-tethered 1,6 enynes bearing a strained ring between the oxygen and the alkyne, such as **97**, reacted in the presence of gold to furnish tricycles **98** (Scheme 24) [108]. The formation of this heterocycle could be explained through a ring expansion via a Wagner–Meerwein transposition from the corresponding cyclopropyl gold carbene. In one of the cases, tetrahydropyran **99** was obtained as a subproduct.

As a particular substructure in the family of enynes, it is important to mention the dienynes in which both double bonds are conjugated. Alternatively to dienynes bearing the second double bond as pendant in the alkene (Scheme 16) or alkyne chain (Schemes 17 and 19), the characteristic distance between the two double bonds in conjugated systems gives rise to new reaction pathways and

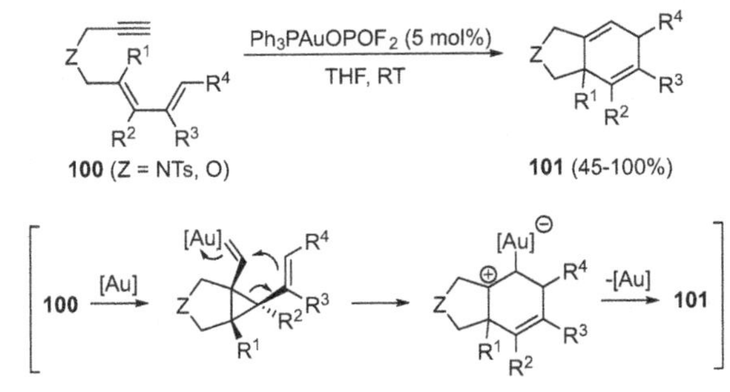

Scheme 25 Formation of bicyclo[4.3.0]nonanes from dienynes

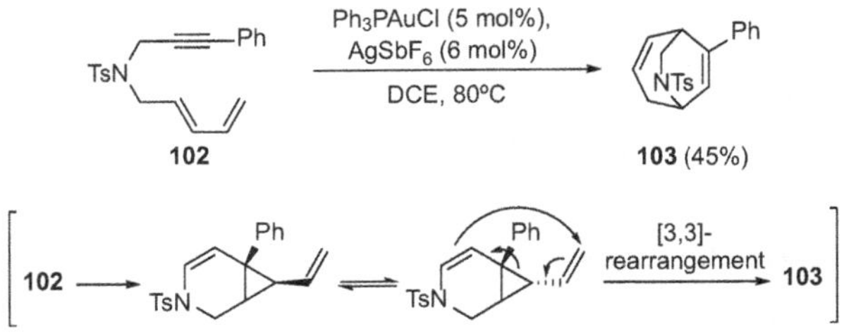

Scheme 26 Formation of fused bicyclo[3.2.2]nonanes from dienynes

complementary ways of trapping the gold carbene. The groups of Fürstner and Chung explored those substrates for the preparation of bicyclo[4.3.0]nonanes **101** (Scheme 25) [109, 110]. The reaction resembles a formal [4+2] cycloaddition, and it would occur through the initial formation of the cyclopropyl gold carbene through the 5-*exo-dig* pathway. The gold carbene would then evolve via cyclopropyl ring opening to a cyclohexyl cation, which after elimination would generate bicycles **101**.

Chung et al. reported a different gold-catalysed reaction of these substrates to form bicyclo[3.2.2]nonanes **103** (Scheme 26) [111]. The divergent reactivity compared to the previous case could be due to the different substitution in the substrate used. Although the reaction worked with gold, platinum catalysis probed to be a better choice in this case. The proposed reaction mechanism would start with a skeletal rearrangement to form the bicyclo[4.1.0]heptane, which then would suffer a [3,3]-sigmatropic rearrangement to furnish the final product **103**.

Yu et al. studied the reactivity of more complex substrates, dienediynes. Under gold catalysis, dienediynes **104** afforded tricyclo[9.3.0.0^{2,7}]tetradecanes **105** in a diastereoselective way (Scheme 27) [112]. The transformation would imply the formation of a cyclopropyl gold carbene followed by a [3,3]-sigmatropic rearrangement that would yield a cyclic allene. C–H activation of the terminal carbon of the pendant chain and cyclisation with the central carbon of the strained allene would lead to the formation of a new cycle in the structure, which after two consecutive 1,2-H shifts would form tricycles **105**.

[3,3]-Rearrangements are very common pathways in enyne chemistry, and we have summarised the most important examples involving synthesis of heterocycles using gold catalysis as a separate section in this chapter (see below Sect. 5).

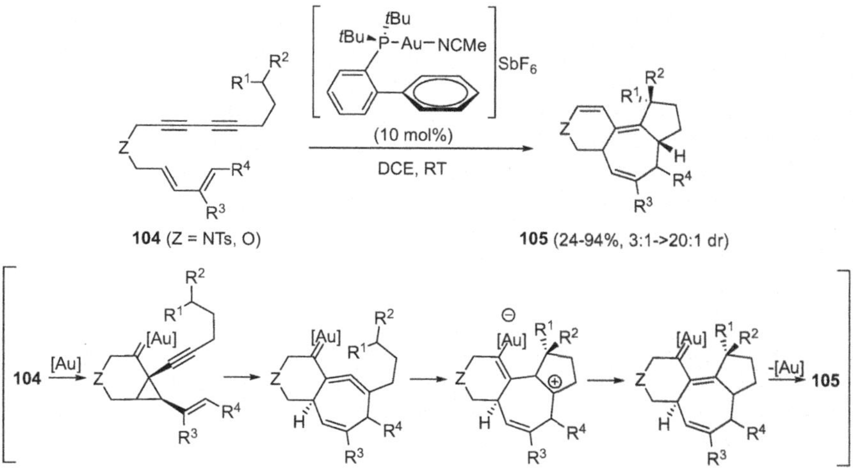

Scheme 27 Formation of fused tricyclo[9.3.0.0^{2,7}]tetradecanes from dienediynes

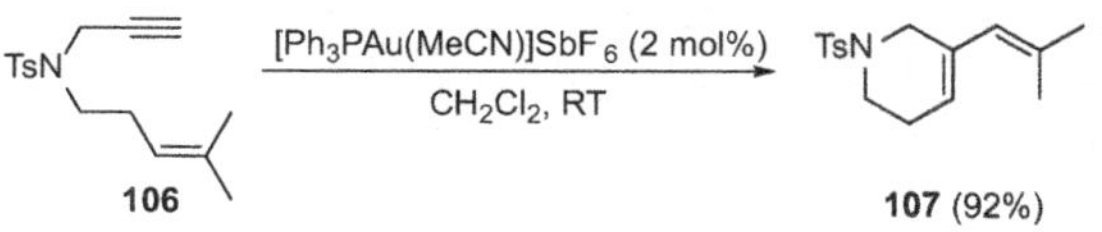

Scheme 28 Formation of 1,2,3,6-tetrahydropyridines from 1,7-enynes

3.2.2 Skeletal Rearrangement of 1,*n*-Enynes with $n \neq 6$

The examples of formation of heterocycles through gold-catalysed reactions of 1,7-enynes are very limited. Echavarren's group has explored the reactivity of mainly carbon-tethered 1,7-enynes, but they also described the synthesis of 1,2,3,6-tetrahydropyridine **107**, resulting from a single cleavage rearrangement, staring from 1,7-enyne **106** (Scheme 28) [113].

Regarding the cyclisations of 1,5-enynes, even though there are a great number of examples of cyclisation involving carbon-tethered 1,5-enynes [45, 114–118], to the best of our knowledge just one example of the synthesis of heterocycles through a gold-catalysed reaction of 1,5-enynes has been described in the absence of nucleophiles (internal or external). However, according to the authors, this methodology does not imply any rearrangement as described in Sect. 3.1. The gold-catalysed cyclisation of *N*-alkenyl-alkynylamides **108** gives rise to the formation of 2-pyridinones **109** via a simple 6-*endo* cyclisation (Scheme 29) [119]. The enantioselective version for the synthesis of chiral heterobiaryl compounds has also been explored [120]. Similar reactions have appeared in the literature using catalysts different than gold, but in those cases, the authors postulate a mechanism based on an initial aza-Claisen rearrangement [121–123]. As mentioned before, the [3,3]-rearrangement of enynes can be exploited for the synthesis of heterocyclic structures as will be explained in Sect. 5.

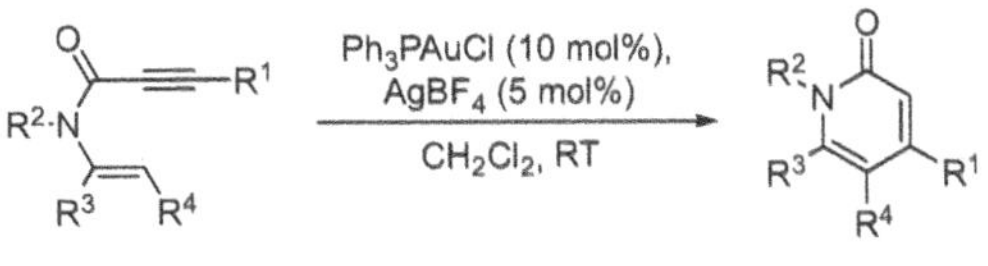

Scheme 29 Formation of 2-pyridinones from 1,5-enynes

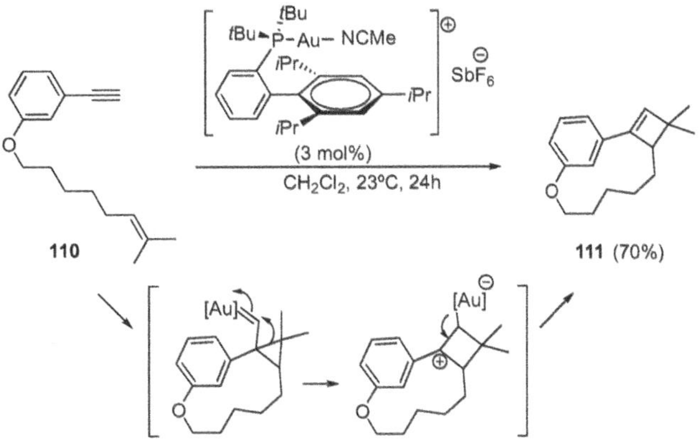

Scheme 30 Formation of macrocycles fused to a cyclobutene moiety from 1,*n*-enynes

The use of 1,*n*-enynes with $n > 8$ has been reported in the literature for the synthesis of macrocyclic structures such as **111** (Scheme 30). An aromatic spacer between alkene and alkyne and a gold catalyst bearing a sterically hindered biphenylphosphine ligand were required in order to favour the reaction [124, 125]. Since the final products of this process are the macrocycles fused to a cyclobutene ring, the transformation could be considered as a formal [2+2] cycloaddition. However, the proposed mechanism involves cyclopropyl gold carbenes and their rearrangement to form the cyclobutenes.

3.2.3 Skeletal Rearrangement of Alkynylfurans

A case that deserves special attention is the one in which the olefinic part of the enyne is an analogue of an alkene rather than an alkene itself. Probably, the most explored analogue is the furan ring. Alkynylfurans have demonstrated to be excellent precursors for the synthesis of phenol derivatives, being gold the most active metal for this transformation (Scheme 31), although platinum has been also explored [126–128]. Hashmi et al. were the ones who first observed this reactivity, and consequently, these authors have explored profusely this methodology [129–132].

The proposed mechanism for this process is shown in Scheme 31. After the coordination of the gold to the alkyne, a 5-*exo* cycloisomerisation would take place to form cyclopropyl gold carbene **114**, which is similar to carbene **10** obtained in

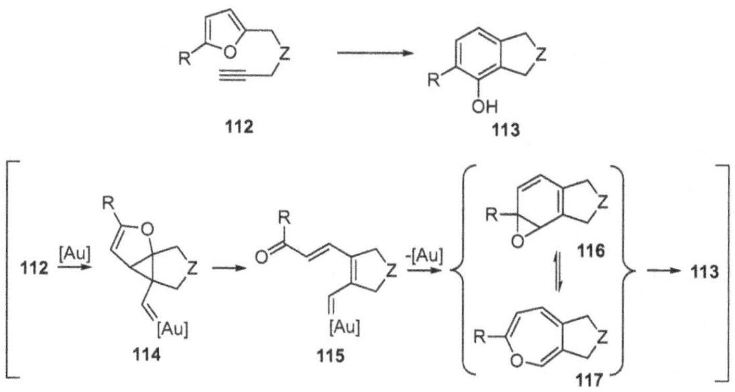

Scheme 31 Mechanism of formation of phenols from alkynylfurans

the cyclisation of enynes (Scheme 3, Sect. 3.1). This intermediate would evolve through a rearrangement to form conjugated gold carbene **115**. After cyclisation, the oxepine **117** would be obtained. This oxepine is in a valence tautomeric equilibrium with the corresponding arene oxide **116**. The ring opening of this arene oxide would form the corresponding bicyclic phenol **113**.

This mechanistic proposal was supported by some experimental evidences. For example, the use of alkynylfurans with R=H in structures like **112** afforded mixtures of two different phenols corresponding to the two possible ring openings of the epoxide ring [132]. Isotopic labelling experiments match the proposed mechanism, and the intermediates oxepine **117** and arene oxide **116** have been identified in the reaction media by NMR techniques [133–135]. The isolation of arene oxides **116** appeared as a challenge due to its rapid isomerisation to the corresponding oxepines **117**. Nevertheless, these compounds can be trapped via in situ Diels–Alder reaction by the addition of *N*-phenyltriazolindione (Scheme 32) [134].

The scope of this reaction has been extensively explored, and phenols fused with tetrahydrofurans, pyrrolidines, pyrans or piperidines have been obtained [136–142]. In the first examples described, Hashmi et al. studied the reactivity of alkynylfurans **120**, obtaining bicycles **121** with good yields (a, Scheme 33) [132]. The synthesis of phenols fused to piperidines **123** was achieved in excellent yields using as a catalyst a gold(III) complex bearing a pyridine derivative as a ligand (b, Scheme 33) [143]. Heterocycles **125** such as dihydroindoles, dihydrobenzofurans, chromans and tetrahydroquinolines, in which there is a variation in the position of the heteroatom, can also be synthesised. In these examples, the stabilising effect of the heteroatom helps to solve the selectivity problems observed for unsubstituted furans (c, Scheme 33) [144]. Finally, as an interesting example, the reaction of furandialkynes such as **126** could be mentioned. In this case, the formed *ortho*-alkynylfuran evolves through cyclisation into benzofuran **127** (d, Scheme 33). However, this reaction is not selective, and in most of the cases, mixtures of the benzofurans and the corresponding phenols are obtained [130].

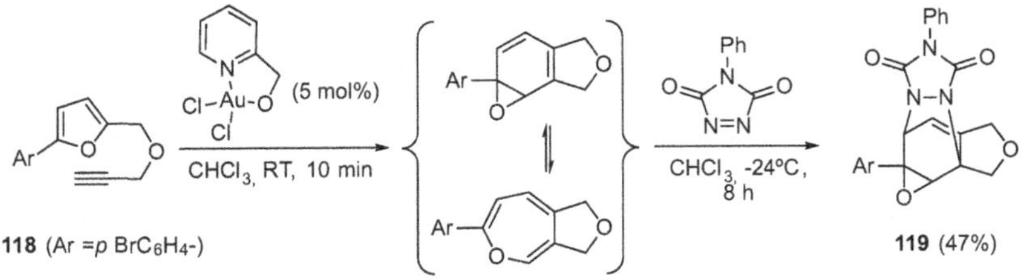

Scheme 32 Formation of a Diels–Alder adduct from alkynylfurans

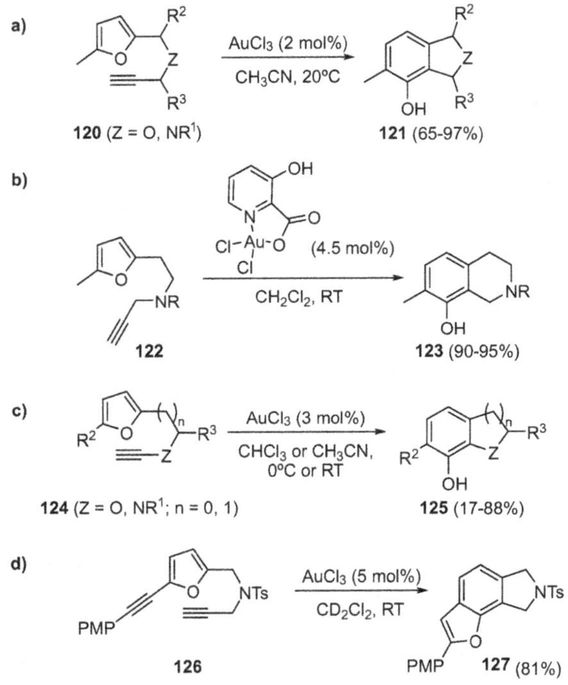

Scheme 33 Formation of phenols fused to different heterocycles from alkynylfurans

Before concluding this section, it is worth mentioning that Hashmi's group has developed a methodology based on this reactivity in which the furan ring is prepared in situ from a 2-alkynylallyl alcohol moieties [145].

3.2.4 Skeletal Rearrangement of Enynes in the Presence of Nucleophiles

Cycloisomerisation of enynes has also been performed in the presence of external and internal nucleophiles, which are incorporated in the skeleton of the final molecules. There are different possible pathways in which the nucleophilic addition could occur, depending on where the nucleophilic attack happens: either in one of

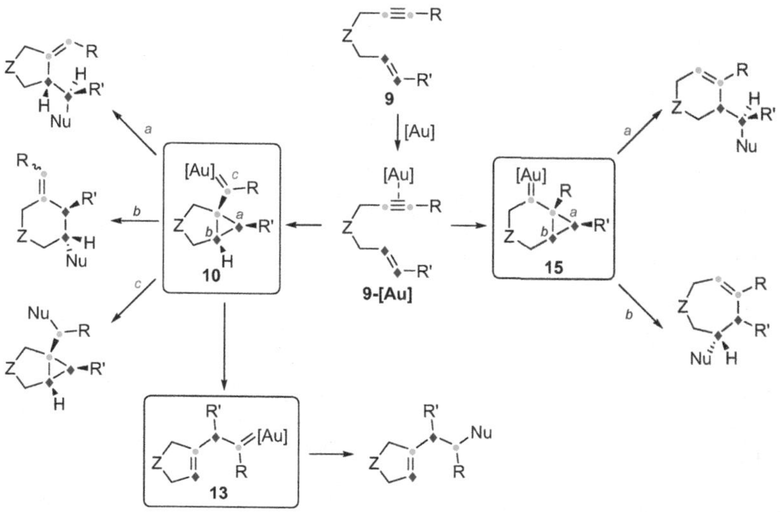

Scheme 34 Possible pathways for the reactivity of 1,6-enynes in the presence of nucleophiles

the positions of the cyclopropyl ring on intermediates **10** or **15** (Scheme 3, Sect. 3.1) or in one of the carbene carbons on intermediates **10** or **13** (Scheme 3, Sect. 3.1). A summary of the possible reaction pathways of the reaction of 1,6-enynes with external nucleophiles is presented in Scheme 34.

The heterocyclic cores that could be formed in the presence of nucleophiles are similar to the ones already discussed, although more complex structures are normally obtained when an extra nucleophile is involved in the reaction. The reactivity of enynes in the presence of nucleophiles has been widely studied [47, 146–150] and already reviewed [41, 62, 151]. Therefore, in this section only a few interesting examples showing appealing reactivity will be outlined.

A particular case of high relevance for this chapter is the cycloisomerisation of 1,6-enynes bearing a nucleophilic group in their structure, since the intramolecular nucleophilic addition opens new possibilities for the creation of heterocycles. The most widely explored nucleophiles are alcohols. For example, enyne **128**, with a carbon-based skeleton but bearing a propargylic alcohol, reacts in the presence of gold to afford oxygenated heterocycle **129** with an excellent yield (a, Scheme 35) [76]. The reaction occurs through a cyclopropyl gold carbene similar to **10** (Scheme 3, Sect. 3.1), which then undergoes cyclopropyl ring opening by the attack of the alcohol group. When enyne **130**, with the alcohol pendant from the alkene chain, is treated with a gold catalyst, compounds like **131** can be obtained through a similar pathway (b, Scheme 35) [76]. Depending on the substitution of the alkene, two different routes are possible from enyne **132** (c, Scheme 35) [51]. Products **133** or **134** can be obtained by attack of the alcohol nucleophile to the carbon of the cyclopropyl gold carbene intermediate where the developing positive charge is best stabilised, reinforcing the carbocationic nature of these intermediates as shown in the resonance forms in Scheme 2, Sect. 3.1.

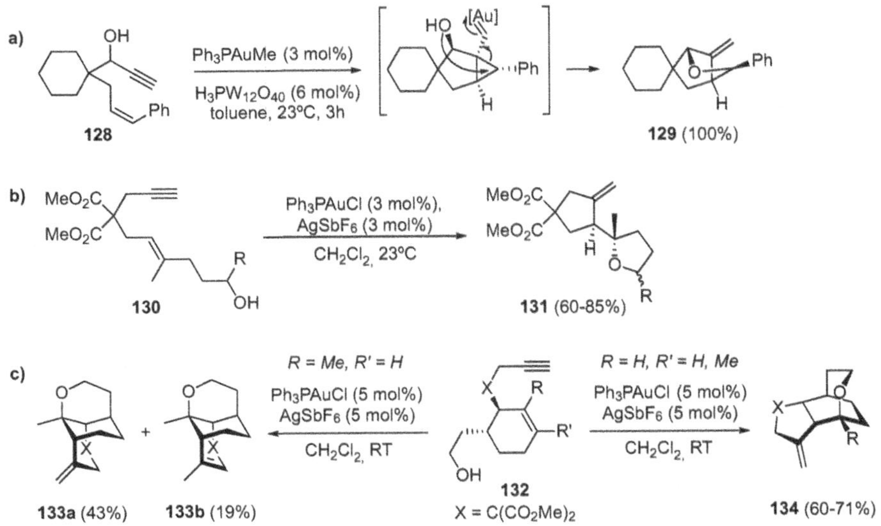

Scheme 35 Formation of heterocycles from 1,6-enynes with a pendant hydroxyl group

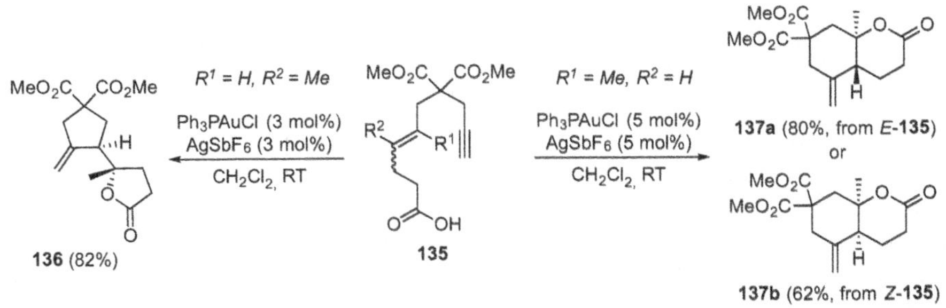

Scheme 36 Formation of heterocycles from 1,6-enynes with a pendant carboxylic acid

Apart from alcohols, the use of other nucleophiles is plausible. Fürstner et al. performed the reaction with carboxylic acids as nucleophiles obtaining a variety of products [51]. For instance, the reaction of enyne **135** upon treatment with gold allows the preparation of two different structures **136** or **137** depending on the substitution of the alkene in the starting material (Scheme 36).

Toste et al. have successfully developed the enantioselective variant of this methodology. These authors extended the scope to different nucleophiles such as amines, phenols or alkenylphenols for the synthesis of a variety of heteropolycyclic scaffolds with good yields and enantiomeric excesses (Scheme 37) [152].

Indoles, one of the most versatile heterocycles [153], are also commonly used as external nucleophiles to give structures as the ones displayed in Scheme 34. Interestingly, Echavarren et al. observed that, when the tether between alkene and alkyne in 1,6-enynes is an ether, the gold-catalysed reaction in the presence of an external indole led to a more complex structure (Scheme 38) [150]. Thus, reaction

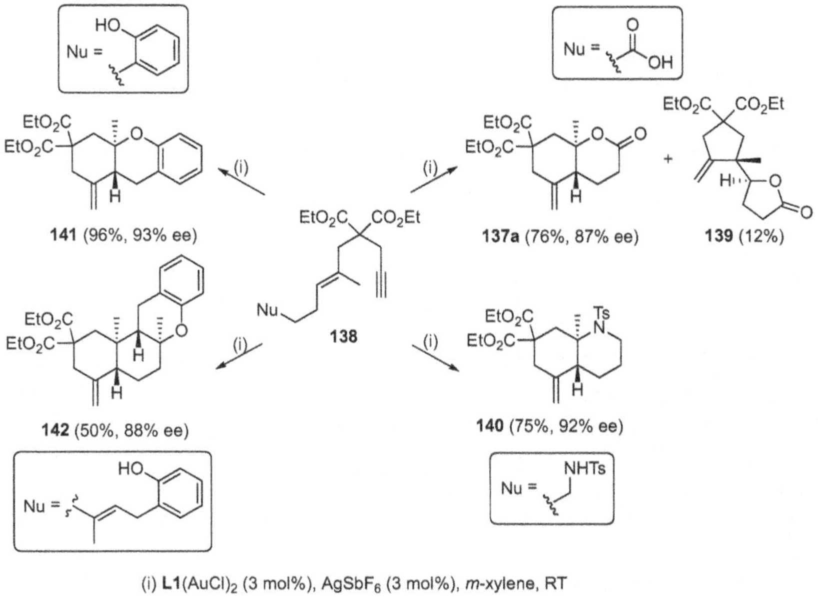

Scheme 37 Enantioselective formation of heterocycles from 1,6-enynes with a pendant nucleophile

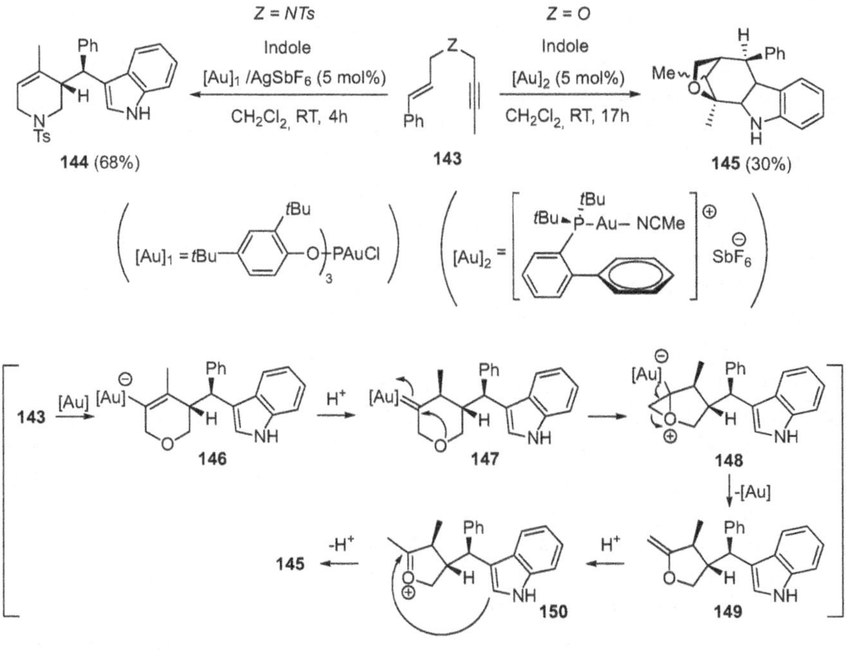

Scheme 38 Formation of heterocycles from 1,6-enynes in the presence of indole

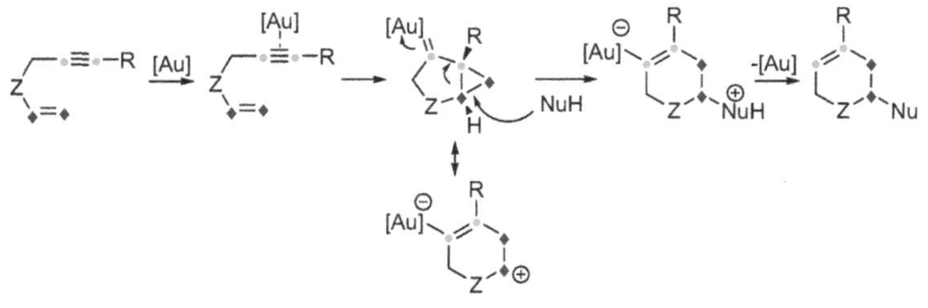

Scheme 39 Pathways for the reaction of 1,5-enynes in the presence of nucleophiles

of enyne **143** with Z=NTs led to tetrahydropyridine **144**, expected product of cycloisomerisation/nucleophilic addition, whereas enyne **143** with Z=O afforded polycycle **145** as the only product. Attainment of this polycyclic structure could be explained by the initial formation of intermediate **146**, which would be similar to the one leading to **144**. However, in this case, protonation of the alkenyl–gold moiety could occur, giving gold carbene **147**. A subsequent ring contraction and proton-catalysed cyclisation would explain the formation of the final product.

Although examples in the absence of nucleophiles are scarce, as shown in Sect. 3.2.2, the cyclisation of 1,5-enynes in the presence of nucleophiles is particularly appealing in the synthesis of heterocycles. Similarly to the case of 1,6-enynes, upon treatment with a gold catalyst, 1,5-enynes evolve through the formation of cyclopropyl gold carbenes, which can also react with nucleophiles in the way depicted in Scheme 39. A cationic resonant form of the cyclopropyl gold carbene could be envisaged, which would also explain the reactivity and the formation of six-membered cycles.

In most of the examples of reactivity of 1,5-enynes described in the literature, Z is a carbon tether. Thus, the formation of heterocycles occurs when the nucleophile is incorporated in the structure of the enyne and the reaction is intramolecular [154]. Some examples of heterocyclic scaffolds prepared with this methodology are shown in Scheme 40 [155–157]. The carbocycle formed from the enyne cycloisomerisation in these cases is a cyclohexene, and the heterocycle comes from the intramolecular nucleophilic attack into the carbocationic intermediate.

Variations in the substitution pattern of the alkene of the starting material have shown to have a profound effect in the reactivity. For example, Gagosz et al. observed that the gold-catalysed reaction of enynes completely substituted in the terminal position of the alkene in the presence of nucleophiles yielded cyclopentene rings as the only products [158]. Later on, these authors applied this methodology to the synthesis of heterocycles by performing a gold-catalysed cyclisation of enynes **157** that afforded bicyclo[4:3:0]nonanes **158** (Scheme 41) [159]. The reaction proceeds through the same type of cyclopropyl gold carbene, but the nucleophilic attack occurs in this case in a different position of the cyclopropyl ring, which carbocationic character is stabilised by the presence of the substituents.

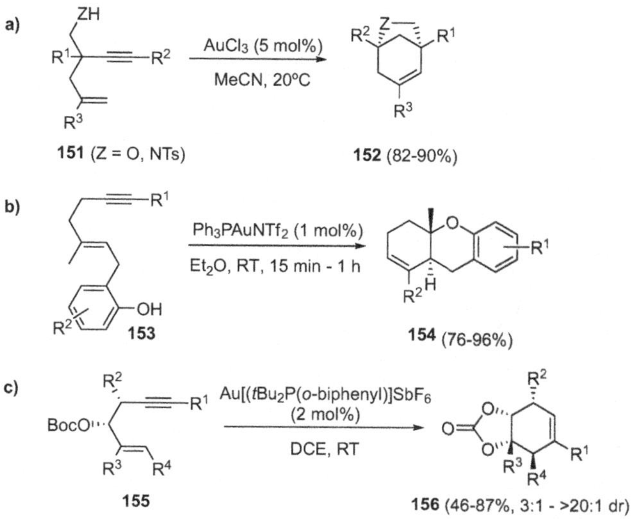

Scheme 40 Formation of heterocycles from 1,5-enynes with a pendant nucleophile

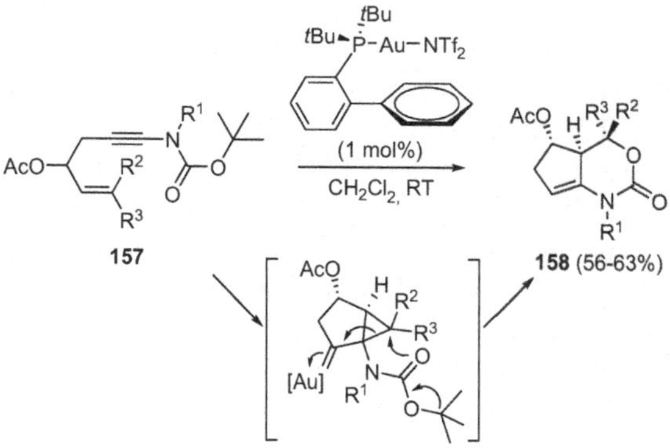

Scheme 41 Formation of bicyclo[4:3:0]nonanes from 1,5-enynes with a pendant nucleophile

As a remarkable case of nucleophilic additions, it is important to mention the carbonyl groups. The oxygen from the carbonyl group can act as a nucleophile attacking to one of the intermediate gold carbenes and forming different oxaheterocycles. Scheme 42 shows an overview of the main cycloisomerisation/ nucleophilic attack pathways that could take place. The pathway followed depends mainly on the substitution on the alkene. When the initial 1,6-enyne bears a terminal alkene, the intermediate cyclopropyl gold carbene **10** would rearrange through the double cleavage pathway to gold carbene **13**, which would then convert into **159** through a nucleophilic attack of the oxygen from the carbonyl compound. A Prins rearrangement of **159** would form cation **160**, which after cyclopropanation would furnish tricycle **161**. On the other hand, substituted alkenes evolve directly through the nucleophilic attack to the cyclopropyl ring giving oxonium cation **162**.

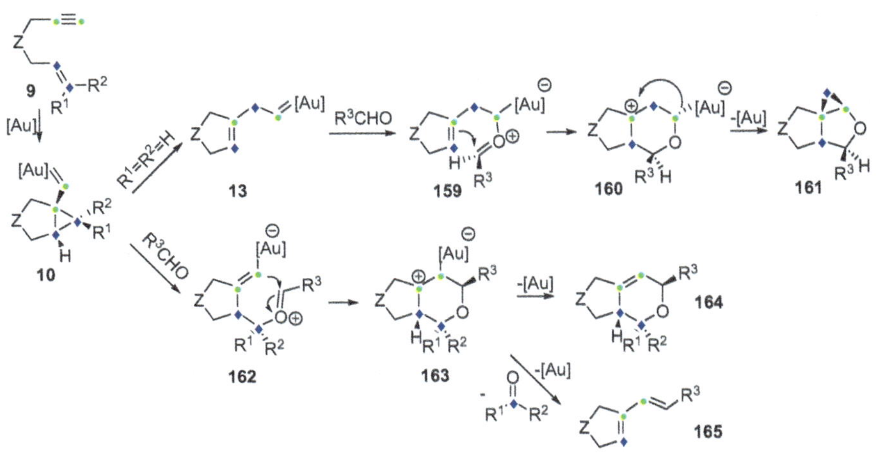

Scheme 42 Pathways for the reaction of 1,6-enynes in the presence of carbonyl compounds

Again a Prins rearrangement would form compound **163** which can turn into **164** by demetalation or fragment to afford **165**.

The groups of Helmchen and Echavarren have explored the synthesis of oxaheterocycles using this strategy [160–162]. Tricycles **167** were obtained as the only products (a, Scheme 43), while bicycles **169** were always obtained along with cyclopentenes **170** (b, Scheme 43). Interestingly, enynes **168b**, with nitrogen or oxygen tethers, only formed product **170**. This procedure can be extended to 1,7-enynes. Surprisingly, Helmchen et al. found that when enynes **171**, which are substituted in the terminal position of the alkene, are submitted to gold catalysis, the products formed were dioxolanes **172** (c, Scheme 43). The mechanism of this transformation, in which two molecules of the aldehyde are incorporated in the final skeleton of the product, is still unclear.

Furthermore, Echavarren et al. performed this reaction intramolecularly, with the carbonyl group incorporated in the enyne skeleton [107]. The reaction of enynes **173** led to the formation of a mixture of products, oxa-tricyclo[6.2.1.$0^{2,6}$]undecane **174** and cyclopentene **175** (Scheme 44), being the first one the major product. Again, the enyne **173b** tethered with a -NTs group gave only product **175b**. The mechanism that explains the formation of these products is similar to the one already explained for the intermolecular nucleophilic addition.

This intramolecular approach has been employed in the total synthesis of some interesting compounds such as Englerins A and B or Orientalol F [163–165]. The example of (−)-Englerin A, an inhibitor of renal cancer, is presented in Scheme 45.

This reactivity has been also studied in the case of 1,5-enynes. Echavarren et al. have explored the reaction in its inter- and intramolecular version [160, 166]. In both cases, the basic core of bicyclo[3:3:0]octane was obtained through a mechanism similar to the one with 1,6-enynes (Scheme 46). In the intermolecular process, the conjugated triene **181b** was the only product when the terminal position of the alkyne was substituted with an aromatic group.

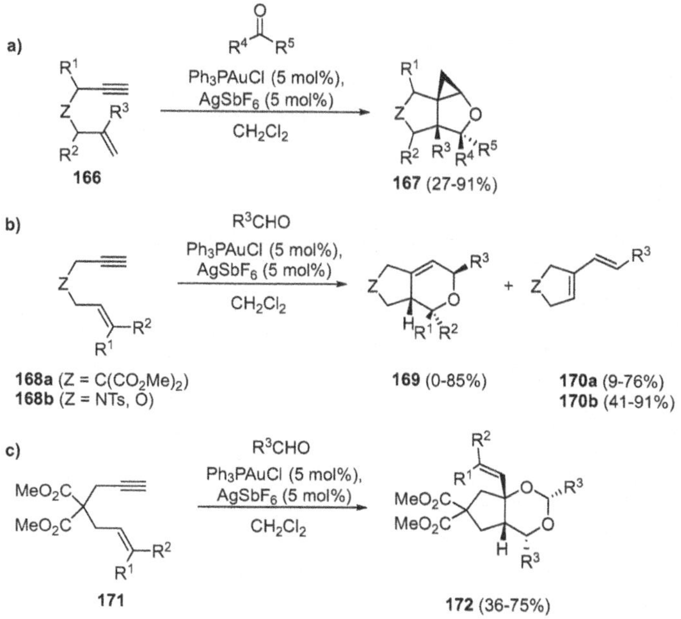

Scheme 43 Formation of oxaheterocycles from 1,6-enynes in the presence of carbonyl compounds

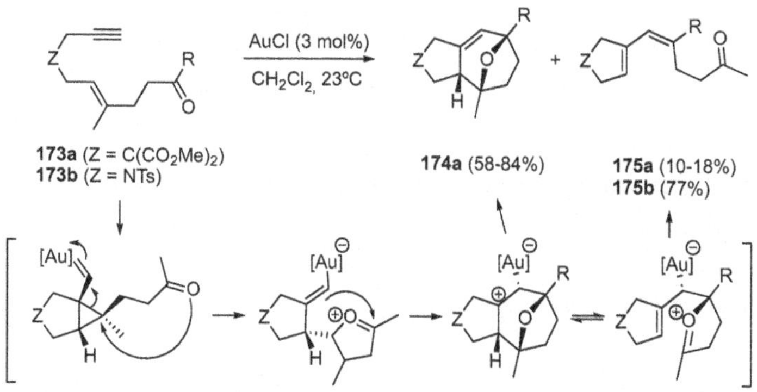

Scheme 44 Formation of oxa-tricyclo[6.2.1.0^{2,6}]undecane from 1,6-enynes with a pendant carbonyl moiety

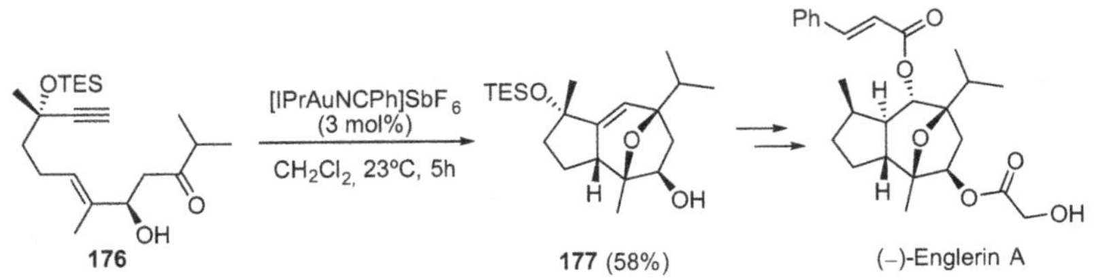

Scheme 45 Total synthesis of (−)-Englerin A from a 1,6-enyne with a pendant carbonyl moiety

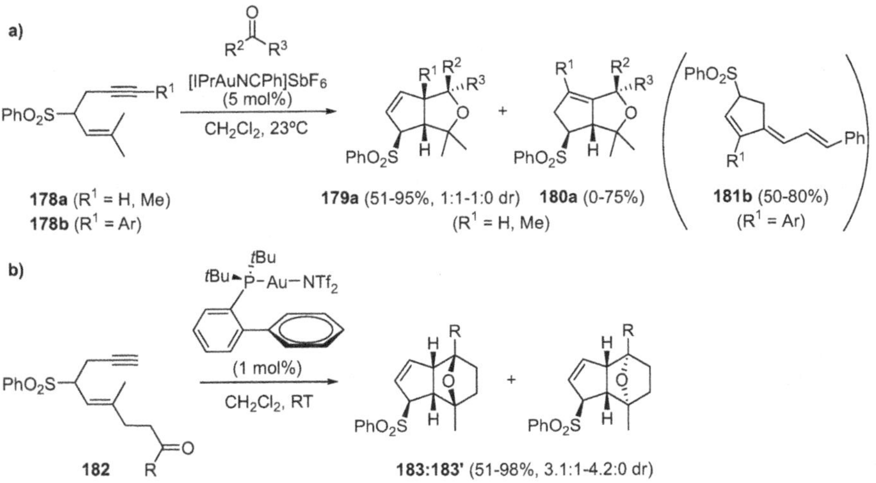

Scheme 46 Formation of bicyclo[3:3:0]octane from 1,5-enynes with a carbonyl moiety

4 Gold-Catalysed Rearrangement of Enynes Bearing a Propargylic Ester

4.1 Mechanistic Insights

In addition to the skeletal rearrangement reviewed in the previous sections, enynes can suffer other types of rearrangement depending on their substitution pattern. Special attention deserves the case in which the enyne structure contains a propargylic ester. The rearrangement of this particular moiety is well known, and it occurs through the nucleophilic attack of the carboxylic oxygen of the ester to one of the carbons of the metal-coordinated alkyne [167, 168]. There are two possible routes for the rearrangement depending on the carbon of the alkyne involved (Scheme 47). On one hand, if the attack takes place in the carbon of the alkyne that is closer to the ester moiety, the process is called a 1,2-acyloxy migration, and it leads to the formation of alkenyl–gold carbene **185** through a 5-*exo-dig* cyclisation and a subsequent ring opening. On the other hand, the attack to the distant carbon of

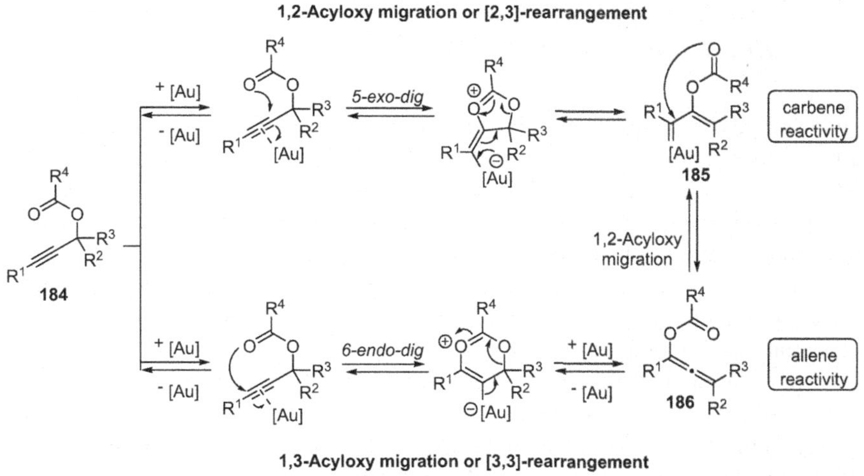

Scheme 47 Possible pathways in the rearrangement of propargylic esters

the alkyne, which is known as 1,3-acyloxy migration, would rearrange to allenic structure **186** through a 6-*endo-dig* cyclisation. DFT calculations suggest that the allene formation could also be explained through two consecutive 1,2-acyloxy migrations from the initial propargylic ester [169].

In the case of enynes, subject of this chapter, there is an additional alkene in the structure that can react with either of these intermediates. The migration pathway followed will determine the intermediate formed, affecting the ulterior reactivity and giving rise to a variety of possible pathways and heterocyclic structures.

Before exploring the different examples reported using this methodology, a comment about nomenclature should be made. Two different ways have appeared in the literature for naming these transformations: 1,2-acyloxy migrations could also be called [2,3]-rearrangements, and 1,3-acyloxy migrations are also named [3,3]-rearrangements. For simplicity and consistency on the text, we will always use the first nomenclature in this chapter.

In this section, the examples would be presented according to the type of rearrangement occurring: 1,2- or 1,3-acyloxy migration. However, in some cases, the reactivity observed could be explained through both pathways, and very similar final structures or intermediates have been reported in the literature to occur through one or the other route. We have organised these examples according to the mechanistic proposal reported by the authors in the references cited.

4.2 Formation of Heterocycles by Rearrangements of Enynes with a Propargylic Ester Moiety

A wide range of examples have appeared in the literature exploring the reactivity of carbon-tethered enynes bearing a propargylic ester [170], and both pathways, 1,2- and 1,3-acyloxy migrations [171], have been observed. However, we will focus here in the examples where the alkyne and alkene are connected through a heteroatom, and therefore, a heterocycle is formed in the reaction.

As mentioned before, if the reaction takes place through a 1,2-acyloxy migration, the intermediate formed is the corresponding carbene **185** (Scheme 47), which will define the outcome of the process.

Toste et al. applied this methodology to enynes **187** bearing an oxygen tether (Scheme 48) [172]. The reaction of these enynes in the presence of a gold(I) catalyst bearing a chiral ligand led to the formation of 4*H*-chromenes **188** with good yields and excellent enantiomeric excesses. The mechanism that explains the formation of this heterocycle starts with a 1,2-acyloxy migration to form the corresponding gold carbene **189**, which evolves through the nucleophilic attack of the oxygen to form oxonium cation **190**. From this intermediate, the authors propose three different pathways: firstly, a 1,2-migration of the allyl group followed by a [3,3]-sigmatropic rearrangement; alternatively, a direct conversion to **188** via a formal 1,4-sigmatropic rearrangement; and finally, the formation of an intermediate allyl cation. According to experimental evidences, the third pathway seems to be more plausible.

The reaction of dienynes **191** under gold catalysis to obtain hexahydroazocines or tetrahydroxocines **192** was reported by the group of She (Scheme 49) [173]. The

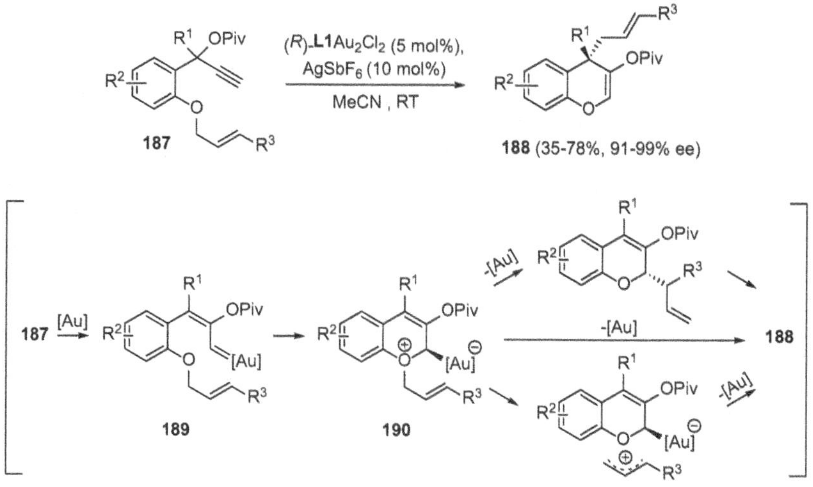

Scheme 48 Formation of 4*H*-chromene from enynes via a 1,2-acyloxy migration. For structure of **L1**, see Fig. 3

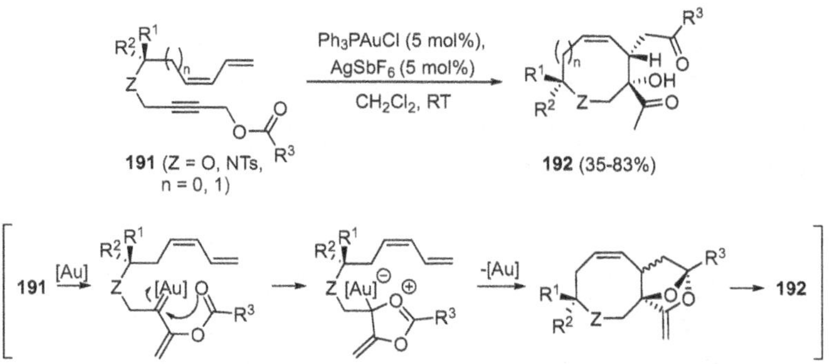

Scheme 49 Formation of hexahydroazocines or tetrahydroxocines from dienynes via a 1,2-acyloxy migration

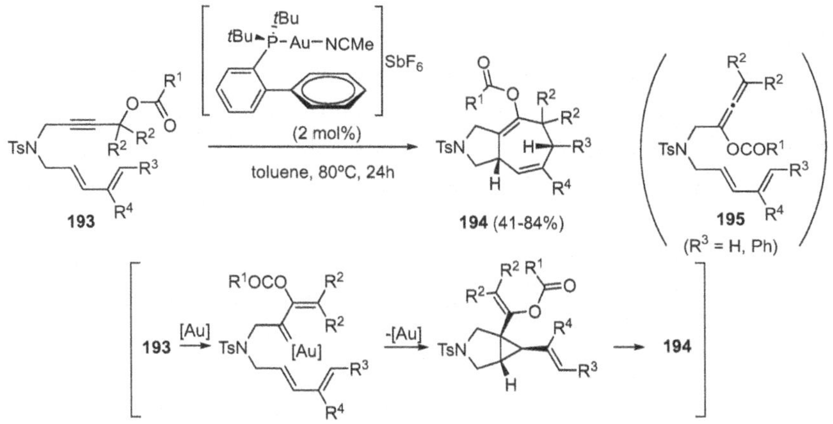

Scheme 50 Formation of bicyclo[5.3.0]decanes from dienynes via a 1,2-acyloxy migration

formation of nine-membered heterocycles was also explored. The mechanistic proposal for this transformation would imply a 1,2-acyloxy migration to form the corresponding gold carbene. Nucleophilic attack of the carboxylic oxygen and subsequent [3+2] cycloaddition would form an enol ketal and regenerate the gold catalyst. This enol ketal would hydrolyse in the presence of the silica gel during column chromatography to generate products **192**.

Chan et al. also studied this reactivity in dienynes such as **193** to obtain bicyclo [5.3.0]decanes **194** (Scheme 50) [174]. Allenes **195** were obtained in the reaction mixture when the substituent in the terminal position of the alkene was H or Ph. The authors postulate a mechanism based on an initial 1,2-acyloxy migration to obtain the corresponding carbene, which would evolve through a cyclopropanation and a subsequent [3,3]-sigmatropic rearrangement to form the final products. The formation of products **195** would be explained through a competitive but reversible reaction pathway based on a 1,3-acyloxy migration.

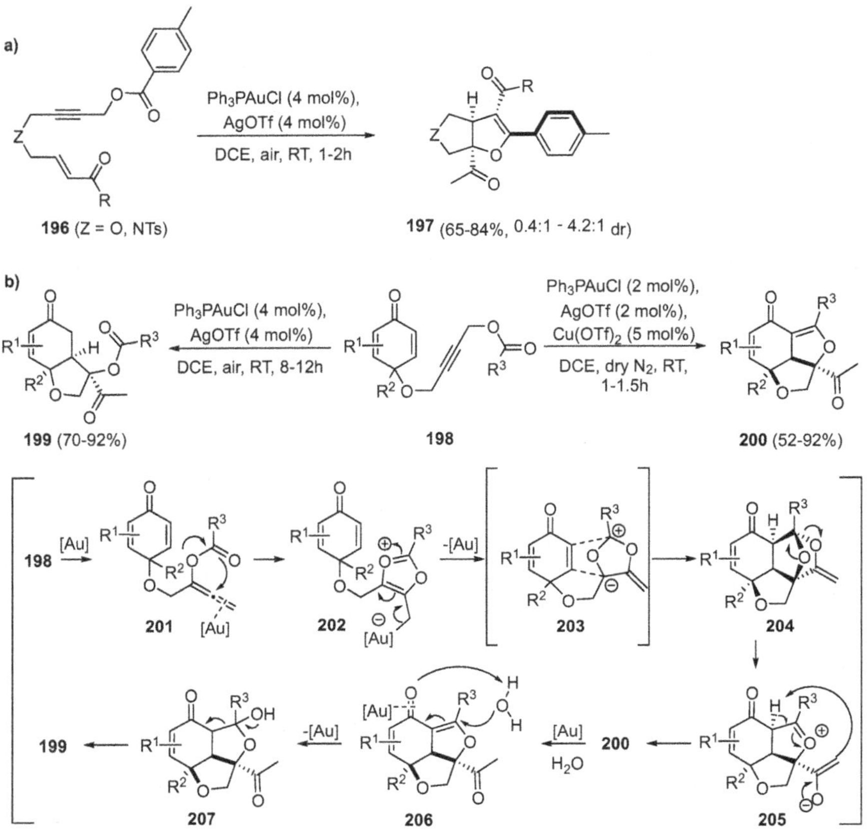

Scheme 51 Formation of polycycles from enynes via a 1,3-acyloxy migration

On the other hand, when the process occurs through a 1,3-acyloxy migration, intermediate allenic species will form. As an example, Wang et al. explored the gold-catalysed reaction of 1,6-enynes **196** to give bicyclo[3.3.0]octanes **197** (a, Scheme 51) [175]. They mainly explored this reactivity in enynes **198** in which the alkene comes from a cyclohexadienone (b, Scheme 51) and discovered that variations in the reaction conditions could alter the product obtained. Thus, when the reaction was run in open air, tetrahydrobenzofuranones **199** were obtained in a stereospecific way (b, Scheme 51, left). Alternatively, when the reaction was carried out in a strictly dry nitrogen atmosphere, the only products observed were tricycles **200**, also obtained with diastereomeric control. In this case, the addition of $Cu(OTf)_2$ as a cocatalyst accelerated the process. The mechanism proposed by the authors is initiated by a 1,3-acyloxy migration to form the allenic species **201**. The attack of the carboxylic oxygen to the central carbon of the allene, followed by the allylic elimination of the gold catalyst, would lead to the formation of a dipolar intermediate, which upon [3+2] cycloaddition with the enone double bond would afford intermediate **204**. Fragmentation of this intermediate triggered by strain

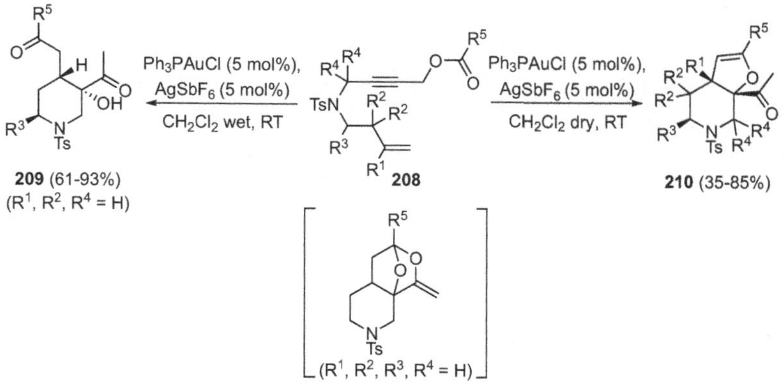

Scheme 52 Formation of piperidines from enynes via a 1,3-acyloxy migration

release would furnish **205**, which will undergo an intramolecular proton transfer to afford the tricyclic product **200**. In the absence of water, the reaction would stop at this point, but in the presence of moisture and under the catalytic conditions, a nucleophilic attack of the water would form hemiketal species **207** that, after a final retro-aldol process, would yield the bicyclic product **199**. It is worth noting that products **197** do not undergo this final ring opening in the presence of water due to the apparent loss of planar geometry on their corresponding enone units.

In a related example, She et al. studied the gold-catalysed reaction of 1,7-enynes **208** to form piperidines (Scheme 52) [176]. The presence of water also affected the reaction outcome in this case. Piperidines **209** were obtained under wet conditions, whereas hexahydro-furo[2,3-c]pyridines **210** were the only products when dry dichloromethane was used. The authors postulate a cyclic acetal as a common intermediate for the formation of both products. The mechanism of this transformation was studied by DFT computational methods by the group of Bi [177]. The calculations suggest that the 1,3-acyloxy migration pathway is preferred over the 1,2-acyloxy migration.

Remarkably, some of the intermediates in the mechanisms shown in Schemes 49, 51 and 52 are very similar despite the different acyloxy migrations proposed by the authors. This suggests that these processes could take place through any of the two migration pathways explained. Probably, the different substituents or reaction conditions would favour one or the other manifolds.

In a different gold-catalysed reaction of 1,7-enyne esters, Chan et al. obtained azabicyclo[4.2.0]oct-5-enes [178]. Reaction of enyne **211** under gold catalysis furnished azabicyclo[4.2.0]oct-5-ene **212** in a regio-, diastereo- and enantio-selective way (Scheme 53). The proposed reaction mechanism starts with a 1,3-acyloxy migration to form an allenene in which the alkene moiety is activated by gold coordination and then undergoes a [2+2] cycloaddition. The corresponding products were obtained as single diastereomers irrespective of whether the starting enyne was a single isomer or a diastereomeric mixture.

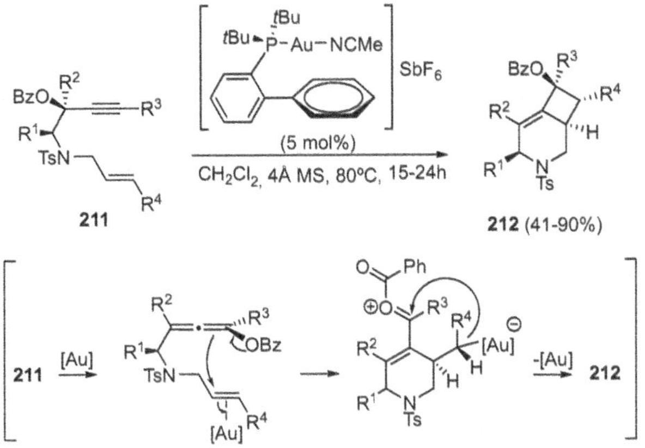

Scheme 53 Formation of azabicyclo[4.2.0]oct-5-enes from enynes via a 1,3-acyloxy migration

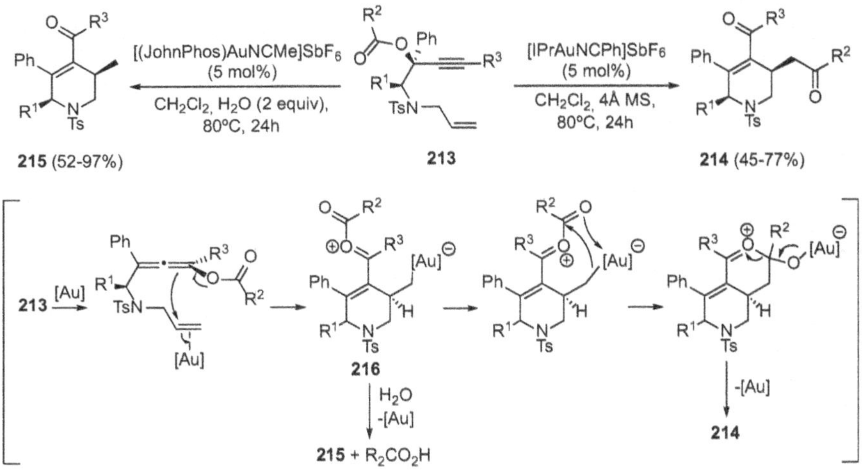

Scheme 54 Formation of tetrahydropyridines from enynes via a 1,3-acyloxy migration

Using the same premise, Chan's group also reported a synthesis of tetrahydropyridines from 1,7-enynes [179]. Interestingly, the choice of the catalyst had a high impact in the product obtained, since the reaction of enyne **213** in the presence of [IPrAuNCPh]SbF_6 led to tetrahydropyridines **214**, whereas the use of [(JohnPhos) AuNCMe]SbF_6 yielded tetrahydropyridines **215** (Scheme 54). The formation of these products could be explained through an initial 1,3-acyloxy migration followed by a 6-*exo-trig* cyclisation to form intermediate **216**. From this intermediate, two divergent pathways could be envisaged. The direct hydrolysis of the 1,7-enyne ester **216** would lead to 1,2,3,6-tetrahydropyridines **215**. On the other hand, a 1,5-acyl migration, triggered by a bond rotation, would explain the formation of 1,2,3,6-tetrahydropyridines **214**. The authors attribute this distinction to the different

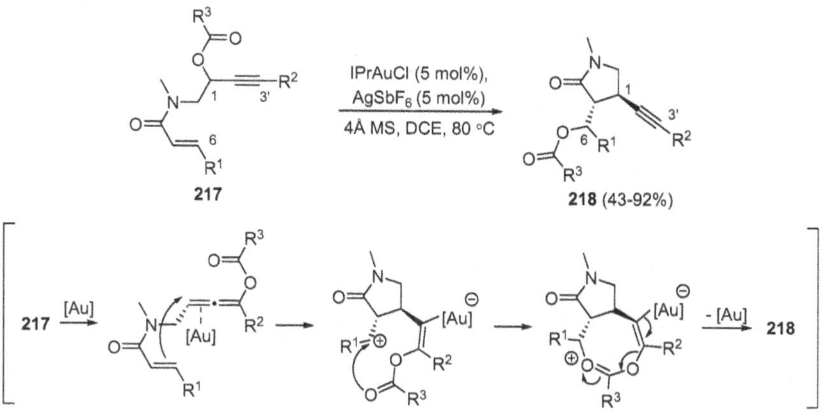

Scheme 55 Formation of pyrrolidinones from enynes via a 1,3-acyloxy migration

electronic properties of the ligands in both catalysts (NHC versus phosphine), since both reactions in presence or absence of water gave the same major products.

It is worth noting that the same catalyst used for the formation of tetrahydropyridines **215** ([(JohnPhos)AuNCMe]SbF_6) from the enynes with terminal alkenes **213** (Scheme 54) gave a different product, namely, azabicyclo[4.2.0]oct-5-ene **212** in the reaction of the substituted derivatives **211** (Scheme 53). This deviation in the reactivity starting from similar substrates could be due to the different properties of the substituents in enynes **211** ($R^2{\neq}Ph$) and **213** ($R^3{\neq}Ph$).

Other interesting examples of reactions of enynes based on an initial 1,3-acyloxy migration are the ones reported by Hashmi et al. describing the transformation of enynes **217** into pyrrolidinones **218** through a formal 1,6-acyloxy migration (Scheme 55) [180]. This process was also explored with phosphates and carbonates instead of esters, occurring in those cases through a formal 1,6-phosphatyloxy or carbonate migration [181]. Experimental evidences and DFT studies were carried out to corroborate the reaction mechanism. The transformation occurs initially through a 1,3-acyloxy migration to form the corresponding allene, which will evolve via the nucleophilic attack of the alkene moiety to form the pyrrolidinone skeleton. A subsequent 1,5-acyloxy migration would take place through the formation of an eight-membered cycle that would finally afford the product **218**.

As it has been stated in Sect. 3.2.3, analogues of alkenes can also exhibit this kind of reactivity. For instance, cases in which indole has been used as the alkene partner can be encountered on the literature [182, 183]. An appealing example of this in which the alkene is replaced by an indole moiety was described by Zhang's group [184]. The reaction of alkynyl-indole **219** in the presence of a gold(I) catalyst gave 2,3-indoline-fused cyclobutanes **220** (Scheme 56). The formation of these products could be explained through a mechanism based on an initial 1,3-acyloxy migration leading to the corresponding allenic intermediate **221**, which subsequently evolves to give oxonium cation **222**. Cyclisation through the attack of the C3 position of the indole to the oxonium cation results in spirocycle **223**, which

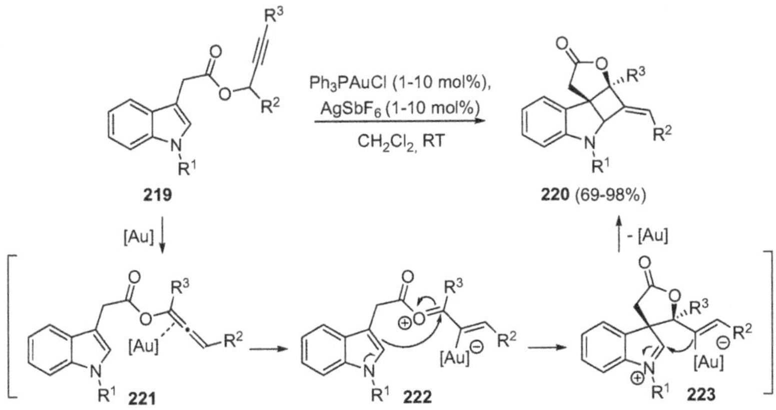

Scheme 56 Formation of 2,3-indoline-fused cyclobutanes from alkynyl-indoles via a 1,3-acyloxy migration

after intramolecular trapping of the iminium with the alkenyl–gold(I) would yield products **220**.

Interestingly, the reaction with platinum instead of gold furnished 2,3-indoline-fused cyclopentanes as products [185].

5 Gold-Catalysed [3,3]-Rearrangements of Enynes

A particularly appealing case amongst the reactions of enynes is when the distance between alkene and alkyne is such that a [3,3]-rearrangement could occur [186–188]. This creates new branches in the general mechanism for the synthesis of heterocycles from enynes. The most straightforward example is the one of 1,5-enynes. The possibility of a [3,3]-rearrangement has been exploited in 1,5-enynes **224** for the gold-catalysed synthesis of furans **225** (a, Scheme 57) [189, 190] and pyrroles **226** (b, Scheme 57) [191]. In the latter case, dihydropyridines appeared as secondary product in some examples. The proposed mechanism would start with a formal Claisen rearrangement, which is proposed to proceed in two steps via the formation of six-membered cycle **227** that then evolves to form the allenyl ketone **228**. This intermediate undergoes a 5-*exo-dig* cyclisation through the attack of the oxygen or nitrogen atoms to the central carbon of the allene to afford the final five-membered cycles.

The formation of intermediate **227** has been explored and exploited for the preparation of tetrahydropyrans [192, 193]. Toste et al. discovered that oxygen-tethered 1,5-enynes **229** react in the presence of nucleophiles, internal or external, forming tetrahydropyrans **230** (a, Scheme 58) [194]. They also discovered that reaction of disubstituted alkenes **231** was stereoselective (b, Scheme 58). The

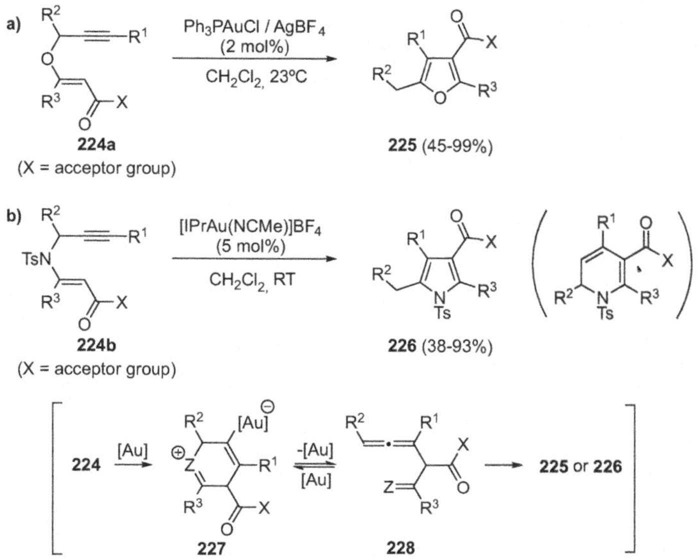

Scheme 57 Formation of furans and pyrroles from 1,5-enynes via formal Claisen rearrangement

mechanism proposed by these authors involves the formation of a six-membered cycle similar to **227** and the trapping of this cyclic intermediate by the nucleophile. On the contrary, in the absence of nucleophiles or carbonylic substituents on the alkene, oxygen-tethered enynes evolve through the classic Claisen rearrangement furnishing acyclic allenols [195].

The groups of Toste and Lee reported almost simultaneously an example based on the same reactivity of oxa-1,5-enynes but with a silicon tether between alkene and alkyne [196, 197]. Toste et al. discovered that the preparation of cyclic or acyclic vinylsilanes could be modulated by the choice of nucleophile. For instance, cyclic silanes **234** were obtained with the use of methanol as a nucleophile, whereas acyclic vinylsilanes **235** were the major product when phenol was used (Scheme 59). The mechanism of the transformations in Schemes 58 and 59 is controversial since the final products can also be explained through the formation of bicyclo[3.1.0]hexanes as depicted in Scheme 39 (Sect. 3.2.4). However, DFT calculations suggest that silacycles **234** are likely to come from an initial [3,3]-rearrangement [198], forming intermediate **236** that would suffer the attack of the nucleophile in the alkenylic carbon or in the silicon atom depending on the nucleophilicity of the alcohol employed.

The Claisen rearrangement of 1,5-enynes to allenes has also been employed in the synthesis of dihydropyridines [199]. An example of this was reported by Xu et al., who prepared pyridines **238** via a gold-catalysed reaction of 1,5-enynes **237** in the presence of tosylamine (Scheme 60) [200]. The reaction would proceed through the initial [3,3]-rearrangement, followed by isomerisation to a dialkenyl

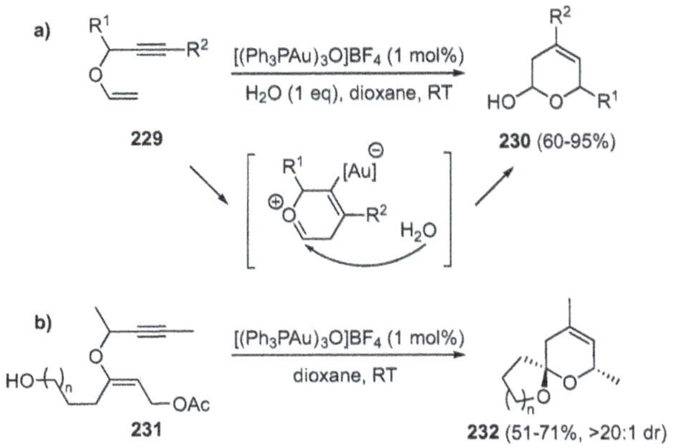

Scheme 58 Formation of tetrahydropyrans from oxa-1,5-enynes via formal Claisen rearrangement

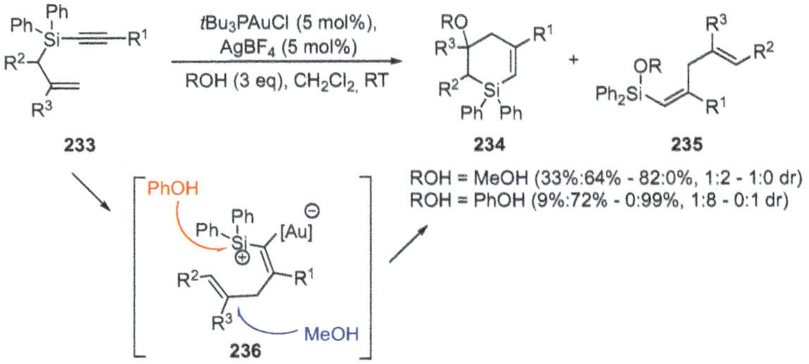

Scheme 59 Formation of sila-cyclohexenes from sila-1,5-enynes via a [3,3]-rearrangement

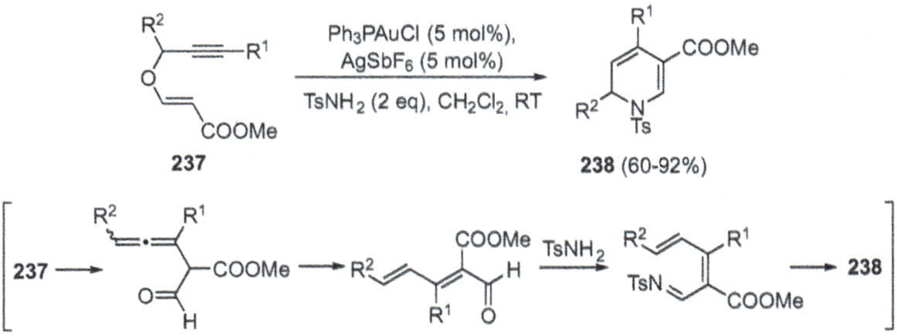

Scheme 60 Formation of dihydropyridines from 1,5-enynes via a Claisen rearrangement

ketone, formation of an imine by reaction with the tosylamine and a final 6π-electrocyclisation.

The examples explained so far imply a formal Claisen rearrangement, in the sense that this [3,3]-rearrangement occurs in two steps, while the classic Claisen rearrangement is a concerted process. In the following examples, this classic [3,3]-

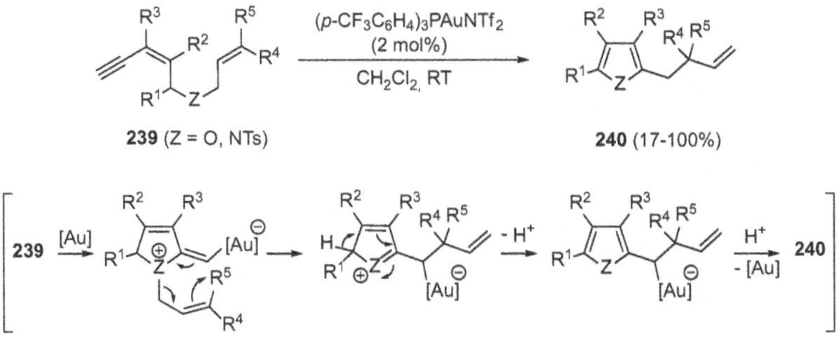

Scheme 61 Formation of pyrroles and furans from enynes via a Claisen rearrangement

sigmatropic rearrangement will be explored in reactions in which the gold catalyst promotes the formation of an intermediate that then evolves through the mentioned rearrangement. Gagosz et al. explored the synthesis of furans and pyrroles from different starting materials in a reaction involving also a Claisen rearrangement as the key step [201, 202]. The reaction of dienynes **239** under gold catalysis gave functionalised pyrroles or furans **240** (Scheme 61). In this example, the mechanistic proposal would start with a 5-*exo-dig* cyclisation to form a vinyl gold intermediate that would be followed by a classic Claisen rearrangement [203]. Proton loss and protodemetalation would lead to the final products.

This sequence of cyclisation/[3,3]-sigmatropic rearrangement has been exploited for the synthesis of diverse heterocyclic structures. In all the cases, the reaction occurs through a nucleophilic attack to one of the carbons of the alkyne moiety to form a vinyl gold intermediate that is prone to undergo a [3,3]-sigmatropic rearrangement. Structures such as indoles [204], isoindoles [205], isoxazoles [206] and isochromenes [207] have been obtained with this methodology (Scheme 62).

Blum et al. reported another example (Scheme 63) that resembles the synthesis of isochromenes **248** shown in Scheme 62d. Starting from enynes **249** and using a mixture of gold and palladium catalysts, they prepared isocoumarins **250** (Scheme 63) [208]. Hashmi et al. studied the mechanism of this reaction and discovered that, unlike the reaction of enynes **247**, the process would occur through an intermolecular transfer of the allyl fragment, rather than a Claisen rearrangement [209].

Prasad et al. applied the same sequence of reactions to the synthesis of isochromanones and isoquinolones (Scheme 64) [210]. In this case, an intermediate isomerisation step is necessary after the cyclisation to get the 1,5-diene moiety that would react through the [3,3]-sigmatropic rearrangement.

In a related and intriguing example, Kumar and Waldmann et al. applied the same principle of cyclisation/rearrangement cascade process for the synthesis of chiral cyclic gold(I) aminocarbene complexes **254** from 1,7-enynes **253** (Scheme 65) [211]. This reaction requires a stoichiometric amount of gold. The transformation would proceed through the formation of the corresponding gold–

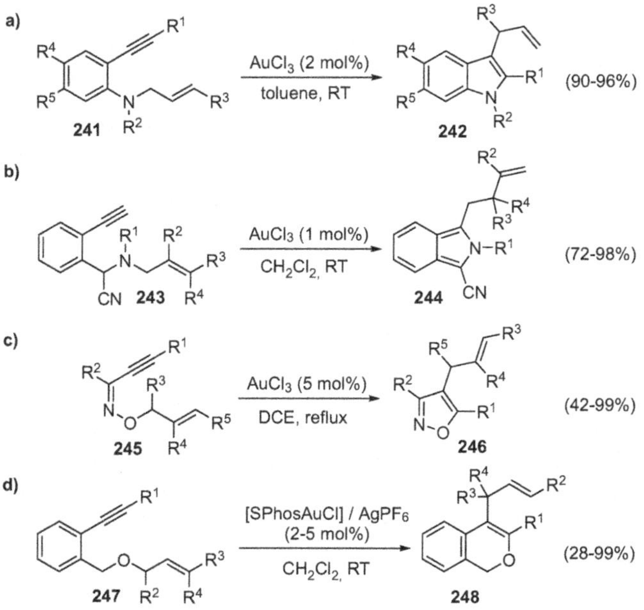

Scheme 62 Formation of different heterocycles from enynes via a Claisen rearrangement

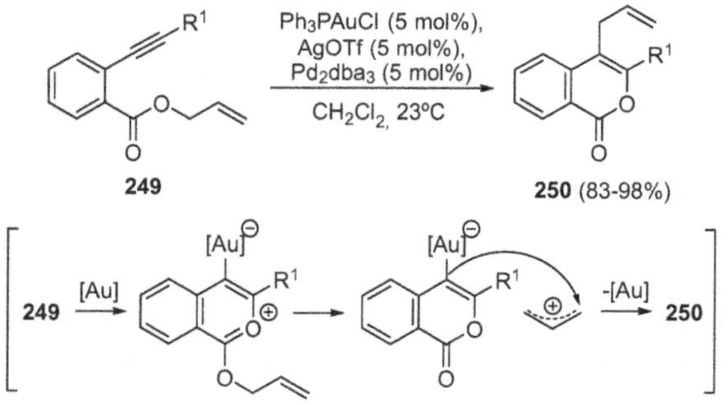

Scheme 63 Formation of isocoumarins from enynes

alkyne complex and its 5-*endo-dig* cyclisation by the attack of the nitrogen to the external carbon of the alkyne. The obtained vinyl gold complex would undergo a [3,3]-sigmatropic rearrangement. The formed iminium cation would evolve to the final product through an isomerisation and a final 1,2-Si shift of the TMS group.

Finally, an attractive example was developed in the group of Yeh for the synthesis of pyrrolidines [212]. 1,6-enynes **255**, bearing a hydroxyl group, were treated with a gold catalyst to obtain pyrrolidines **256** as the only diastereoisomers (Scheme 66). The reaction would start with the nucleophilic attack of the alcohol to the distant carbon of the alkyne to form a nine-membered cycle. This cyclic intermediate would then evolve through a Claisen rearrangement to form the pyrrolidine ring, which after protodemetalation would furnish pyrrolidines **256**.

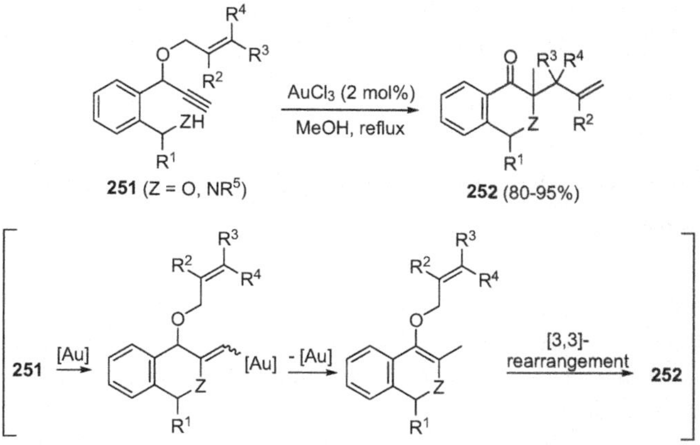

Scheme 64 Formation of isochromanones and isoquinolones from enynes via a Claisen rearrangement

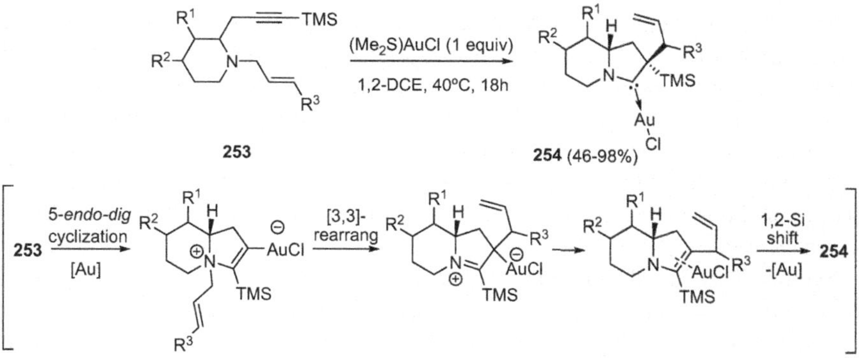

Scheme 65 Formation of gold(I) carbenes via a [3,3]-sigmatropic rearrangement

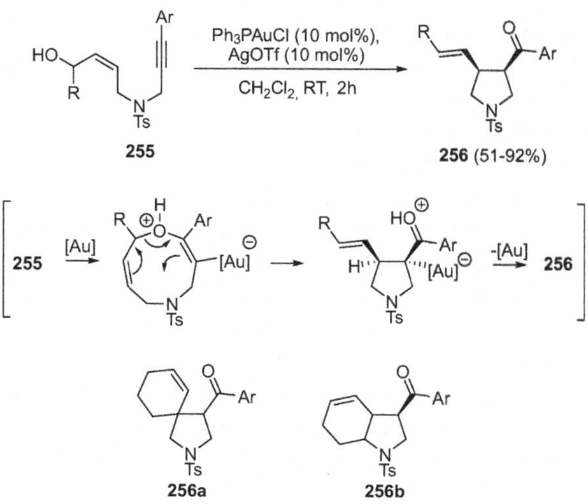

Scheme 66 Formation of pyrrolidines from 1,6-enynes via a Claisen rearrangement

This methodology has been also applied to the synthesis of spiro- and cyclohexenyl-fused pyrrolidines **256a** and **256b**, respectively [213, 214].

6 Special Cases of Gold-Catalysed Rearrangement of Enynes

Some examples of gold-catalysed reaction of enynes to form heterocycles that have appeared in the literature deserve a special comment due to their particular mechanisms. Most of the cases are related to the main mechanisms already described in this chapter but with some peculiarities.

Davies et al. described the synthesis of dihydrothiophenes **259** through the gold (I)-catalysed reaction of propargylic ester **257** and sulphur-tethered enyne **258** (Scheme 67) [215]. The mechanism of this reaction would start with a 1,2-acyloxy migration of the propargylic ester to form the corresponding gold carbene. This carbene would react with the sulphur in the enyne furnishing an ylide, which after a 1,2-migration of the propargylic fragment and nucleophilic attack of the sulphur to the gold activated alkyne would generate the dihydrothiophene core. A final [3,3]-sigmatropic rearrangement would form product **259**. This reaction is a remarkable combination of the acyloxy migration of propargylic esters and a [3,3]-rearrangement described in Sects. 4 and 5 in this chapter, and an alternative method for the synthesis of the less studied substituted thiophene core.

Syntheses of dihydrothiophenones and dihydropyranones were performed by Davies group starting from 1,7- or 1,8-enynes **260** tethered with a sulphoxide group (Scheme 68) [216]. The reaction would start with the activation of the alkyne moiety by coordination with the gold(III) catalyst. The gold–alkyne complex will suffer a nucleophilic attack from the oxygen of the sulphoxide group to form a cyclic intermediate, which ring opening would form a gold carbene. The attack of the sulphide to the carbene would form an allyl sulphonium ylide, which will evolve through a [2,3]-rearrangement of the allyl group to generate the final products.

Using a similar strategy, Yang and Tang et al. developed a synthesis of dihydrofuranone rings from oxygen-tethered 1,7-enynes (Scheme 69) [217]. The reaction needed of an external oxidant, in this case a pyridine *N*-oxide, and the yield proved to be better in the presence of $Yb(OTf)_3$ as additive. This process would proceed through the formation of an α-oxo-gold carbene from the activated 1,7-enyne in the presence of the gold catalyst and the oxidant. This carbene would evolve through the nucleophilic attack of the oxygen to form a cyclic oxonium ylide. From this intermediate, the authors propose two possible pathways: Path *a* would lead directly to the final product through a [2,3]-sigmatropic rearrangement, whereas path *b* is a stepwise process consisting of tautomerisation, 1,4-allyl migration and Claisen rearrangement. Different evidences point to one or the other mechanism depending on the substituents of the starting materials.

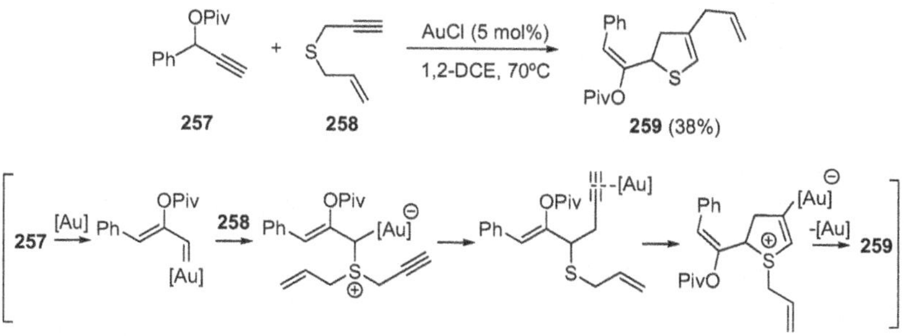

Scheme 67 Formation of dihydrothiophenes from 1,6-enynes and propargylic esters

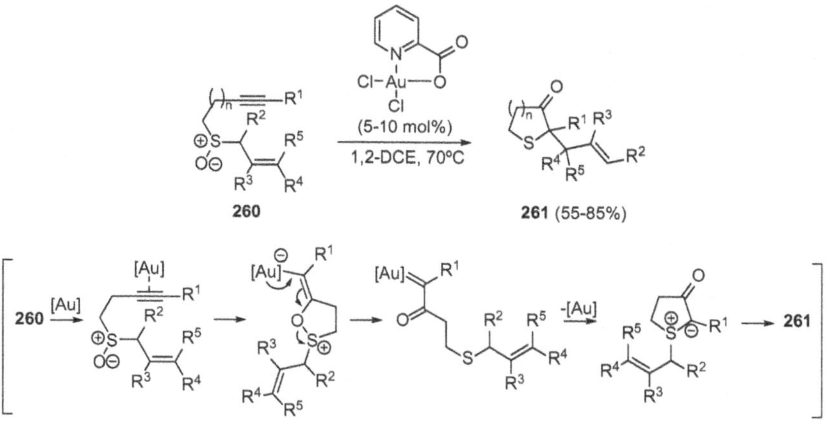

Scheme 68 Formation of dihydrothiophenones or dihydropyranones from enynes

An intriguing example based on the skeletal rearrangement of 1,6-enynes was described by the Malacria, Gandon, Fensterbank et al. These authors synthesised vinylidene-tetrahydrofurans **265** from oxygen-tethered 1,6-enynes **264** (Scheme 70) [218]. In a similar way to the skeletal rearrangements the mechanism would imply the formation of cyclopropyl gold carbenes. The corresponding vinyl gold cation, in resonance with the cyclopropyl gold carbene, can evolve through a 1,5-shift to give the final products. A 1,5-hydride shift is favoured when the R substituent is a donor group, as in the case of **265a**. Besides, if that donor group is also a good leaving group, it could also migrate to the cationic carbon, as in the case of **265b**. Although the authors explored different tethers and substituents in the reaction, the migration process could not be fully controlled in all the cases.

Based also in the skeletal rearrangement of enynes, a sequence of cycloisomerisation/cycloaddition was developed by Liu et al. These authors explored the reaction of 1,6-enynes using nitrones as a coupling partner (a, Scheme 71) [219] and of 1,5-enynes with nitrosobenzene derivatives (b, Scheme 71) [220]. In the reaction of 1,6-enynes **266**, 1,2-oxazepanes **267** were obtained diastereoselectively. The enantioselective version was also explored. The reaction would occur through

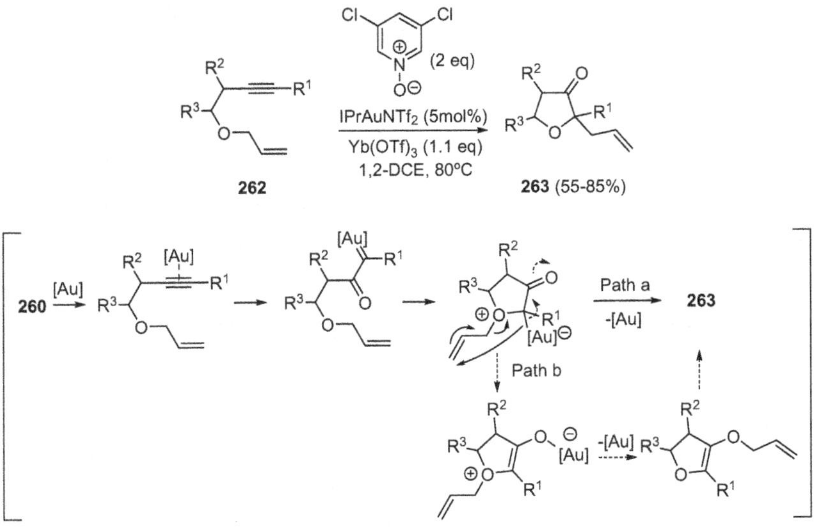

Scheme 69 Formation of dihydrofuranones from enynes

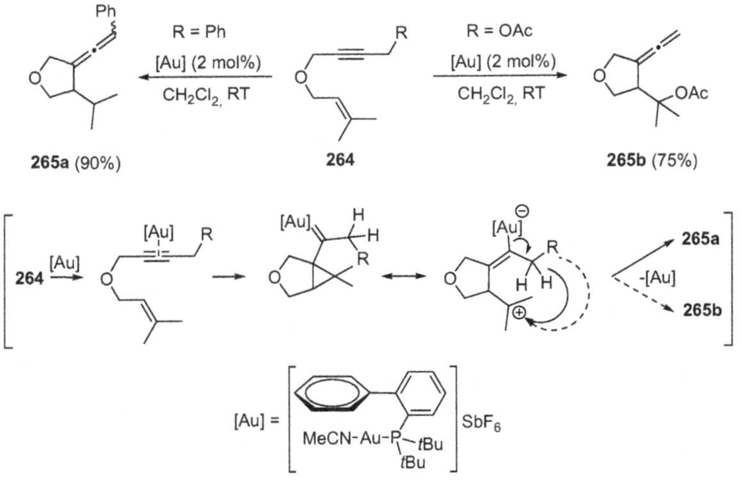

Scheme 70 Formation of dihydrofurans from 1,6-enynes

the formation of the cyclopropyl gold carbene, in resonance with the vinyl gold cation, and a subsequent [2+2+3] cycloaddition. Likewise, the reaction of 1,5-enynes afforded isoxazolidines **269** as single diastereoisomers, together with diazene oxide **271** and in some cases cyclopentene **270** as a secondary product. Heterocycles **269** would be also explained as a result of the formation of the alkenyl–gold cation. A [3+2] cycloaddition would form a gold carbene, which would then be oxidised by the nitrosobenzene to the corresponding ketone, forming diazene oxide **271** in the process.

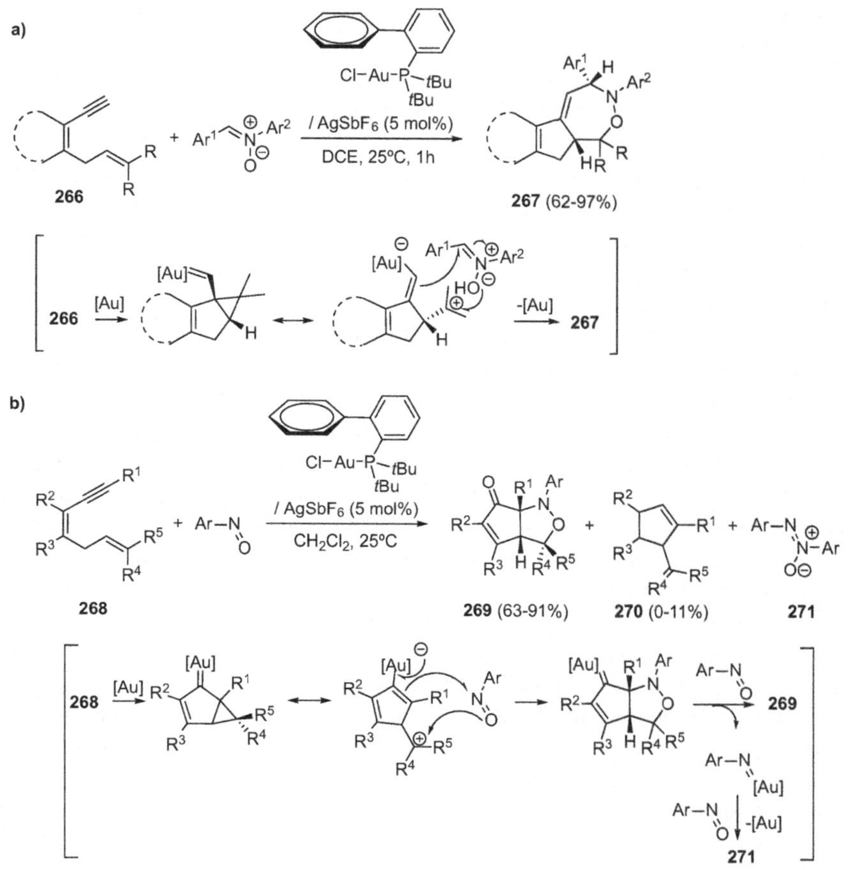

Scheme 71 Formation of oxazepanes and isoxazolidines from enynes via cycloaddition reactions

7 Conclusions

The research focused on the synthesis of heterocyclic structures through gold-catalysed rearrangement of enynes has advanced considerably since the first examples appeared in 2004. Many different branches have been explored so far and, as it has been demonstrated, are still an active area of research. The attractiveness of this methodology lies in the possibility of constructing heterocycles with high complexity from relatively simple structures. However, this is not the only appeal.

Formation of different heterocyclic structures is possible from the same starting materials modulating different factors. It has been described how in some cases the change in the electronic properties of the ligands of the catalyst could lead to different products. Similarly, variations in the substituents of the enyne or in the reaction conditions can also favour one or other reaction pathways.

Gold catalysis has evolve to be a breakthrough in the area, and many catalysts, from the simple ones such as AuCl to complexes bearing really complicated ligands, have been discovered, including those for the study of the asymmetric versions of the reaction, which allow the attainment of enantiopure heterocyclic structures.

Furthermore, it has been shown that this approach is not only applicable for simple enynes. Substitution of the alkene by analogous structures and introduction of additional functionalities in the starting materials such as internal nucleophiles, alkenes or carbonyl groups open new branches in the reaction mechanism for the synthesis of novel heterocycles.

Gold-catalysed enyne rearrangement has been most commonly employed for the preparation of oxygen and nitrogen heterocycles, but examples of sulphur and silica heterocycles have also been described and reviewed in this chapter.

A comment should be made about the mechanisms of the processes. Much effort has been put in understanding the basic mechanism of the skeletal rearrangement of enynes and all its variations and all these studies have contributed to the development of new approaches for the synthesis of functionalised heterocycles. With this deep understanding and the constant development of new more potent and selective gold catalysts, the future of this methodology for the synthesis of carbo- and heterocycles is guaranteed to move forward in exciting ways.

This methodology has already proved to be important in the synthesis of natural products, and new total syntheses of heterocycle-containing natural products are likely to be developed in the future from enynes or derivatives as starting points. Besides, novel natural-product-like heterocyclic structures will be designed using this methodology, since variations in the reaction conditions or starting materials are bound to open new reaction pathways in which just the imagination of the scientist will be the limit.

References

1. Corma A, Leyva-Pérez A, Sabater MJ (2011) Chem Rev 111(3):1657–1712
2. Gomtsyan A (2012) Chem Heterocycl Compd 48(1):7–10
3. Hashmi ASK, Rudolph M (2008) Chem Soc Rev 37(9):1766–1775
4. Hashmi ASK (2007) Chem Rev 107(7):3180–3211
5. Gorin DJ, Toste FD (2007) Nature 446(7134):395–403
6. Brooner REM, Widenhoefer RA (2013) Angew Chem Int Ed 52(45):11714–11724
7. Schmidbaur H, Schier A (2010) Organometallics 29(1):2–23
8. Frenking G, Fröhlich N (2000) Chem Rev 100(2):717–774
9. Ciancaleoni G, Biasiolo L, Bistoni G, Macchioni A, Tarantelli F, Zuccaccia D, Belpassi L (2013) Organometallics 32(15):4444–4447
10. Savjani N, Roşca D-A, Schormann M, Bochmann M (2013) Angew Chem Int Ed 52(3): 874–877
11. Cinellu MA (2012) Modern gold catalyzed synthesis. Wiley-VCH, Weinheim, pp 153–173
12. Cinellu MA (2012) Modern gold catalyzed synthesis. Wiley-VCH, Weinheim, pp 175–199
13. Zhu Y, Day CS, Jones AC (2012) Organometallics 31(21):7332–7335

14. Brown TJ, Widenhoefer RA (2011) J Organomet Chem 696(6):1216–1220
15. Brown TJ, Dickens MG, Widenhoefer RA (2009) J Am Chem Soc 131(18):6350–6351
16. Brown TJ, Dickens MG, Widenhoefer RA (2009) Chem Commun (42):6451–6453
17. Flügge S, Anoop A, Goddard R, Thiel W, Fürstner A (2009) Chem Eur J 15(34):8558–8565
18. Shibata T, Toshida N, Takagi K (2002) J Org Chem 67(21):7446–7450
19. Méndez M, Muñoz MP, Nevado C, Cárdenas DJ, Echavarren AM (2001) J Am Chem Soc 123(43):10511–10520
20. Jimenez-Nunez E, Echavarren AM (2007) Chem Commun (4):333–346
21. García-Mota M, Cabello N, Maseras F, Echavarren AM, Pérez-Ramírez J, Lopez N (2008) ChemPhysChem 9(11):1624–1629
22. Chatt J, Duncanson LA (1953) J Chem Soc (Resumed) (0):2939–2947
23. Dewar MJS (1951) Bull Soc Chim Fr 18:C71–C79
24. Zuccaccia D, Belpassi L, Rocchigiani L, Tarantelli F, Macchioni A (2010) Inorg Chem 49(7):3080–3082
25. Wu J, Kroll P, Dias HVR (2009) Inorg Chem 48(2):423–425
26. Shapiro ND, Toste FD (2008) Proc Natl Acad Sci U S A 105(8):2779–2782
27. Sanguramath RA, Hooper TN, Butts CP, Green M, McGrady JE, Russell CA (2011) Angew Chem Int Ed 50(33):7592–7595
28. Soriano E, Marco-Contelles J (2011) In: Soriano E, Marco-Contelles J (eds) Topics in current chemistry. Springer, Berlin
29. Eisenstein O, Hoffmann R (1981) J Am Chem Soc 103(15):4308–4320
30. Vergara E, Cerrada E, Casini A, Zava O, Laguna M, Dyson PJ (2010) Organometallics 29(11):2596–2603
31. Vicente J, Chicote M-T, Abrisqueta M-D, Jones PG (2000) Organometallics 19(13): 2629–2632
32. Yam VW-W, Choi SW-K (1996) J Chem Soc Dalton Trans (22):4227–4232
33. Hashmi ASK (2014) Acc Chem Res 47(3):864–876
34. Gómez-Suárez A, Nolan SP (2012) Angew Chem Int Ed 51(33):8156–8159
35. Hashmi ASK, Braun I, Nösel P, Schädlich J, Wieteck M, Rudolph M, Rominger F (2012) Angew Chem Int Ed 51(18):4456–4460
36. Brown TJ, Widenhoefer RA (2011) Organometallics 30(21):6003–6009
37. Grirrane A, Garcia H, Corma A, Álvarez E (2011) ACS Catal 1(12):1647–1653
38. Hooper TN, Green M, Russell CA (2010) Chem Commun 46(13):2313–2315
39. Trost BM (1990) Acc Chem Res 23(2):34–42
40. Fensterbank L, Malacria M (2014) Acc Chem Res 47(3):953–965
41. Michelet V (2014) In: Knochel P (ed) Comprehensive organic synthesis II, 2nd edn. Elsevier, Amsterdam, pp 1483–1536
42. Obradors C, Echavarren AM (2014) Acc Chem Res 47(3):902–912
43. Wang Y, Zhang L (2013) Catalytic cascade reactions. Wiley, Hoboken, pp 145–177
44. Zhang L, Sun J, Kozmin SA (2006) Adv Synth Catal 348(16–17):2271–2296
45. Luzung MR, Markham JP, Toste FD (2004) J Am Chem Soc 126(35):10858–10859
46. Mamane V, Gress T, Krause H, Fürstner A (2004) J Am Chem Soc 126(28):8654–8655
47. Nieto-Oberhuber C, Muñoz MP, Buñuel E, Nevado C, Cárdenas DJ, Echavarren AM (2004) Angew Chem Int Ed 43(18):2402–2406
48. Obradors C, Echavarren AM (2014) Chem Commun 50(1):16–28
49. Reeds JP, Healy MP, Fairlamb IJS (2014) Catal Sci Technol 4(10):3524–3533
50. Soriano E, Marco-Contelles J (2009) Acc Chem Res 42(8):1026–1036
51. Fürstner A, Morency L (2008) Angew Chem Int Ed 47(27):5030–5033
52. Nieto-Oberhuber C, López S, Muñoz MP, Cárdenas DJ, Buñuel E, Nevado C, Echavarren AM (2005) Angew Chem Int Ed 44(38):6146–6148
53. Echavarren AM, Nevado C (2004) Chem Soc Rev 33(7):431–436
54. Dorel R, Echavarren AM (2015) J Org Chem 80(15):7321–7332

55. Nunes dos Santos Comprido L, Klein JEMN, Knizia G, Kästner J, Hashmi ASK (2015) Angew Chem Int Ed 54(35):10336–10340
56. Seidel G, Mynott R, Fürstner A (2009) Angew Chem Int Ed 48(14):2510–2513
57. Fürstner A, Davies PW (2007) Angew Chem Int Ed 46(19):3410–3449
58. Hussong MW, Rominger F, Krämer P, Straub BF (2014) Angew Chem Int Ed 53(35): 9372–9375
59. Joost M, Estévez L, Mallet-Ladeira S, Miqueu K, Amgoune A, Bourissou D (2014) Angew Chem Int Ed 53(52):14512–14516
60. Fedorov A, Moret M-E, Chen P (2008) J Am Chem Soc 130(28):8880–8881
61. Lee SI, Chatani N (2009) Chem Commun (4):371–384
62. Jiménez-Núñez E, Echavarren AM (2008) Chem Rev 108(8):3326–3350
63. Lloyd-Jones GC (2003) Org Biomol Chem 1(2):215–236
64. Nieto-Oberhuber C, López S, Jiménez-Núñez E, Echavarren AM (2006) Chem Eur J 12(23): 5916–5923
65. Escribano-Cuesta A, Perez-Galan P, Herrero-Gomez E, Sekine M, Braga AAC, Maseras F, Echavarren AM (2012) Org Biomol Chem 10(30):6105–6111
66. Nakai H, Chatani N (2007) Chem Lett 36(12):1494–1495
67. Amijs CHM, Ferrer C, Echavarren AM (2007) Chem Commun (7):698–700
68. López S, Herrero-Gómez E, Pérez-Galán P, Nieto-Oberhuber C, Echavarren AM (2006) Angew Chem Int Ed 45(36):6029–6032
69. Nieto-Oberhuber C, López S, Muñoz MP, Jiménez-Núñez E, Buñuel E, Cárdenas DJ, Echavarren AM (2006) Chem Eur J 12(6):1694–1702
70. Witham CA, Mauleón P, Shapiro ND, Sherry BD, Toste FD (2007) J Am Chem Soc 129(18): 5838–5839
71. Cabello N, Jiménez-Núñez E, Buñuel E, Cárdenas DJ, Echavarren AM (2007) Eur J Org Chem 2007(25):4217–4223
72. Ferrer C, Raducan M, Nevado C, Claverie CK, Echavarren AM (2007) Tetrahedron 63(27): 6306–6316
73. Cheong PH-Y, Morganelli P, Luzung MR, Houk KN, Toste FD (2008) J Am Chem Soc 130(13):4517–4526
74. Simonneau A, Jaroschik F, Lesage D, Karanik M, Guillot R, Malacria M, Tabet J-C, Goddard J-P, Fensterbank L, Gandon V, Gimbert Y (2011) Chem Sci 2(12):2417–2422
75. Bastin S, Barthes C, Lugan N, Lavigne G, César V (2015) Eur J Inorg Chem 2015(13): 2216–2221
76. Nieto-Oberhuber C, Muñoz MP, López S, Jiménez-Núñez E, Nevado C, Herrero-Gómez E, Raducan M, Echavarren AM (2006) Chem Eur J 12(6):1677–1693
77. Trost BM, Gutierrez AC, Ferreira EM (2010) J Am Chem Soc 132(26):9206–9218
78. Lee SI, Kim SM, Choi MR, Kim SY, Chung YK, Han W-S, Kang SO (2006) J Org Chem 71(25):9366–9372
79. Lee YT, Kang YK, Chung YK (2009) J Org Chem 74(20):7922–7934
80. Jullien H, Brissy D, Sylvain R, Retailleau P, Naubron J-V, Gladiali S, Marinetti A (2011) Adv Synth Catal 353(7):1109–1124
81. Nevado C, Ferrer C, Echavarren AM (2004) Org Lett 6(18):3191–3194
82. Fürstner A, Stelzer F, Szillat H (2001) J Am Chem Soc 123(48):11863–11869
83. Fürstner A, Szillat H, Stelzer F (2000) J Am Chem Soc 122(28):6785–6786
84. Schröder F, Tugny C, Salanouve E, Clavier H, Giordano L, Moraleda D, Gimbert Y, Mouriès-Mansuy V, Goddard J-P, Fensterbank L (2014) Organometallics 33(15):4051–4056
85. Zhang D-H, Wei Y, Shi M (2012) Chem Eur J 18(23):7026–7029
86. Nieto-Oberhuber C, Pérez-Galán P, Herrero-Gómez E, Lauterbach T, Rodríguez C, López S, Bour C, Rosellón A, Cárdenas DJ, Echavarren AM (2008) J Am Chem Soc 130(1):269–279
87. Sethofer SG, Staben ST, Hung OY, Toste FD (2008) Org Lett 10(19):4315–4318
88. Chen Z, Zhang Y-X, Wang Y-H, Zhu L-L, Liu H, Li X-X, Guo L (2010) Org Lett 12(15): 3468–3471

89. Chao C-M, Beltrami D, Toullec PY, Michelet V (2009) Chem Commun (45):6988–6990
90. Pradal A, Chao C-M, Toullec PY, Michelet V (2011) Beilstein J Org Chem 7:1021–1029
91. Teller H, Corbet M, Mantilli L, Gopakumar G, Goddard R, Thiel W, Fürstner A (2012) J Am Chem Soc 134(37):15331–15342
92. Yavari K, Aillard P, Zhang Y, Nuter F, Retailleau P, Voituriez A, Marinetti A (2014) Angew Chem Int Ed 53(3):861–865
93. Aillard P, Voituriez A, Dova D, Cauteruccio S, Licandro E, Marinetti A (2014) Chem Eur J 20(39):12373–12376
94. Barreiro EM, Boltukhina EV, White AJP, Hii KK (2015) Chem Eur J 21(6):2686–2690
95. Deschamps NM, Elitzin VI, Liu B, Mitchell MB, Sharp MJ, Tabet EA (2011) J Org Chem 76(2):712–715
96. Teller H, Fürstner A (2011) Chem Eur J 17(28):7764–7767
97. Wang W, Yang J, Wang F, Shi M (2011) Organometallics 30(14):3859–3869
98. Qian D, Hu H, Liu F, Tang B, Ye W, Wang Y, Zhang J (2014) Angew Chem Int Ed 53(50): 13751–13755
99. Perez-Galan P, Herrero-Gomez E, Hog DT, Martin NJA, Maseras F, Echavarren AM (2011) Chem Sci 2(1):141–149
100. Kim SM, Park JH, Choi SY, Chung YK (2007) Angew Chem Int Ed 46(32):6172–6175
101. Hudlicky T, Reed JW (2010) Angew Chem Int Ed 49(29):4864–4876
102. Kim SY, Park Y, Son S, Chung YK (2012) Adv Synth Catal 354(1):179–186
103. Jiménez-Núñez E, Raducan M, Lauterbach T, Molawi K, Solorio CR, Echavarren AM (2009) Angew Chem Int Ed 48(33):6152–6155
104. McGonigal PR, de León C, Wang Y, Homs A, Solorio-Alvarado CR, Echavarren AM (2012) Angew Chem Int Ed 51(52):13093–13096
105. Couty S, Meyer C, Cossy J (2009) Tetrahedron 65(9):1809–1832
106. Couty S, Meyer C, Cossy J (2006) Angew Chem Int Ed 45(40):6726–6730
107. Jiménez-Núñez E, Claverie CK, Nieto-Oberhuber C, Echavarren AM (2006) Angew Chem Int Ed 45(33):5452–5455
108. Simonneau A, Harrak Y, Jeanne-Julien L, Lemière G, Mouriès-Mansuy V, Goddard J-P, Malacria M, Fensterbank L (2013) ChemCatChem 5(5):1096–1099
109. Kim SM, Park JH, Chung YK (2011) Chem Commun 47(23):6719–6721
110. Fürstner A, Stimson CC (2007) Angew Chem Int Ed 46(46):8845–8849
111. Kim SY, Park Y, Chung YK (2010) Angew Chem Int Ed 49(2):415–418
112. Cai P-J, Wang Y, Liu C-H, Yu Z-X (2014) Org Lett 16(22):5898–5901
113. Cabello N, Rodríguez C, Echavarren AM (2007) Synlett 2007(11):1753–1758
114. Speck K, Karaghiosoff K, Magauer T (2015) Org Lett 17(8):1982–1985
115. López-Carrillo V, Huguet N, Mosquera Á, Echavarren AM (2011) Chem Eur J 17(39): 10972–10978
116. Gagosz F (2005) Org Lett 7(19):4129–4132
117. Mézailles N, Ricard L, Gagosz F (2005) Org Lett 7(19):4133–4136
118. Zhang L, Kozmin SA (2004) J Am Chem Soc 126(38):11806–11807
119. Imase H, Noguchi K, Hirano M, Tanaka K (2008) Org Lett 10(16):3563–3566
120. Shibuya T, Nakamura K, Tanaka K (2011) Beilstein J Org Chem 7:944–950
121. Xin X, Wang D, Wu F, Li X, Wan B (2013) J Org Chem 78(8):4065–4074
122. Xin X, Wang D, Li X, Wan B (2013) Tetrahedron 69(48):10245–10248
123. Cacchi S, Fabrizi G, Filisti E (2008) Org Lett 10(13):2629–2632
124. Homs A, Obradors C, Lebœuf D, Echavarren AM (2014) Adv Synth Catal 356(1):221–228
125. Obradors C, Leboeuf D, Aydin J, Echavarren AM (2013) Org Lett 15(7):1576–1579
126. Echavarren AM, Mendez M, Munoz MP, Nevado C, Martin-Matute B, Nieto-Oberhuber C, Cardenas DJ (2004) Pure Appl Chem 76(3):453–463
127. Martin-Matute B, Nevado C, Cardenas DJ, Echavarren AM (2003) J Am Chem Soc 125(19): 5757–5766

128. Martín-Matute B, Cárdenas DJ, Echavarren AM (2001) Angew Chem Int Ed 40(24): 4754–4757
129. Hashmi ASK, Hamzić M, Rudolph M, Ackermann M, Rominger F (2009) Adv Synth Catal 351(14–15):2469–2481
130. Hashmi ASK, Enns E, Frost TM, Schäfer S, Frey W, Rominger F (2008) Synthesis 2008(17): 2707–2718
131. Hashmi ASK, Frost TM, Bats JW (2001) Org Lett 3(23):3769–3771
132. Hashmi ASK, Frost TM, Bats JW (2000) J Am Chem Soc 122(46):11553–11554
133. Hashmi ASK, Rudolph M, Siehl H-U, Tanaka M, Bats JW, Frey W (2008) Chem Eur J 14(12):3703–3708
134. Hashmi ASK, Kurpejović E, Wölfle M, Frey W, Bats JW (2007) Adv Synth Catal 349(10): 1743–1750
135. Hashmi ASK, Rudolph M, Weyrauch JP, Wölfle M, Frey W, Bats JW (2005) Angew Chem Int Ed 44(18):2798–2801
136. Pažický M, Loos A, Ferreira MJ, Serra D, Vinokurov N, Rominger F, Jäkel C, Hashmi ASK, Limbach M (2010) Organometallics 29(20):4448–4458
137. Carrettin S, Blanco MC, Corma A, Hashmi ASK (2006) Adv Synth Catal 348(10–11): 1283–1288
138. Hashmi ASK, Blanco MC, Kurpejović E, Frey W, Bats JW (2006) Adv Synth Catal 348(6): 709–713
139. Hashmi ASK, Weyrauch JP, Kurpejović E, Frost TM, Miehlich B, Frey W, Bats JW (2006) Chem Eur J 12(22):5806–5814
140. Hashmi ASK, Salathé R, Frey W (2006) Chem Eur J 12(26):6991–6996
141. Hashmi ASK, Wölfle M, Ata F, Hamzic M, Salathé R, Frey W (2006) Adv Synth Catal 348(16–17):2501–2508
142. Hashmi ASK, Haufe P, Schmid C, Rivas Nass A, Frey W (2006) Chem Eur J 12(20): 5376–5382
143. Hashmi ASK, Weyrauch JP, Rudolph M, Kurpejović E (2004) Angew Chem Int Ed 43(47): 6545–6547
144. Hashmi ASK, Rudolph M, Bats JW, Frey W, Rominger F, Oeser T (2008) Chem Eur J 14(22):6672–6678
145. Hashmi ASK, Häffner T, Rudolph M, Rominger F (2011) Chem Eur J 17(29):8195–8201
146. Robertson BD, Brooner REM, Widenhoefer RA (2015) Chem Eur J 21(15):5714–5717
147. Notar Francesco I, Giauffret J, Fontaine-Vive F, Edwards JK, Kiely CJ, Hutchings GJ, Antoniotti S (2014) Tetrahedron 70(51):9635–9643
148. Yang J, Zhang R, Wang W, Zhang Z, Shi M (2011) Tetrahedron Asymmetry 22(23): 2029–2038
149. Chao C-M, Genin E, Toullec PY, Genêt J-P, Michelet V (2009) J Organomet Chem 694(4): 538–545
150. Amijs CHM, López-Carrillo V, Raducan M, Pérez-Galán P, Ferrer C, Echavarren AM (2008) J Org Chem 73(19):7721–7730
151. Muzart J (2008) Tetrahedron 64(25):5815–5849
152. Sethofer SG, Mayer T, Toste FD (2010) J Am Chem Soc 132(24):8276–8277
153. Shiri M (2012) Chem Rev 112(6):3508–3549
154. Pradal A, Chen Q, Faudot dit Bel P, Toullec PY, Michelet V (2012) Synlett 2012(01):74–79
155. Toullec PY, Blarre T, Michelet V (2009) Org Lett 11(13):2888–2891
156. Lim C, Kang J-E, Lee J-E, Shin S (2007) Org Lett 9(18):3539–3542
157. Zhang L, Kozmin SA (2005) J Am Chem Soc 127(19):6962–6963
158. Buzas AK, Istrate FM, Gagosz F (2007) Angew Chem Int Ed 46(7):1141–1144
159. Buzas A, Istrate F, Le Goff XF, Odabachian Y, Gagosz F (2009) J Organomet Chem 694(4): 515–519
160. Escribano-Cuesta A, López-Carrillo V, Janssen D, Echavarren AM (2009) Chem Eur J 15(23):5646–5650

161. Schelwies M, Moser R, Dempwolff AL, Rominger F, Helmchen G (2009) Chem Eur J 15(41):10888–10900
162. Schelwies M, Dempwolff AL, Rominger F, Helmchen G (2007) Angew Chem Int Ed 46(29): 5598–5601
163. Zhou Q, Chen X, Ma D (2010) Angew Chem Int Ed 49(20):3513–3516
164. Molawi K, Delpont N, Echavarren AM (2010) Angew Chem Int Ed 49(20):3517–3519
165. Jimenez-Nunez E, Molawi K, Echavarren AM (2009) Chem Commun (47):7327–7329
166. Huguet N, Echavarren AM (2012) Synlett 2012(01):49–53
167. Kazem Shiroodi R, Gevorgyan V (2013) Chem Soc Rev 42(12):4991–5001
168. Wang S, Zhang G, Zhang L (2010) Synlett 2010(05):692–706
169. Correa A, Marion N, Fensterbank L, Malacria M, Nolan SP, Cavallo L (2008) Angew Chem Int Ed 47(4):718–721
170. Marco-Contelles J, Soriano E (2007) Chem Eur J 13(5):1350–1357
171. Petrović M, Scarpi D, Fiser B, Gómez-Bengoa E, Occhiato EG (2015) Eur J Org Chem 2015(18):3943–3956
172. Uemura M, Watson IDG, Katsukawa M, Toste FD (2009) J Am Chem Soc 131(10): 3464–3465
173. Zhao C, Xie X, Duan S, Li H, Fang R, She X (2014) Angew Chem Int Ed 53(40): 10789–10793
174. Rao W, Sally, Berry SN, Chan PWH (2014) Chem Eur J 20 (41):13174–13180
175. Cai S, Liu Z, Zhang W, Zhao X, Wang DZ (2011) Angew Chem Int Ed 50(47):11133–11137
176. Zheng H, Huo X, Zhao C, Jing P, Yang J, Fang B, She X (2011) Org Lett 13(24):6448–6451
177. Liu Y, Yang X, Liu L, Wang H, Bi S (2015) Dalton Trans 44(12):5354–5363
178. Rao W, Susanti D, Chan PWH (2011) J Am Chem Soc 133(39):15248–15251
179. Rao W, Sally, Koh MJ, Chan PWH (2013) J Org Chem 78 (7):3183–3195
180. Hashmi ASK, Yang W, Yu Y, Hansmann MM, Rudolph M, Rominger F (2013) Angew Chem Int Ed 52(4):1329–1332
181. Yang W, Yu Y, Zhang T, Hansmann MM, Pflästerer D, Hashmi ASK (2013) Adv Synth Catal 355(10):2037–2043
182. Ferrer C, Amijs CHM, Echavarren AM (2007) Chem Eur J 13(5):1358–1373
183. Ferrer C, Echavarren AM (2006) Angew Chem Int Ed 45(7):1105–1109
184. Zhang L (2005) J Am Chem Soc 127(48):16804–16805
185. Zhang G, Catalano VJ, Zhang L (2007) J Am Chem Soc 129(37):11358–11359
186. Martín-Castro AM, Tortosa M (2014) In: Knochel P (ed) Comprehensive organic synthesis II, 2nd edn. Elsevier, Amsterdam, pp 912–977
187. Tejedor D, Mendez-Abt G, Cotos L, Garcia-Tellado F (2013) Chem Soc Rev 42(2):458–471
188. Henrion G, Gagosz F (2009) PATAI'S chemistry of functional groups. Wiley, Chichester
189. Gille A, Rehbein J, Hiersemann M (2011) Org Lett 13(8):2122–2125
190. Suhre MH, Reif M, Kirsch SF (2005) Org Lett 7(18):3925–3927
191. Saito A, Konishi T, Hanzawa Y (2010) Org Lett 12(2):372–374
192. Xiong Y, Schaus SE, Porco JA (2013) Org Lett 15(8):1962–1965
193. Matousova E, Ruzicka A, Kunes J, Kralova J, Pour M (2011) Chem Commun 47(33): 9390–9392
194. Sherry BD, Maus L, Laforteza BN, Toste FD (2006) J Am Chem Soc 128(25):8132–8133
195. Sherry BD, Toste FD (2004) J Am Chem Soc 126(49):15978–15979
196. Horino Y, Luzung MR, Toste FD (2006) J Am Chem Soc 128(35):11364–11365
197. Park S, Lee D (2006) J Am Chem Soc 128(33):10664–10665
198. Ariafard A, Asadollah E, Ostadebrahim M, Rajabi NA, Yates BF (2012) J Am Chem Soc 134(40):16882–16890
199. Harschneck T, Kirsch SF (2011) J Org Chem 76(7):2145–2156
200. Wei H, Wang Y, Yue B, Xu P-F (2010) Adv Synth Catal 352(14–15):2450–2454
201. Istrate FM, Gagosz F (2011) Beilstein J Org Chem 7:878–885
202. Istrate FM, Gagosz F (2007) Org Lett 9(16):3181–3184

203. Hong YJ, Tantillo DJ (2011) Organometallics 30(21):5825–5831
204. Majumdar KC, Hazra S, Roy B (2011) Tetrahedron Lett 52(50):6697–6701
205. Heugebaert TSA, Stevens CV (2009) Org Lett 11(21):5018–5021
206. Ueda M, Sato A, Ikeda Y, Miyoshi T, Naito T, Miyata O (2010) Org Lett 12(11):2594–2597
207. Ackermann M, Bucher J, Rappold M, Graf K, Rominger F, Hashmi ASK (2013) Chem Asian J 8(8):1786–1794
208. Shi Y, Roth KE, Ramgren SD, Blum SA (2009) J Am Chem Soc 131(50):18022–18023
209. Hashmi ASK, Lothschütz C, Döpp R, Ackermann M, De Buck Becker J, Rudolph M, Scholz C, Rominger F (2012) Adv Synth Catal 354(1):133–147
210. Prasad KR, Nagaraju C (2013) Org Lett 15(11):2778–2781
211. Kolundžić F, Murali A, Pérez-Galán P, Bauer JO, Strohmann C, Kumar K, Waldmann H (2014) Angew Chem Int Ed 53(31):8122–8126
212. Yeh M-CP, Lin M-N, Chang W-J, Liou J-L, Shih Y-F (2010) J Org Chem 75(17):6031–6034
213. Liang C-J, Jiang X-Y, Yeh M-CP (2014) Synthesis 46(16):2220–2224
214. Yeh M-CP, Pai H-F, Hsiow C-Y, Wang Y-R (2010) Organometallics 29(1):160–166
215. Davies PW, Albrecht SJC (2008) Chem Commun (2):238–240
216. Davies PW, Albrecht SJC (2009) Angew Chem Int Ed 48(44):8372–8375
217. Fu J, Shang H, Wang Z, Chang L, Shao W, Yang Z, Tang Y (2013) Angew Chem Int Ed 52(15):4198–4202
218. Harrak Y, Simonneau A, Malacria M, Gandon V, Fensterbank L (2010) Chem Commun 46(6):865–867
219. Gawade SA, Bhunia S, Liu R-S (2012) Angew Chem Int Ed 51(31):7835–7838
220. Chen C-H, Tsai Y-C, Liu R-S (2013) Angew Chem Int Ed 52(17):4599–4603

Top Heterocycl Chem (2016) 46: 175–226
DOI: 10.1007/7081_2015_5005

Published online: 3 February 2016

C–H Functionalisation of Heteroaromatic Compounds via Gold Catalysis

Nanna Ahlsten, Xacobe C. Cambeiro, Gregory J.P. Perry, and Igor Larrosa

Abstract In this chapter, examples of the C–H functionalisation of heteroarenes using Au catalysis are presented. The majority of examples to date describe the hydroheteroarylation of multiple bonds or Friedel–Crafts-type substitution reactions with heteroaromatic nucleophiles. These reactions are redox neutral and take advantage of the Lewis acidity of Au complexes. The reactivity of [Au(I)] and [Au (III)] in C–H activation of heteroaromatics is also discussed, and examples where this mode of reactivity has been used in oxidative couplings as well as redox-neutral reactions for the functionalisation of heteroarene C–H bonds are presented.

Keywords Alkenes • Alkynes • Allenes • C–H activation • C–H functionalisation • Catalysis • Cross-coupling • Gold • Heteroarenes • Hydroarylation

Contents

N. Ahlsten, X.C. Cambeiro, G.J.P. Perry, and I. Larrosa (✉)
University of Manchester, Oxford Road, M13 9PL Manchester, UK
e-mail: igor.larrosa@manchester.ac.uk

1 Introduction

The ability to directly functionalise C–H bonds is of paramount interest in today's scientific community as it allows for the development of short and atom economical synthesis from readily available building blocks while minimising the generation of harmful waste [1]. One of the greatest challenges in C–H functionalisation is to selectively target a given C–H bond among many others in a chosen molecule. Heterocycles are a privileged class of compounds in C–H transformations as many possess both nucleophilic sites and acidic C–H bonds. This allows for selective C–H functionalisation through electrophilic aromatic substitution (S_EAr)-type reactivity or via deprotonative pathways. The rich coordination and organometallic chemistry of gold [2, 3] has given way to a huge rise in its use in catalytic C–H functionalisations [4–7]. While many of the transformations are not unique to heterocycles, but can be applied to electron-rich aromatics in general, here we discuss the use of gold in the functionalisation of heteroaromatic C–H bonds specifically.

Most well-known is the ability of gold to act as a Lewis acid in the activation of π-bonds [8–10], increasing their electrophilicity and therefore their tendency to react with electron-rich arenes in hydroarylation reactions. This reactivity will be the topic of the first section, describing the reaction of heteroarenes with alkenes, allenes and alkynes (Scheme 1, left). In a similar fashion, gold can also promote Friedel–Crafts-type reactivity through coordination of a given heteroatom. In general, coordination improves the leaving group capability of this heteroatom, thus activating it towards nucleophilic attack (Scheme 1, bottom).

The C–H auration of aromatics is one of the earliest known examples of reactivity between gold and organic molecules; however, it is only recently that this mode of activation has been exploited in synthetically useful catalytic transformations. The formation of an aryl-gold intermediate is particularly attractive as it provides access to cross-coupling-type reactivity (Scheme 1, right). This will be the focus of the last section in this chapter.

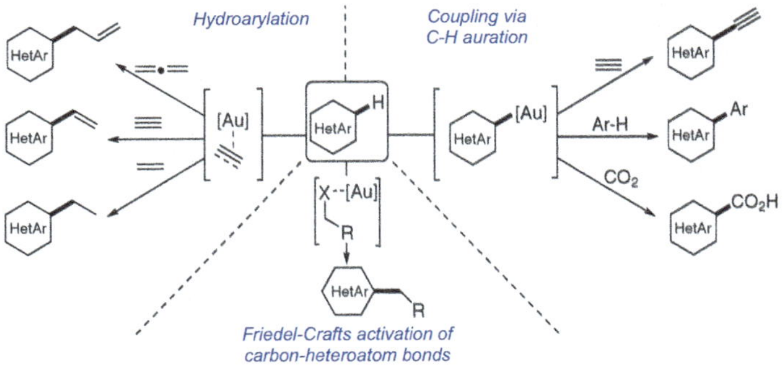

Scheme 1 Summary of Au-catalysed C–H functionalisations of heteroarenes

In some cases, a definite line between whether one type of reactivity (Lewis acid type activation or C–H auration) is favoured over the other is hard to interpret conclusively, for example, Au-catalysed hydroarylations may proceed via Friedel-Crafts-type reactivity promoted by the activation of a π-bond or via auration of the aryl unit. Nevertheless, both mechanisms are related to the pronounced Lewis acidity of Au, a property that has allowed for the development of the vast set of transformations for the functionalisation of heterocycles introduced in the coming sections [11–15].

2 Hydroarylation

As a carbophilic Lewis acid, one of the main characteristics of Au complexes is their ability to activate C–C π-bonds. The reactivity of Au catalysts with alkynes [10, 16–22], alkenes [23] and allenes [24, 25] allows the addition of nucleophiles across the π-system [26, 27]. This mode of activation has been applied using heteroatom and carbon nucleophiles, including heterocycles in which case it allows for the formal C–H functionalisation of the heterocycle. Both [Au(I)] and [Au(III)] sources have commonly been employed, either as salts or in combination with phosphines or *N*-heterocyclic carbenes and often in the presence of Ag salts as halide abstractors to enhance the electrophilicity of the catalyst.

In principle, two mechanistic scenarios are possible for the hydroarylation of heterocycles and other electron-rich arenes. The more commonly invoked mechanism involves electrophilic activation of a π-bond coordinated to Au through a σ-donation/π-backdonation-type interaction (**I** in Scheme 2a). The detailed physical aspects and origin of the π-Lewis acidity of Au is outside the scope of this chapter, but have been described excellently elsewhere [28]. The coordination leads to an umpolung of electron-rich unsaturated bonds, or an increased reactivity of electron-deficient bonds, thus promoting a Friedel–Crafts-type addition of an arene or

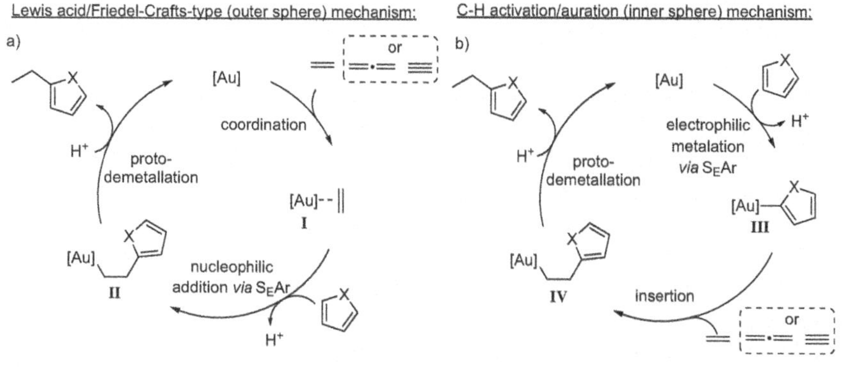

Scheme 2 Alternative mechanisms for the Au-catalysed hydroarylation: (**a**) via activation of the π-bond and (**b**) via activation of the heterocycle C–H bond

heterocyclic nucleophile. Protodemetallation of the formed σ-bonded Au intermediate (**II**) affords the hydroarylation product, but other more complex pathways such as rearrangements, reactions with other unsaturated moieties or elimination of adjacent leaving groups are often possible.

The alternative mechanism is based on the ability of [Au(III)] to promote C–H activation (auration) of electron-rich arenes via an S_EAr-type mechanism (Scheme 2b). The formed [Ar-Au(III)] species (**III**) subsequently inserts or conjugatively adds to a π-system. Protodeauration of the formed intermediate (**IV**) affords the corresponding hydroarylation product. Although some mechanistic data has been provided to support it (vide infra), the feasibility of such a mechanism has not been conclusively established ([29] and references therein).

Furthermore, the interpretation of Ag-additives as mere halide scavengers is not necessarily straightforward since some transformations are readily catalysed by other Lewis acids – including Ag – or by Brønsted acids which may be generated in situ from Au/Ag complexes [30, 31] ([32] and references therein). Thus, the importance of control reactions has been pointed out when elucidating the mechanistic role of Au [33]. Furthermore, the gold species formed by Ag-halide abstraction has been shown not always to be the expected cationic gold formed from precipitation of equimolar Ag-halide [34]. In this sense, Gagosz catalyst, PPh_3AuNTf_2, which does not require activation by Ag salts, has found use in many protocols [35]. Nevertheless, Au catalysts have in many cases proved more efficient, more selective and milder catalysts than alternative metals or acids. Importantly, some of the transformations shown in this chapter, including complex multi-bond forming reactions, are unique to gold, and their discovery has initiated a wide, new field of research in catalysis.

2.1 Hydroarylation of Enones and Alkenes

Enones are activated Michael acceptors that readily participate in intermolecular hydroarylation reactions. Usually, the conditions are very mild and the reactions high yielding. In most cases, [Au(III)] sources have been used, but there are also examples of [Au(I)] catalysts in this transformation. The gold catalyst has been suggested to act either as a Lewis acid activating the enone and promoting the attack of the free arene or through the activation of the arene via gold C–H auration followed by a conjugate addition to the enone. However, the often noted ability of Ag salts and Brønsted acids to catalyse this reaction complicates mechanism elucidation, as Ag salts have often been used to produce more electrophilic gold species and the Au/Ag salts can be sources of Brønsted acids. Nonconjugated alkenes are significantly less reactive than enones, and only a handful of hydroarylation examples have been described with heterocycles.

In 2000, Hashmi and co-workers reported the first observation of a Au-catalysed hydroarylation reaction [36]. While investigating the cyclisation of allenyl ketones **1** to furans **2**, they observed that in the presence of [Au(III)] salts, mono- and

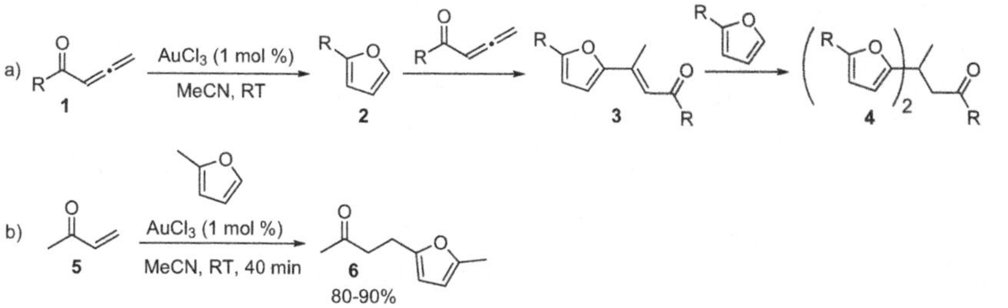

Scheme 3 (**a**) First observation of a hydroarylation reaction of allenes and enones; (**b**) hydroarylation of methyl vinyl ketone

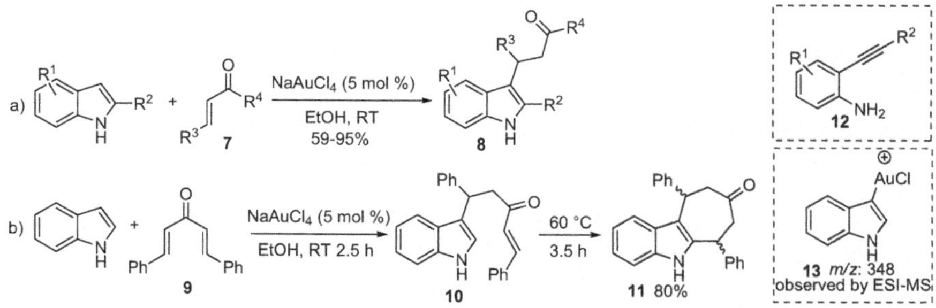

Scheme 4 (**a**) Hydroarylation of enones; (**b**) double hydroarylation of dibenzylideneacetone

bis-hydroarylation products **3** and **4** resulting from reaction of the formed furans with unreacted allene were readily formed (Scheme 3a). The observation that Au could promote hydroarylation of the intermediate enones was further investigated with the use of other allenyl ketones or enones (Scheme 3b) [33].

NMR experiments in which a stoichiometric amount of methyl vinyl ketone (**5**) was mixed with $AuCl_3$ showed no obvious coordination of the enone to $AuCl_3$. However, mixing 2-methylfuran with $AuCl_3$ resulted in changes in the spectra indicative of a Au-furan interaction, possibly C–H auration. The authors however did not exclude that Au was acting as a Lewis acid or a Brønsted acid source, in particular as it was found that the hydroarylation of methyl vinyl ketone could also be catalysed by *p*-TsOH (3 mol %) in equal efficiency [33].

Following Hashmi's report, several groups described similar systems [37, 38] with other heterocycles. Unprotected indoles react with various enones **7** in the presence of $NaAuCl_4$ in EtOH to give C3-alkylated products **8** (Scheme 4a) [39]. When the Michael acceptor was dibenzylideneacetone (**9**), a consecutive hydroarylation of the second double bond at the indole C2 position in **10** gave place to a 7-membered ring **11** via a formal 7-*endo*-trig cyclisation (Scheme 4b). Products **8** could also be obtained under similar conditions starting from alkynyl anilines **12** via alkyne hydroamination [40]. A mechanism via C–H auration was first favoured based on the observation of [indolyl-AuCl]$^+$ adduct **13** by mass

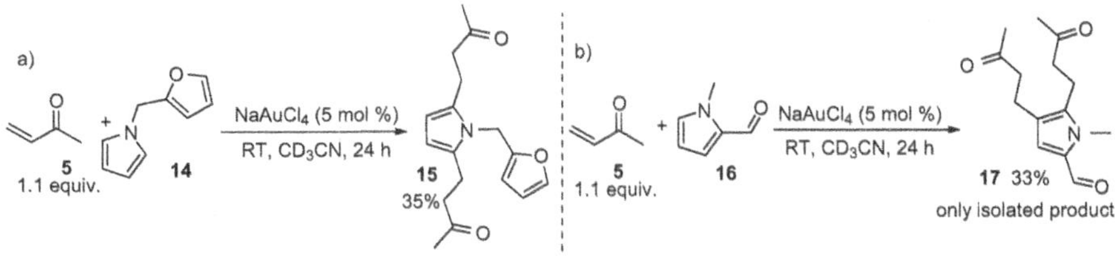

Scheme 5 Bis- vs. mono-hydroarylation of methyl vinyl ketone with pyrroles

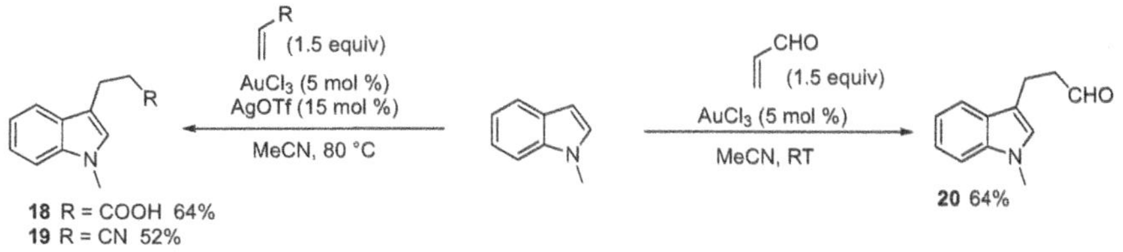

Scheme 6 Indole hydroarylation of acrylic acid, acrylonitrile and acrolein

spectrometry in a solution containing a stoichiometric quantity of $NaAuCl_4$; however, Lewis acid activation of the enone was later proposed [40].

Attempts to use pyrroles in these reactions demonstrate the higher nucleophilicity of these substrates compared to furans. The *N*-methylene-linked furan substrate **14** which allows for intramolecular competition between the two heterocyclic scaffolds afforded twofold alkylation of the pyrrole moiety to give **15** (Scheme 5a) [41]. Even for pyrrole **16** containing a deactivating aldehyde group in C2, the doubly C2, C3 hydroarylated product **17** was favoured exclusively over mono-alkylation as a result of the electron-donating effect of the alkyl groups (Scheme 5b).

Other Michael acceptors also undergo hydroarylation. For example, *N*-methylindole affords alkylated products **18** and **19** with acrylic acid or acrylonitrile in the presence of 5 mol% $AuCl_3$ and 15 mol % AgOTf at 80°C (Scheme 6) [42]. Also α,β-unsaturated aldehydes give **20** by selective conjugate addition over addition to the carbonyl.

The low nucleophilicity of 7-azaindoles renders them significantly less reactive substrates than regular indoles, and the hydroarylation of these substrates require both high temperatures and long reaction times. Arcadi and co-workers found the C3 vs. N-alkylation of these with enones to be substrate dependent; 7-azaindole **21** afforded mainly N-alkylation with methyl vinyl ketone in the presence of several [Au(III)] halides (Cl, Br), while 6-substituted-7-azaindole **22** gave C3-alkylation with $NaAuCl_4$ as the sole effective catalyst (Scheme 7a) [43]. Introducing substitution at the enone β-carbon also favoured hydroarylation. The authors invoked coordination of [Au(III)] through the pyridine nitrogen with concomitant release of HCl (**V** → **VI**) as a possible cause for the observed selectivity (Scheme 7b). Reaction of **VI** with the enone would favour N-alkylation product **24**, while **22** would be

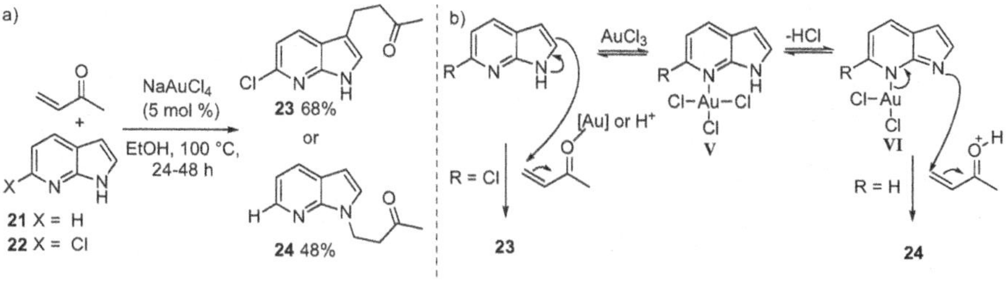

Scheme 7 (**a**) Hydroarylation of enones with 7-azaindoles; (**b**) proposed origin of C3- vs. *N*-selectivity

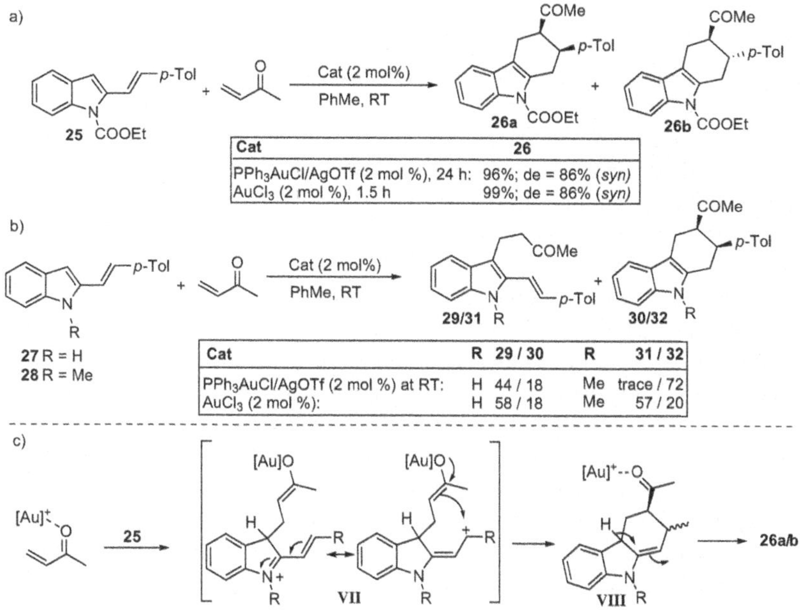

Cat	26
$PPh_3AuCl/AgOTf$ (2 mol %), 24 h:	96%; de = 86% (*syn*)
$AuCl_3$ (2 mol %), 1.5 h	99%; de = 86% (*syn*)

Cat	R	29 / 30	R	31 / 32
$PPh_3AuCl/AgOTf$ (2 mol %) at RT:	H	44 / 18	Me	trace / 72
$AuCl_3$ (2 mol %):	H	58 / 18	Me	57 / 20

Scheme 8 (**a**) and (**b**) [4 + 2] addition vs. hydroarylation under [Au(I)] and [Au(III)] catalysis; (**c**) proposed mechanism

less prone to form **VI** and reacts at C3 to give **23** (Scheme 7b). The authors note that *p*-TsOH was also found to catalyse both processes, indicating that the conjugate addition step may be an acid-catalysed process.

Rossi and co-workers found that 2-vinyl indoles either can undergo hydroarylation with enones or take part in [4 + 2] cycloaddition type reactions, depending on the protecting group and the catalyst [44]. Thus, ethyl carbamate-protected indole **25** exclusively afforded Diels–Alder adducts **26** with [Au(I)], [Au(III)] and a range of other Lewis acids, $AuCl_3$ giving the fastest reactions at ambient conditions (Scheme 8a). On the other hand, unprotected indole **27** gave mixtures of hydroarylated **29** and cycloaddition product **30**, and *N*-methyl indole **28** afforded

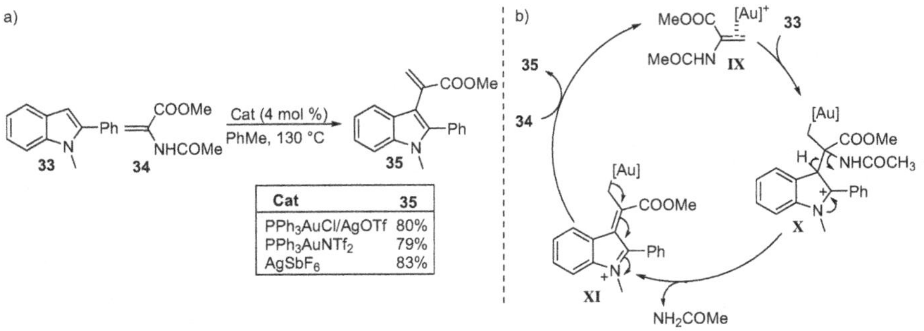

Cat	35
$PPh_3AuCl/AgOTf$	80%
PPh_3AuNTf_2	79%
$AgSbF_6$	83%

Scheme 9 Formation of indolylacrylates and proposed mechanism

exclusively **32** in the presence of $PPh_3AuCl/AgOTf$, but gave **31** as the major product with $AuCl_3$ (Scheme 8b). The cycloaddition was proposed to occur by a stepwise or pseudo-concerted mechanism in which intermediate **VII** quickly cyclises to **VIII** when the protecting group is electron-withdrawing, but gives competing hydroarylation with unprotected or *N*-methylindoles (Scheme 8c). The authors speculated that the differences in reactivity of [Au(I)] and [Au(III)] with **28** could relate to competing mechanisms, e.g. C–H auration for [Au(III)] vs. a Lewis acid pathway for [Au(I)].

When the same group attempted the hydroarylation of 2-acetamidoacrylate **34** with cationic [Au(I)] or [Ag(I)] sources, addition occurred at the α- rather than the β-position and was followed by an elimination of acetamide to give vinylated products **35** (Scheme 9a) [45]. Similar yields were obtained in the presence of [Au(I)] or [Ag(I)] catalysts. Based on NMR experiments showing no obvious coordination of the amidoacrylate nitrogen to Au, a mechanism triggered by π-activation of α-amidoacrylate was proposed (Scheme 9b).

The hydroarylation of unactivated alkenes is also possible. For example, Wong, Che and co-workers reported the alkylation of indoles and thiophene with styrenes and aliphatic alkenes **36** and dienes **38** using $PPh_3AuCl/AgOTf$ to give Markovnikov adducts **37** and **39** (Scheme 10) [46]. Microwave irradiation was found essential for less reactive aliphatic substrates and significantly accelerated the hydroarylation of styrenes compared to conventional heating. The hydroarylation of styrenes also proceeded with TfOH, although Au was found significantly more active. In the reaction with dienes **38**, *E*-allylated indoles **39** were obtained exclusively even when *E*/*Z* mixtures or pure *Z*-dienes were used. The same authors have also reported the hydroarylation of styrene using $AuCl_3/AgSbF_6$ in DCE [47].

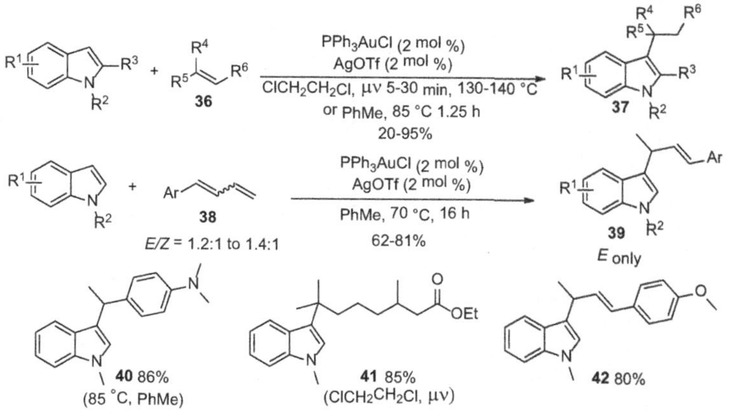

Scheme 10 Hydroarylation of alkenes and conjugated dienes

2.2 *Hydroarylation of Allenes*

Allenes have been shown to participate in both intra- and intermolecular hydroarylation reactions with heterocyclic nucleophiles. The vast majority of examples employ indoles or pyrroles reacting at the expected C3 and C2 positions respectively, although this can be reversed when the most nucleophilic position is blocked.

Au-allene complexes have been shown to exist in several coordination modes, either with η^2 or η^1 coordination. In a computational study, Gandon, Fensterbank and Malacria showed that allene complexes can exist as η^2-coordinated complexes **XII** or slipped versions **XIII–XIV**, σ-bonded η^1-coordinatied planar allyl cations **XV** and bent allenes **XVI** (Fig. 1) [48]. The barrier for interconversion between these structures and the nature of the ground state was highly dependent on the substitution of the allene. For example, with 1,3-dimethylallene, η^1-coordinated complexes of type **XII** were localised as ground states, whereas with heteroatom-substituted allenes ($R^1 = Me$, $R^2 = OAc$) planar **XV** coordination modes were lower in energy and for trisubstituted vinyl allenes, a bent structure such as **XVI** was favoured ($Au = AuPPh_3$).

As an axially chiral molecule ($R \neq H$), reactions involving additions to allenes have the possibility to proceed with transfer of axial to central chirality, and examples of this with heterocyclic nucleophiles are shown in the coming section. Importantly, while complexes **XII–XIV** and **XVI** are non-planar and still retain their chiral information, the interconversion of allene-gold complexes into planar intermediates of type **XV** has consequences for the stereochemical outcome and will result in racemisation of the allene and an erosion of the enantiomeric excess.

After Hashmi's seminal report [36], gold-catalysed intramolecular hydroarylation of allenes with indoles was reported in 2006 by Widenhoefer [49]

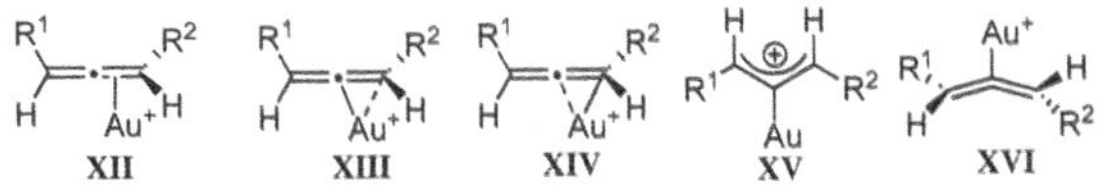

Fig. 1 Allene-Au coordination modes; η^2-coordinated **XII–XIV** and η^1-coordinatied **XV–XVI**

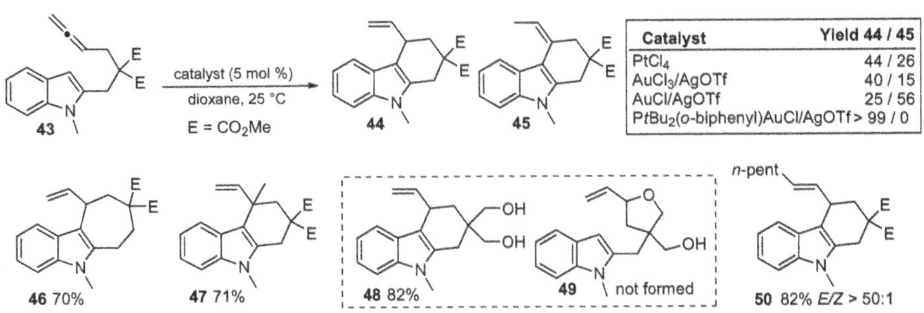

Scheme 11 Formation of tetrahydrocarbazoles from intramolecular allene hydroarylation

for a number of allenyl tethered indoles and by Nelson [50] with a pyrrole substrate in the total synthesis of (−)-Rhazinilam (vide infra).

In Widenhoefer's system, simple Au or Pt salts gave mixtures of the *6-exo*-tetrahydrocarbazole **44** and an isomeric product **45** resulting from migration of the olefin to the internal position (Scheme 11). However, using P*t*-Bu_2(*o*-biphenyl) AuCl/AgOTf as the catalyst afforded **44** as a single isomer. Depending on the length of the tether, 6- (**47**, **48**, **50**) or 7-membered rings (**46**) are obtained, always fully selective for the *exo*-cyclisation, also in the presence of hydroxyl groups capable of engaging in hydroalkoxylation (e.g. **48** vs. **49**).

The possibility of chirality transfer from the allene to the formed tertiary stereogenic carbon was also investigated. When the method was applied to enantiomerically enriched allene **51**, cyclised product **52** was obtained with complete conservation of *ee* (Scheme 12). Such axial to central chirality transfer implicates a mechanism that proceeds via chiral intermediates without racemisation of the allene. The proposed mechanism involves an outer sphere *anti*-attack of the indole on the intermediate **XIX** in which Au is coordinated *cis* to the *n*-pentyl substituent. Cyclisation from the *trans*-intermediate **XVII** would give a Z-alkene, which is not observed, probably because of steric interactions between the indole and the *n*-pentyl substituent in this approach. Three examples of efficient chirality transfer via 6-*endo*-cyclisation of enantioenriched thiophene and benzothiophene substrates with (dppm)Au_2Cl_2 were also recently reported by Ma and co-workers [51].

Gagné similarly used pyrrole **53** and other electron-rich arenes, finding that the best catalyst was formed in situ from gold carrying a less electron-rich triphenylphosphite ligand and $AgSbF_6$ (Scheme 13a) [52]. Studying the mechanism and the scope of this transformation, the same group reported the finding of *gem*-diaurated vinyl-gold species **56** as a resting state in the reaction with dimethoxybenzene

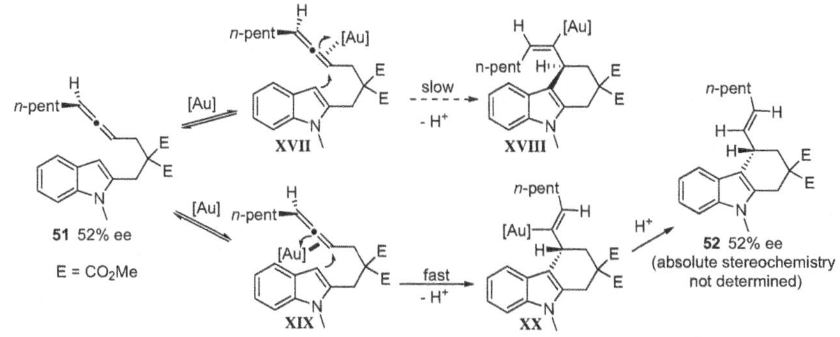

Scheme 12 Chirality transfer from enantiomerically enriched allene **51** into tetrahydrocarbazole **52**

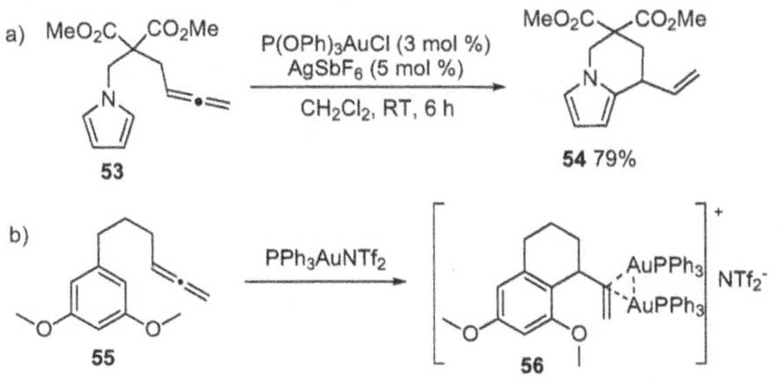

Scheme 13 (**a**) Intramolecular hydroarylation of allenyl pyrroles; (**b**) formation of *gem*-diaurated species in the hydroarylation of a dimethoxybenzene substrate

substrate **55** (Scheme 13b) [53]. This was readily formed in the reaction of vinyl-gold species with 1 equiv. of cationic gold, independent of the nature of the counter ion, and was shown to be more stable towards protodemetallation than the corresponding vinyl-gold intermediate. This and related *gem*-diaurated species are thought to predominantly exist as off-cycle resting states, as demonstrated for the hydroalkoxylation of allenes [54, 55].

Examples so far have shown the 6-*exo*-cyclisation with indoles and pyrroles. However, also intramolecular 6-*endo*- or 5-*endo*-hydroarylations are possible when the length of the tether is shortened. For example, Barluenga used P*t*-Bu$_2$(*o*-biphenyl)AuNTf$_2$ to catalyse the 6-*endo*-hydroarylation of allenylindoles **57** at the C2 position (Scheme 14a) [56]. Equally, Ma and co-workers have shown the 5-*endo*-cyclisation of C3-substituted indoles **59** (Scheme 14b) [57]. This reaction required somewhat forceful temperatures of 110°C for useful yields to form, possibly due to the more difficult attack of the indole on the allene in this more rigid substrate. Fully substituted allenes containing an electron-withdrawing group were crucial for the formation of products. Both reactions are proposed to be initiated by allene complexation, as shown in Scheme 14c for 5-*endo*-hydroarylation. Cyclisations of C2-tethered indole allenols [58, 59] provide carbazoles by loss of water and rearomatisation (vide infra, Scheme 31). In addition,

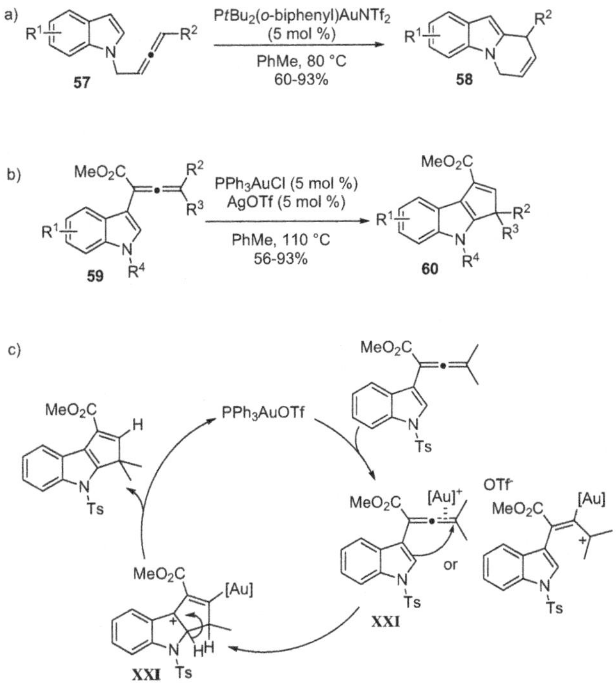

Scheme 14 (**a**) 6-*endo*- and (**b**) 5-*endo*-cyclisation of allenyl indoles; (**c**) proposed mechanism

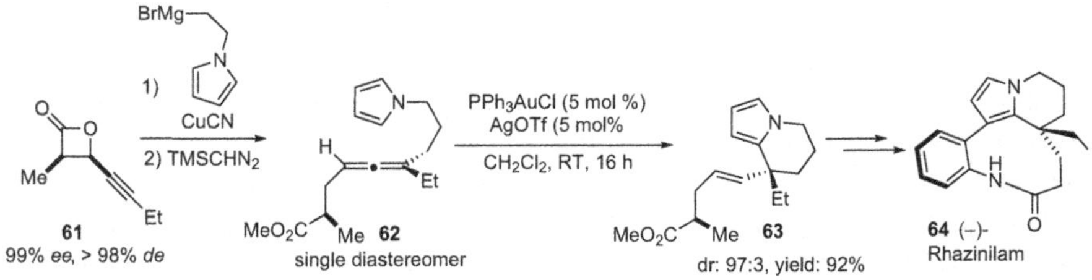

Scheme 15 Application of Au-catalysed hydroarylation with axial to central chirality transfer in the total synthesis of (−)-Rhazinilam

Sanz has reported a formal 6-*endo*-cyclisation of C3-tethered allenylindoles giving regioisomeric products selectively depending on the choice of catalyst. Several mechanistic scenarios were proposed to be operating initiated either by a 3-*exo*- or a 5-*endo-spiro*cyclisation followed by rearrangements [60].

With the common occurrence of heterocycles in natural products, gold-catalysed hydroarylation reactions of allenylpyrroles and indoles have featured as key steps in some natural product synthesis (for the use of Au catalysis in total synthesis, see [61]). Nelson and co-workers utilised PPh_3AuOTf as catalyst to construct the quaternary stereogenic centre of (−)-Rhazinilam **64** (Scheme 15). The allene intermediate **62**, prepared as a single diastereomer from ring opening of alkynyl

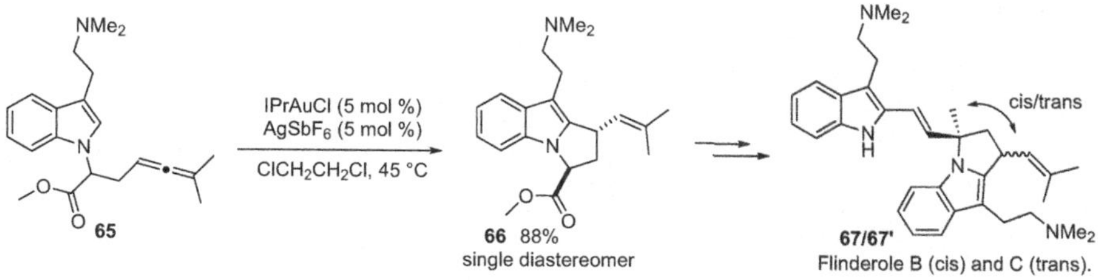

Scheme 16 Diastereoselective hydroarylation in the total synthesis of Flinderoles B and C

β-lactone **61**, was subjected to a variety of catalysts known to promote hydroarylation. It was envisioned that the pendant ester functionality could help direct the catalyst to the same face of the allene, while the second stereogenic centre would allow the authors to assess the absolute configuration of the formed tertiary carbon on the basis of diastereomeric ratio. While Ag salts were ineffective and $(MeCN)_2PdCl_2$ gave erosion of diastereomeric ratio, PPh_3AuCl in combination with AgOTf afforded the cyclised product **63** in high yields with high translation of allene chirality.

Toste and co-workers used an indole–allene hydroarylation in the total synthesis of antimalarial compounds Flinderoles B and C (**67** and **67′**) containing two indole motives, of which one is part of a fused 5-membered ring containing an indole C2-allyl fragment (Scheme 16) [62]. Whereas PPh_3AuCl failed to provide the cyclised product **66** from **65**, the use of a more electron donating and bulky NHC ligand, IPrAuCl (IPr = 1,3-bis-(2,6-diisopropylphenyl)imidazol-2-ylidene), afforded **66** as a single diastereomer, which in further transformations could be converted into diastereomeric Flinderoles B and C.

Following the development of intermolecular addition of oxygen and nitrogen nucleophiles catalysed by IPrAuCl/AgOTf [63], the Widenhoefer group used the same catalyst system in the intermolecular hydroarylation of alkyl or aryl allenes **68** to give *E*-allylated indoles **69** (Scheme 17a) [64]. The regioselectivity of the arylation was sensitive to both electronic and steric factors. Thus, arylation occurred selectively α to the phenyl group to give **70**, but regioisomeric mixtures of **71** formed with disubstituted phenyl methyl allenes. Despite exhaustive substitution, even tetramethyl-substituted allenes underwent the hydroarylation (**72**). The intermolecular hydroarylation of allenyl ethers **73** was also recently reported by Ramana and co-workers, proceeding with very low catalyst loading to give allylated products **74** in generally high yields (Scheme 17b) [65].

Rossi, Vicente and co-workers observed interesting selectivity in the reaction between 2-vinyl indoles **75** and *N*-allenamides **76** [66]. The reaction, proposed to go via the hydroarylation intermediate **XXIII**, could take different pathways depending on the catalyst and the indole protecting group (Scheme 18). Thus, with unprotected or *N*-methylindole and 5 mol% PPh_3AuNTf_2, hydroarylation products **77** and **78** resulting from the protodeauration of **XXIII** were formed. Introducing an electron-withdrawing carbamate group on the indole and thus

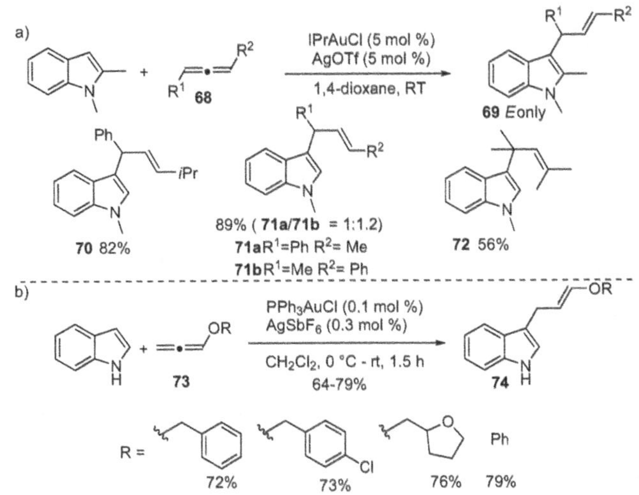

Scheme 17 Intermolecular hydroheteroarylation of (**a**) allenes and (**b**) allenyl ethers

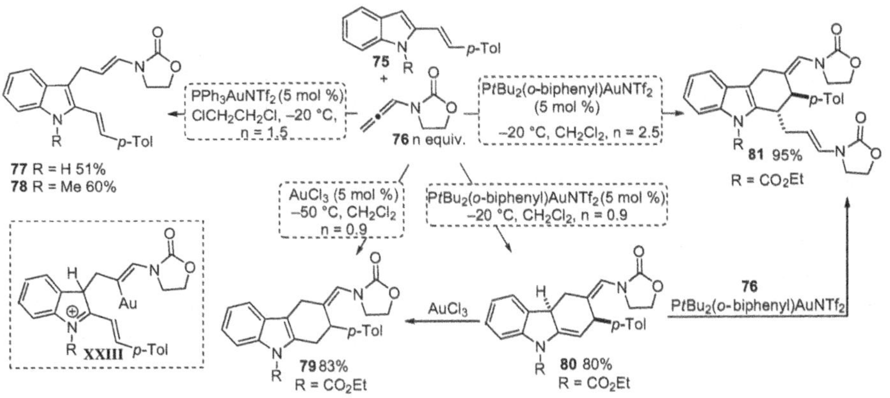

Scheme 18 Catalyst and protecting group-controlled transformations of 2-vinylindoles with allene **76**

increasing the electrophilicity of the putative iminium intermediate could instead promote the formation of products originating from formal [4 + 2] cycloaddition. Thus, in the presence of $AuCl_3$, tetrahydrocarbazole **79** was obtained selectively, while the same substrate in the presence of $PtBu_2$(*o*-biphenyl)$AuNTf_2$ gave the isomeric product **80**. Finally, in the presence of 2.5 equiv of **76**, a second hydroarylation took place to give **81**. Furthermore, **80** could be converted into **79** or **81** by treatment with the appropriate catalyst, supporting their stepwise formation via **80**.

Widenhoefer and co-workers reported a Au-catalysed enantioselective intramolecular hydroarylation of allenyl tethered indoles **82** with dinuclear [Au(I)]-catalyst

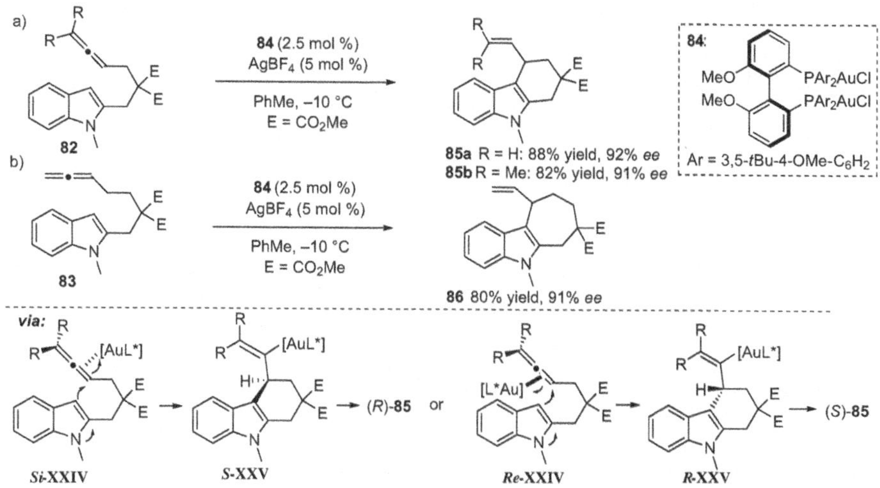

Scheme 19 Intramolecular enantioselective hydroarylation of allenes with indoles

84 in good to high *ee* (Scheme 19a) [67]. The longer tether in **83** afforded **86** with a 7-membered ring (Scheme 19b). The reaction is proposed to proceed with an attack of the nucleophile on Au-coordinated intermediate **XXIV** to give *R*- and *S*-products, respectively.

An enantioselective intermolecular hydroarylation between *N*-methyl indoles and racemic 1,3-diarylallenes **87** was reported by Che and co-workers (Scheme 20) [68]. Dinuclear catalyst **89** was found to give the best results, providing generally high yields and up to 60% *ee*. NMR experiments between **89**/AgOTf and *N*-methylindole showed the appearance of a peak in the ^{31}P spectrum at 43.8 ppm, characteristic of a phosphine-ligated aurated species of the type $R_3PAu(I)Ar$. Nevertheless, a computational comparison showed significantly lower energies (in the gas-phase) for the mechanism in which Au-coordination of the allene precedes the attack of the indole rather than one via indolyl-Au intermediates (Scheme 20b). The reaction is an example of a DYKAT (dynamic kinetic asymmetric transformation) in which the *R*- and *S*-enantiomers of the allene **87** are interconverted via a planar intermediate **XXVII** or **XXVIII**, while one of the diasteromeric Au complexes **XXVI** reacts faster to give enantioenriched product **88** (Scheme 20b and c). Only the *E* configuration in the product is observed, the pathway giving Z-isomers (from coordination of Au_2L^* to the opposite faces of **XXVI**) likely being disfavoured due to steric clashes between the substrate and the incoming nucleophile.

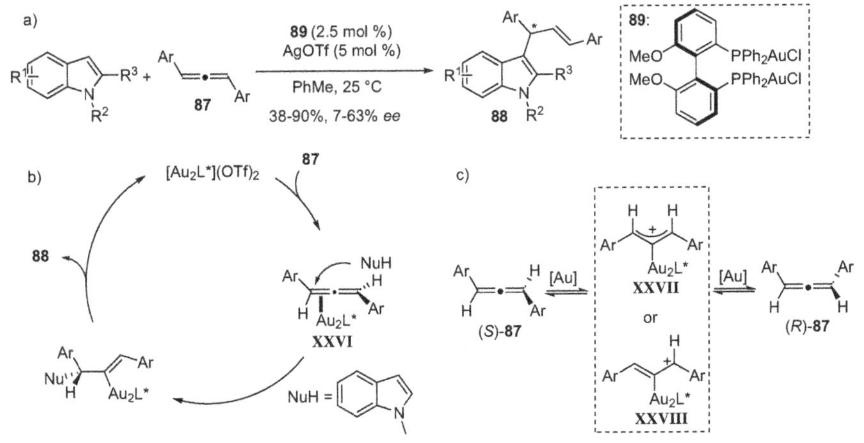

Scheme 20 (**a**) Enantioselective intermolecular hydroarylation; (**b**) proposed mechanism showing the formation of one enantiomer via nucleophilic attack on **XXVI**; (**c**) racemisation of (*R*)-**87** and (*S*)-**87** via planar intermediates **XXVII** or **XXVIII**

2.3 Hydroarylation of Alkynes

Activation of alkynes by [Au(I)] and [Au(III)] catalysts has been extensively explored in the last decade, giving place to numerous useful methodologies based on both intra- and intermolecular attack of nucleophiles on the alkyne, frequently followed by rearrangements. Consistently with this general reactivity, alkynes react with electron-rich, nucleophilic heteroarenes in Friedel-Crafts-type processes, resulting on C–H functionalisation of the heteroarene. The alkenyl-functionalised heteroarenes that would be expected from these transformations can also frequently be activated by Au and suffer a second attack by another unit of the heteroarene or another nucleophile, allowing access to more complex structures through cascade processes. Alternatively, reaction of the alkyne with a nucleophile can give place to a reactive intermediate which is then attacked by the heteroarene. The distinction between these two possibilities is not always trivial, and, in some of the cases presented here, no mechanistic data relevant to this point are available.

Au-catalysed intermolecular hydroarylation reactions between heteroarenes and electron-deficient alkynes were first described in the context of general hydroarylation methodologies. Reetz and co-workers in 2003 [69] reported the reaction of furan and 2-methyfuran with ethyl propiolate to give the corresponding (*Z*)-α-alkenylfuran products **90** and **91** (Scheme 21a). Hydroarylation of the same alkyne substrate with *N*-methylindole or benzo[*b*]furan was reported by He in 2005 [42], but in these cases, the single hydroarylation product could not be obtained. Instead, the corresponding double hydroarylation products **92** and **93** were obtained (Scheme 21b). In all the cases, the bond formation took place between the most nucleophilic position of the heteroarene (i.e. α for furan, β for benzo[*b*]furan and

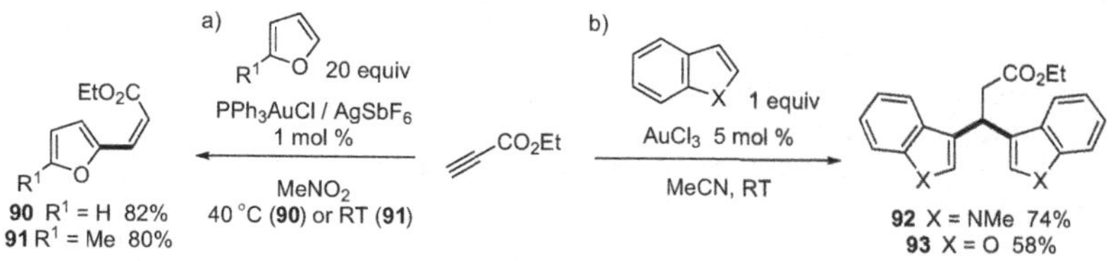

Scheme 21 Divergent outcomes on the Au-catalysed hydroheteroarylation of ethyl propiolate

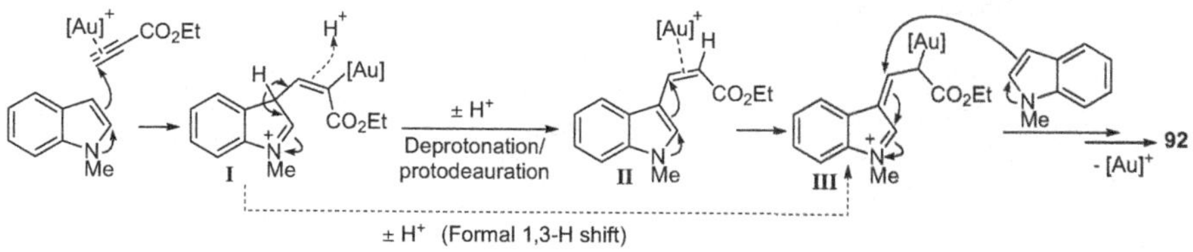

Scheme 22 Possible mechanism for the twofold hydroheteroarylation of ethyl propiolate

indole) and the electrophilic terminal position of the alkyne, as would be expected for a Michael-type addition.

In the reaction leading to double hydroarylation products **92/93**, the authors describe that no single addition product was observed. This led them to propose that the alkenylindole or alkenylpyrrole intermediates are significantly more reactive than ethyl propiolate, which might be explained by the higher polarisation of the double bond induced by the electron-donating heteroaryl group and the electron-withdrawing carboxylate group (Scheme 22).

Similar twofold hydroheteroarylation of nonactivated alkynes was later described with indoles, pyrroles [70] and furans [71] (Scheme 23a–c, respectively). With terminal, alkyl- or arylacetylenes, Markovnikov selectivity was observed, affording 1,1,1-trisubstituted ethane derivatives **94–96**. With indoles, the alkylation took place at the C3 position, while with pyrroles and furans, it took place at the α-position. With indoles and pyrroles, the hydroheteroarylation took place under mild reaction conditions with cationic [Au(I)] complex **97** as the catalyst or the combination of phosphite-ligated complex **98** with $AgSbF_6$. In the case of furans, the Cl-bridged dinuclear Au-complex **99** was used, and the reaction proceeded slowly at 50°C over 7 days. The reaction of indole with prop-1-ynylbenzene **100** (Scheme 23d) provided double hydroarylation product **101**, derived from attack of the indole at the Me-substituted position of the alkyne.

This mode of reactivity is unusual with furans, which generally react with alkynes (either inter- or intramolecularly) to provide phenols (**102**) rather than the simple hydroarylation products (Scheme 24) [16, 19, 72–75]. The reaction proceeds through initial attack of the furan on the Au-coordinated alkyne to form cyclopropylcarbene **IV**, followed by fragmentation and cyclisation of carbene **V** to provide epoxide **VII**. Final rearrangement gives place to the phenol product [76–80].

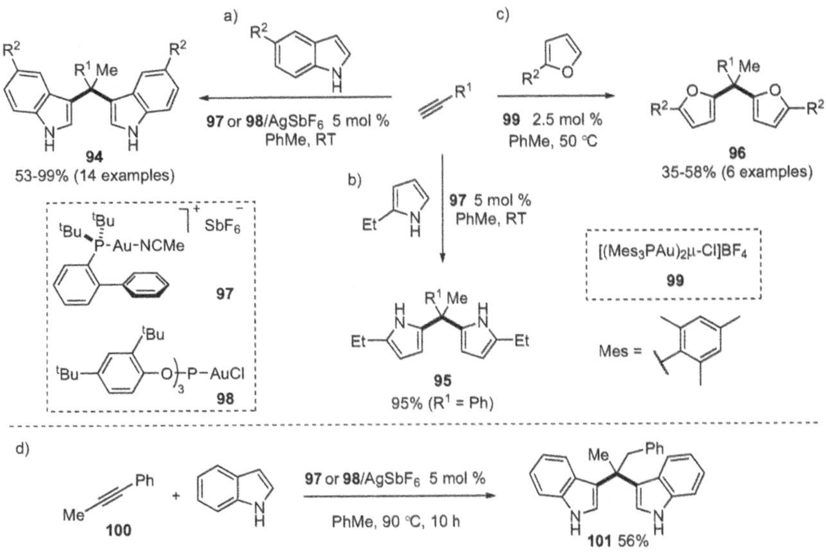

Scheme 23 Twofold hydroheteroarylation of nonactivated alkynes

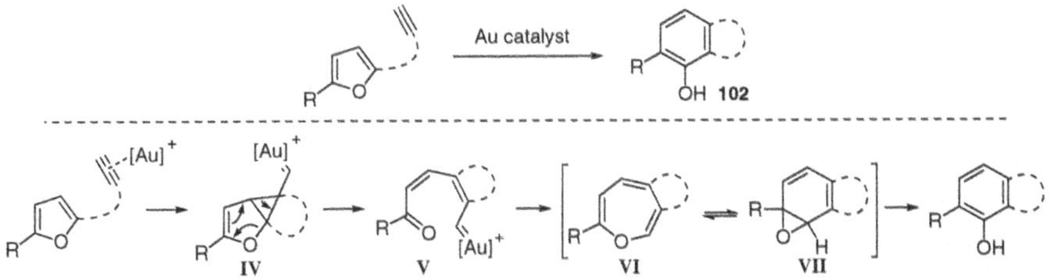

Scheme 24 Phenol synthesis from the reaction of furans and alkynes

This transformation and variations of it have been extensively explored for the preparation of phenols with different substitution patterns [81] as well as more complex polycyclic structures [82]. Other transformations involving the cleavage of the furan unit have been reported for the synthesis of fulvenes [83] and aryl enones [84]. This type of reactivity, not providing overall a C–H functionalisation of the initial heteroarene, falls beyond the scope of this chapter.

Homopropargylic alcohols **103** (with either terminal or internal alkynes) reacted with *N*-methylindole to provide double C3 hydroheteroarylation products **104** (Scheme 25), arising from attack at the distal alkyne position respect of the hydroxyl group [85]. With terminal alkynes ($R^1 = H$), this anti-Markovnikov selectivity is opposite to that observed with nonfunctionalised alkynes. The unusual selectivity has been explained by intramolecular alkoxycyclisation of the homopropargylic alcohol to form Au-coordinated dihydrofuran **IX**, which is

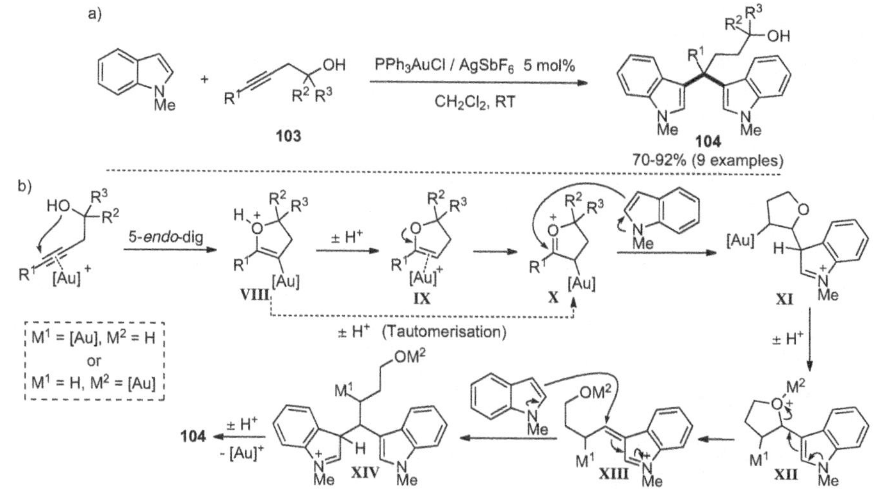

Scheme 25 Distal twofold hydroheteroarylation of homopropargylic alcohols

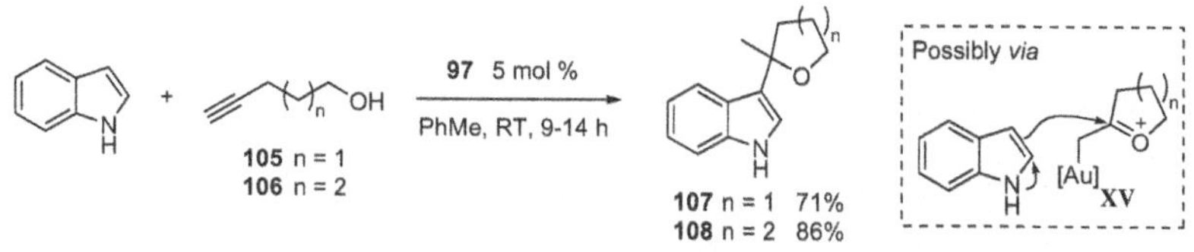

Scheme 26 Formal C3 alkylation of indoles via alkoxycyclisation/hydroarylation

activated for reacting with the indole. Alternatively, the activated species **X** could be directly accessed by tautomerisation, without protodeauration taking place.

Interestingly, this inversion of regioselectivity by the presence of a hydroxyl group on the alkyne was only observed for homopropargylic alcohols. Instead, 5-pentyn-1-ol and 6-hexyn-1-ol (**105** and **106**, respectively, Scheme 26) reacted with indole in the presence of catalyst **97** to give cyclised 3-alkylindoles **107** and **108**, arising from alkoxycyclisation/hydroarylation [70]. The reaction proceeded in these cases through 5-*exo*- or 6-*exo*-alkoxycyclisation which, after tautomerisation or protodemetallation/Au insertion (see Scheme 25), provide intermediate **XV**. Attack of the indole on **XV** provides products **107** and **108** with Markovnikov selectivity. Second addition of the indole to obtain products of the type **104** was not described for substrates **105** and **106**.

Further exploiting the reactivity of alkenylindoles under Au catalysis, the reaction between 2-alkynylindoles **109** with terminal alkynes has been explored. Markovnikov C3 alkenylation provided in this case enyne intermediate **XVI**, which reacted further through a 6-*endo*-dig cyclisation to provide disymmetrically substituted carbazoles **110** (Scheme 27) [86].

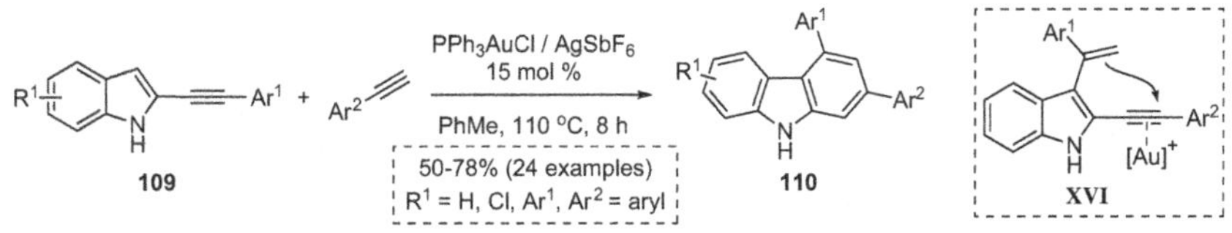

Scheme 27 Synthesis of carbazoles by C3 alkenylation/cyclisation

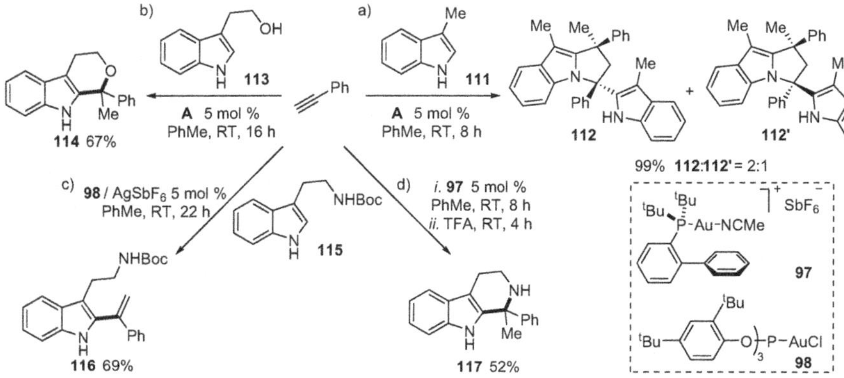

Scheme 28 C2 alkylation of indole derivatives via hydroarylation

In 3-substituted indoles, the alkyne insertion took place at the C2 position. However, in such cases, instead of the twofold hydroarylation (vide supra, Schemes 21b and 23a), reaction of the intermediate alkenylindole with other nucleophiles took place (Scheme 28). Thus, skatole (**111**) reacted with phenylacetylene in a 2 + 2 stoichiometry to afford polycyclic adducts **112** and **112′**, possibly arising from dimerisation of an initially formed alkenylindole product (a related dimerisation has been described in the reaction of 3-substituted indoles with methyl ketones [70, 87]). Tryptophol (**113**) provided cyclised product **114** by hydroarylation/hydroalkoxylation. Finally, Boc-protected tryptamine (**115**) provided the simple hydroarylation product **116**, which could be isolated or converted in situ to tricyclic product **117** upon deprotection of the amine side chain.

Intermolecular C2 hydroindolylation reactions are believed to proceed through an initial C3 attack to give intermediate **XVII** (Scheme 29) followed by migration to give **XVIII**, which after rearomatisation of the indole and protodeauration gives the alkenylindole product **116** [70]. This mechanism was supported by an intramolecular hydroarylation with acyl group migration (vide infra, Scheme 38), although direct C2 attack to give **XVIII** or formation of cyclopropyl carbene intermediate **XIX** cannot be ruled out in all cases.

Au-catalysed intramolecular reactions of electron-rich heteroarenes bearing alkyne functionalities have also been reported, with outcomes depending strongly on factors such as the position of the heteroarene at which the alkyne is linked and the length and structure of the tether.

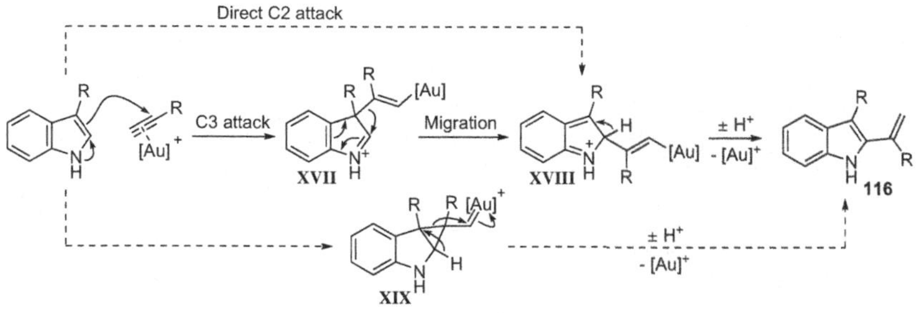

Scheme 29 Possible mechanism for Au-catalysed C2 alkenylation of indoles

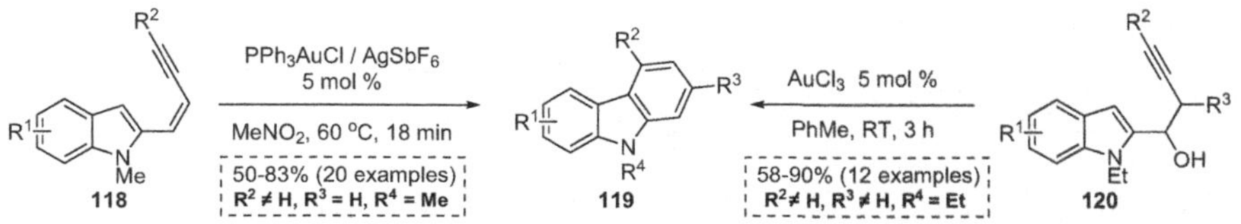

Scheme 30 Preparation of carbazoles by 6-*endo*-cyclisation of alkyne-functionalised indoles

Indole derivatives with an alkyne group tethered at the C2 position have been used for the preparation of carbazoles through 6-*endo*-cyclisations (Scheme 30). *Z*-2-Enynyl indoles **118**, with diverse substituents on the alkyne, cyclised in the presence of $PPh_3AuCl/AgSbF_6$ to afford the corresponding 4-substituted-9*H*-carbazoles **119** ($R^3 = H$) [88]. Homopropargyl alcohol derivatives **120** reacted similarly with $AuCl_3$ to provide, after dehydration, disubstituted carbazoles **119** ($R^3 \neq H$) [89]. This route avoided the troublesome preparation of *Z*-enynyl indoles **118**.

In a related transformation, more densely functionalised carbazoles **122** (Scheme 31a) were prepared from substrates **121**, in situ precursors of allenes through 1,3-migration of the benzoyloxy group (**XX** → **XXI**, Scheme 31b) [90]. Rearomatisation to the final product involved a 1,2-migration of R^3/R^4 in **XXIII** followed by elimination of the $IPrAu^+$ unit. The different tendencies of substituents for 1,2-migration (H > Aryl > Alkyl > Me) allowed to differentiate between R^3 and R^4. Similar reactions starting from allenes have also been reported [58, 59] and used in the synthesis of several naturally occurring carbazole alkaloids [91, 92].

Similarly, by using alkyne-functionalised furans **123** with short, two-atom tethers, the synthesis of benzo[*b*]furans **124** has been achieved (Scheme 32a) [93]. When an *exo*-vinyl group was present in the substrate (**125**), the reaction could be combined with attack by an external nucleophile, to provide diversely functionalised benzo[*b*]furans **126** (Scheme 32b) [94]. The reaction proceeded through an initial 5-*endo*-spirocyclisation followed by 1,2-shift (**XXV** → **XXVII**, Scheme 32c).

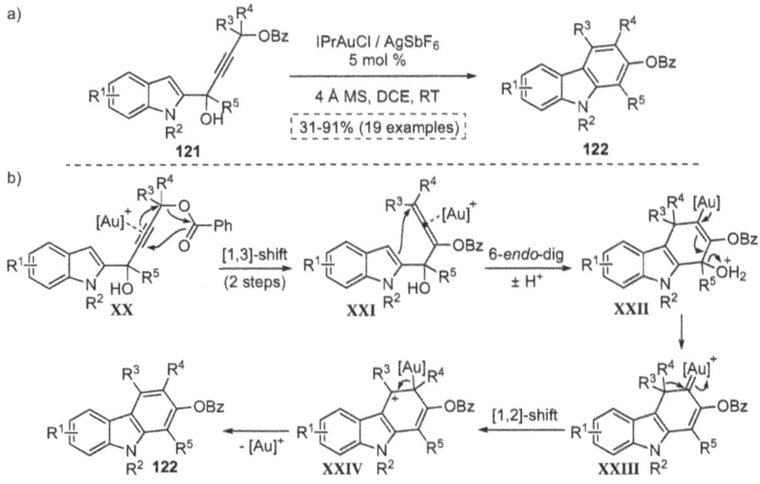

Scheme 31 Synthesis of polysubstituted carbazoles from alkyne-functionalised indoles

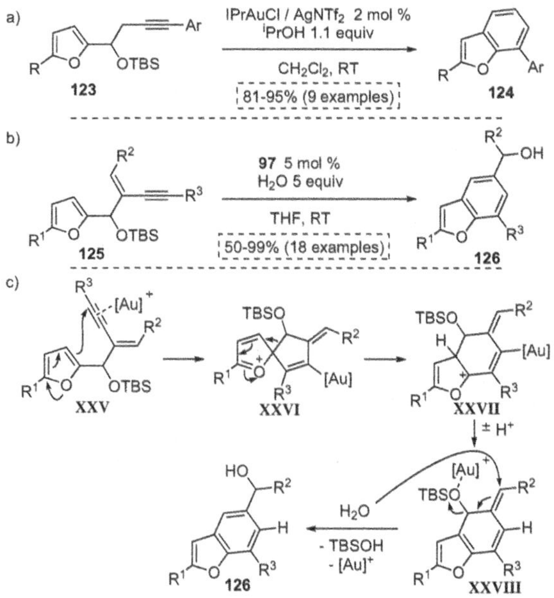

Scheme 32 Synthesis of benzo[*b*]furans by intramolecular hydroarylation of alkynes

A formal [2 + 4] cycloaddition, proceeding through a cascade C2/C3 double hydroarylation, has been reported by Ohno for the preparation of 4,7-disubstituted indoles **128** from pyrroles (Scheme 33) [95]. This process afforded generally moderate yields of the indole product with symmetrically substituted diynes **127**. With nonsymmetrically substituted diynes, the regioselectivity was low. The methodology could be applied also to the preparation of carbazoles from indoles.

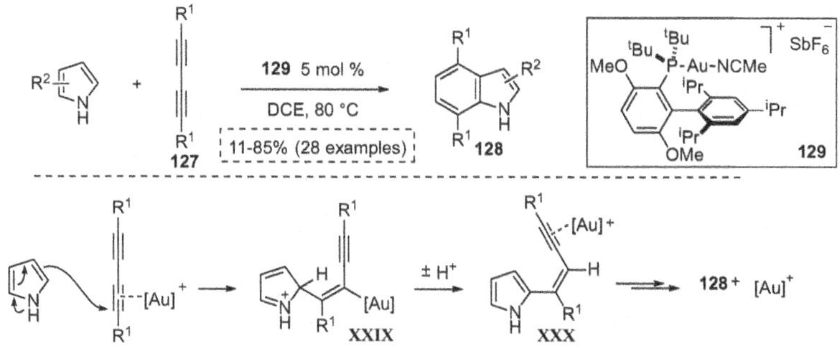

Scheme 33 Au-catalysed formal [4 + 2] cycloaddition of pyrroles or indoles and diynes

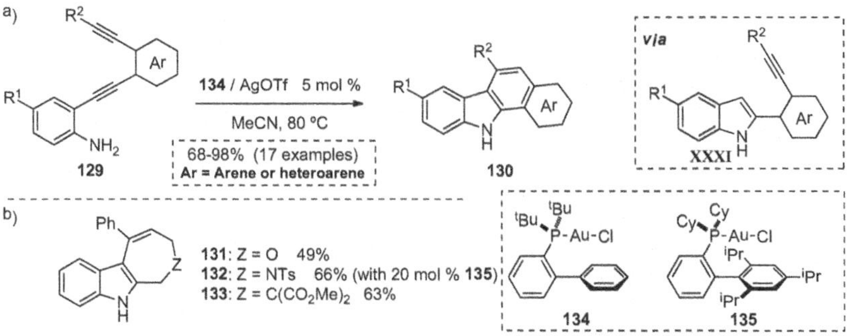

Scheme 34 Cascade Au-catalysed hydroamination/hydroarylation for the preparation of carbazoles, oxepino- and azepinoindoles and cyclohepta[*b*]indole derivatives

Cascade reactions involving in situ formation of alkyne-substituted indoles **XXXI** by a 5-*endo*-aminocyclisation reaction followed by 6-*endo*-carbocyclisation have been described, leading to benzocarbazoles **130** (Scheme 34) (for a related deacylative cyclisation, see [96–99]). The same strategy, with different linkers between the two alkyne units, allowed the preparation of oxepino- and azepinoindoles, as well as cyclohepta[*b*]indole derivatives (**131–133**, respectively) via a 7-*endo*-cyclisation (Scheme 34b) [97].

Au-catalysed cyclisations of indoles with an alkyne tethered through the C3 position have also been reported, giving place to fused 6-, 7- or 8-membered rings. Echavarren reported the cyclisation of propargyl-substituted tryptophan and tryptamine derivatives **136** to proceed under mild conditions in the presence of cationic [Au(I)] catalyst **97**, providing the formal 7-*exo*-dig cyclisation products **137** (Scheme 35) [70, 100]. Interestingly, other catalysts such as AuCl or $AuCl_3$ displayed a completely reversed selectivity, favouring the formal 8-*endo*-dig products **138**, while the combination of PPh_3AuCl/AgOTf provided mixtures of both. The reaction was highly substrate dependent, providing also allenylindoles and

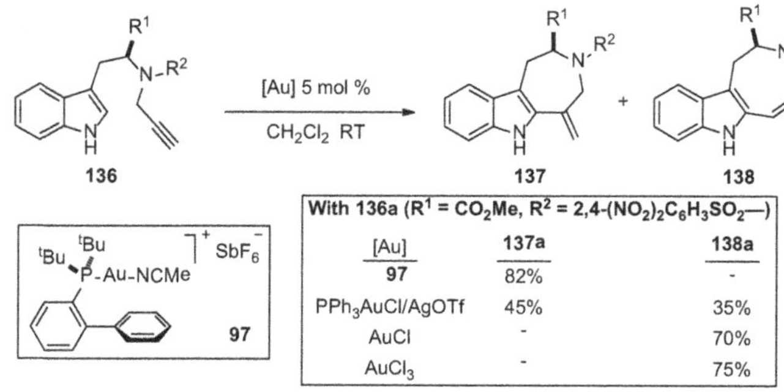

With 136a ($R^1 = CO_2Me$, $R^2 = 2,4-(NO_2)_2C_6H_3SO_2$—)		
[Au]	137a	138a
97	82%	-
$PPh_3AuCl/AgOTf$	45%	35%
AuCl	-	70%
$AuCl_3$	-	75%

Scheme 35 Au-catalysed cyclisation of alkyne-containing tryptophan derivatives **136**

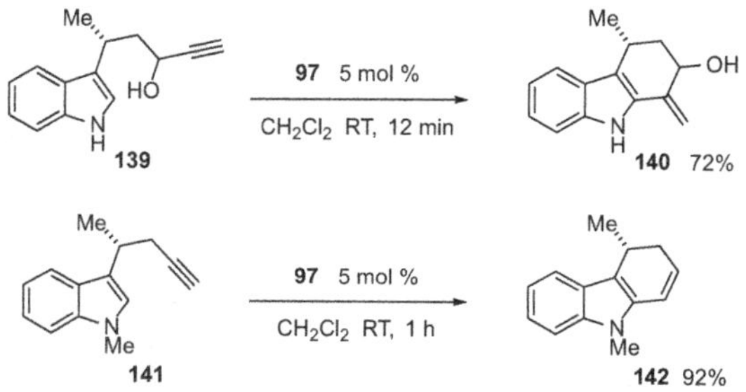

Scheme 36 Au-catalysed cyclisation of alkyne-containing indole derivatives **140** and **142**

bicyclic indolines in some cases. Also, the formal 7-*exo*-products were only accessible with terminal alkynes, while internal ones favoured the formal 8-*endo*.

Similarly, compounds containing the alkyne and the indole separated by shorter 3-centred linkers provided selectively the product from formal 6-*exo*-dig cyclisation (**140**), while with 2-centre linkers, the 6-*endo*-dig product was obtained (**142**, Scheme 36) [70, 100].

Zhang reported the cyclisation of indoles bearing both alkyne and amido functionalities (**143**) to provide tetracyclic products **144** (Scheme 37a) [101]. The reaction was proposed to take place through initial attack of the carbonyl oxygen on the Au-coordinated alkyne (**XXXII**) to provide cyclised intermediate **XXXIII** (Scheme 37b). Attack by the indole provides **XXXIV**, which after migration, deprotonation and protodeauration gives place to **XXXVI**. Finally, a Ferrier rearrangement affords product **144**.

Cyclisations of indoles in which the most nucleophilic C3 position is blocked are proposed to proceed through C3 attack followed by migration or by fragmentation of an intermediate cyclopropane Au carbene rather than by direct C2 attack (see above, Scheme 29). Taking advantage of this mechanism, Hashmi has reported a synthesis of azepinoindolones **146** (Scheme 38), proceeding through 6-*endo*-cyclisation followed by preferential migration of the acyl group in intermediate **XXXVIII** [102].

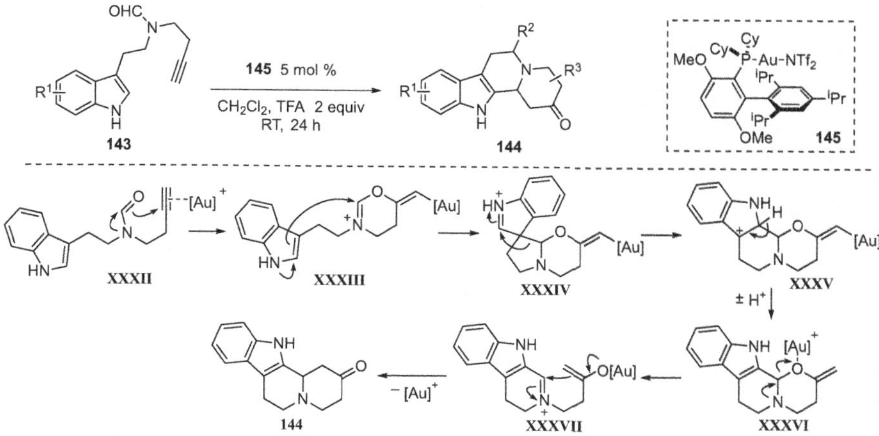

Scheme 37 Cyclisation of indoles functionalised with an *N*-homopropargyl amide unit

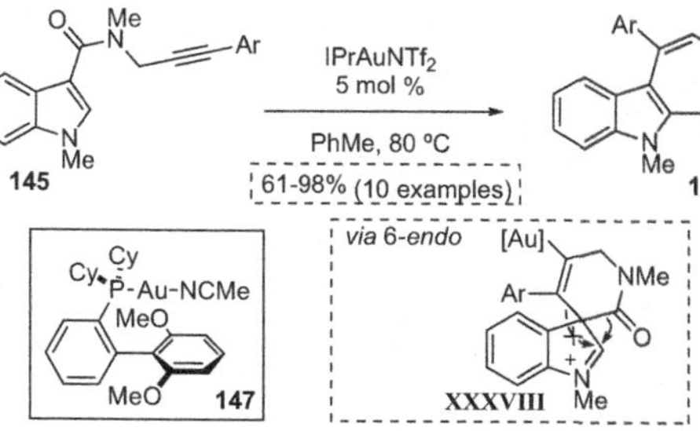

Scheme 38 Preparation of azepinoindolones by Au-catalysed cyclisation and acyl migration

In similar transformations, both intra- and intermolecular alkenylation of alkyne-functionalised pyrroles **148** have been described, catalysed by cationic [Au(I)] complex **97** [103]. In the absence of an external alkyne, the reaction proceeded through an intramolecular 7-*endo*-cyclisation to form bicyclic products of the type **149**, or 6-*exo* in the case of terminal alkynes, to provide product **150** (Scheme 39). Under the same conditions, the reaction could be performed with an external alkyne (1 equiv) to obtain 2-alkenylpyrroles **151** (phenylacetylene derivatives) or **152** (methyl propiolate). The intermolecular reaction took place with terminal arylacetylene or propiolic esters, while alkylacetylenes or internal alkynes did not react, giving place instead to just the intramolecular reaction.

Intramolecular hydroarylation has also been reported with silyl-substituted furans **153**, giving place to 7-membered fused rings **155** through a formal 7-*exo*-cyclisation (Scheme 40) [77, 79]. The presence of the α-silyl group was essential for the simple intramolecular hydroarylation to take place, in contrast with the rearrangement to phenol product **154** obtained with a methyl substituent.

This reactivity has been applied in more complex tandem transformations with in situ formation of the alkyne-heteroarene unit which then cyclises. Enders

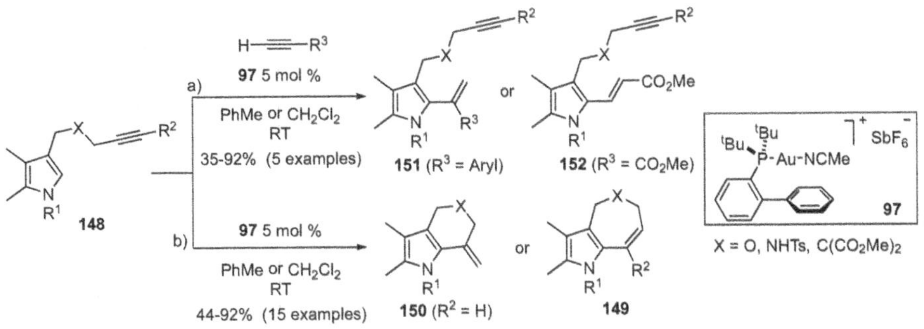

Scheme 39 Intra- and intermolecular alkenylation of pyrroles

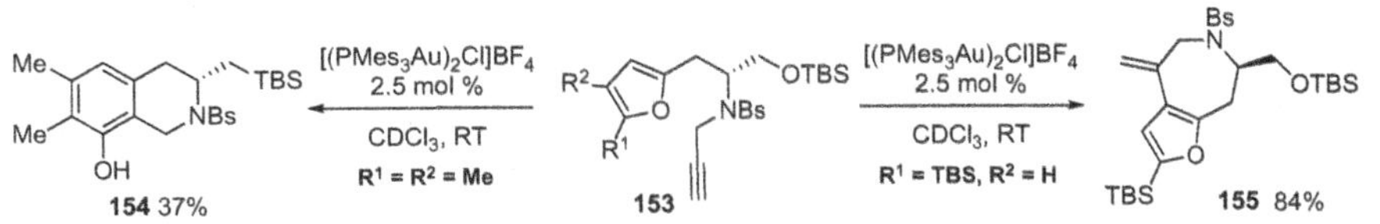

Scheme 40 Intramolecular hydroarylation of 2-silylfuran derivatives versus phenol synthesis. Bs = 4-Bromobenzenesulfonyl

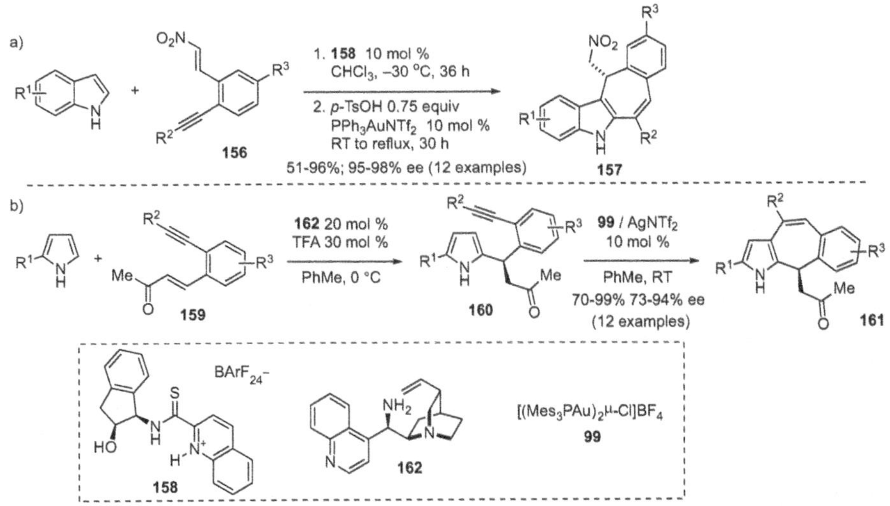

Scheme 41 Tandem difunctionalisation of indoles and pyrroles through formation of an alkynyl-functionalised substrate and Au-catalysed cyclisation

reported the preparation of tetracyclic derivatives **157** via a one-pot sequential organocatalytic conjugate addition/Au-catalysed cyclisation (Scheme 41a) [104]. Formal 7-*endo*-cyclisation products of pyrroles could be accessed via an organocatalytic Michael reaction/Au-catalysed cyclisation sequence (Scheme 41b) [105]. The Michael reaction gave place to relatively rigid C2-tethered alkynylpyrrole compounds **160**, which cyclised to give tricyclic products **161**, proposed to proceed through a 6-*endo*-cyclisation/1,2-migration.

Van der Eycken has described the cyclisation of Ugi reaction products **163** containing an indole unit to obtain indoloazepinones **164**, indoloazocinones **164′** and complex indoline derivatives [106–108]. Similarly, C4-tethered alkynylindoles have been used for C3 alkenylation, giving place to azocinoindoles **166** through Au-catalysed 8-*endo*-cyclisations (Scheme 42) [109].

With pyrroles, post-Ugi cyclisation strategies have been developed for the Au-catalysed preparation of pyrrolopyridinones [110, 111] **168**, pyrrolopyridines [112] **170** and diazoninones [113] **172**, through variations on the structures of the starting materials (Scheme 43). Formation of 6-membered ring products **168** and

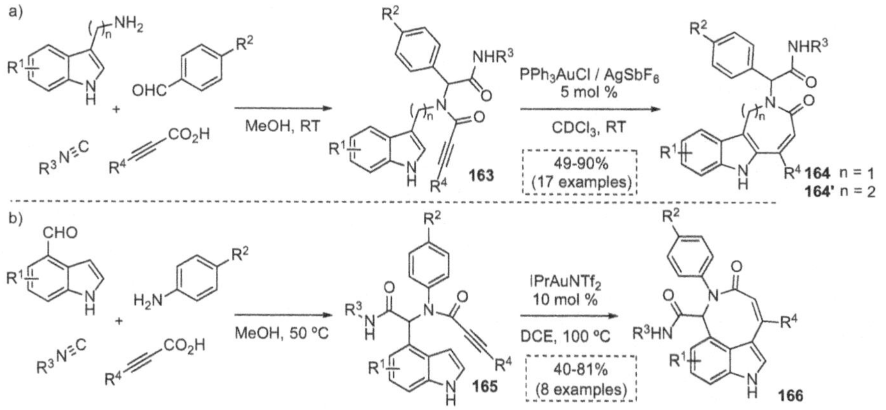

Scheme 42 Cyclisation of indole-containing Ugi four-component reaction products to give complex polycyclic structures with formation of 7- and 8-membered rings

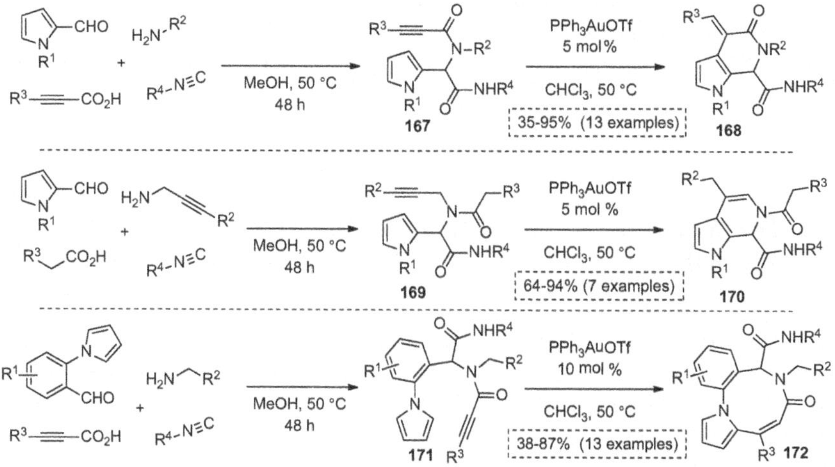

Scheme 43 Cyclisation of pyrrole-containing Ugi four-components reaction products to give complex polycyclic structures with formation of 6- and 9-membered rings

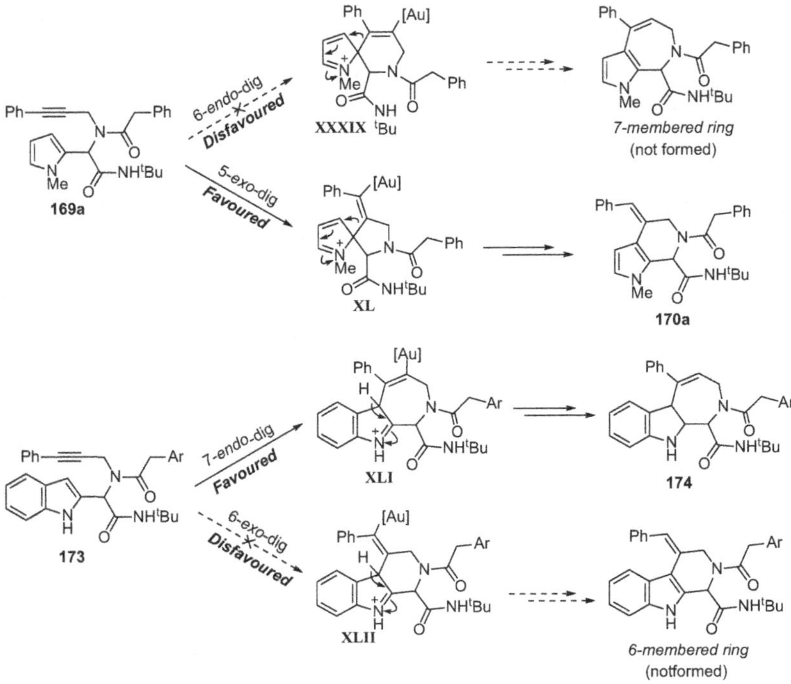

Scheme 44 Comparison of the cyclisation selectivities of C2-tethered alkynylpyrroles and indoles

170 contrasts with the 7-membered rings obtained with related indoles (Scheme 44). This is explained by a C2-spirocyclisation/migration taking place in the case of pyrroles (**167** and **169**). In such case, the 5-*exo*-cyclisation (to **XL**) is favoured over 6-*endo*-(**XXXIX**), which would lead after migration to the 7-membered ring. In indoles **173**, the reaction takes place at the most nucleophilic C3 position, in which case the 7-*endo*-cyclisation (to **XLI**) is favoured over the alternative 6-*exo*-(**XLII**). Product **172** is proposed to arise directly from **171** via a 9-*endo*-cyclisation (Scheme 43).

2.4 Tandem and Domino Reactions Involving Alkenes and Alkynes

A combined enone-alkyne double annulation from enediynones **175** was reported by Liu and co-workers (Scheme 45a) [114]. Polycyclic-fused indole scaffolds **176** were formed in this carbonyl-yne cyclisation/Friedel-Crafts/indole-yne cascade cyclisation in the presence of $NaAuCl_4$ according to the proposed mechanism in Scheme 45b. Although AgOTf could catalyse the first two steps (**I** → **III**) of the

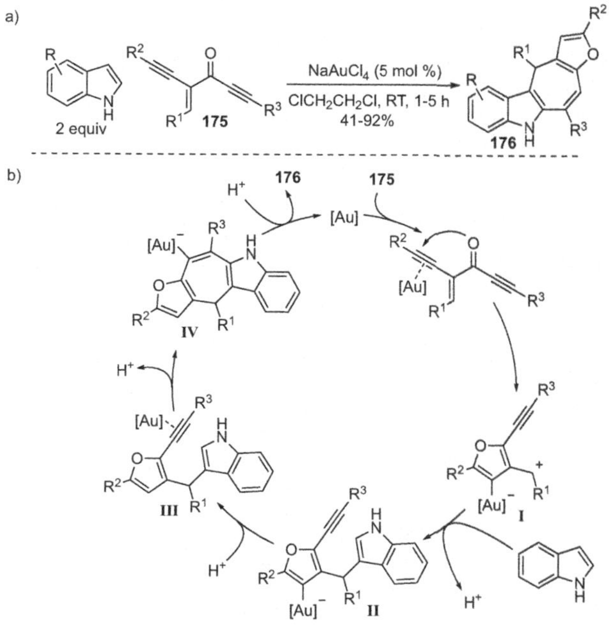

Scheme 45 Au-catalysed tandem oxycyclisation/arylation/cyclisation of enediynones **175**

catalytic cycle in reduced yield, it failed to promote the final alkyne hydroarylation (**III** → **IV**).

A related intermolecular hydroarylation/annulation of enynones **177** giving fused tricyclic structures **178** was reported by Carbery and co-workers (Scheme 46a) [115]. A variety of indoles with functional groups such as boronic esters, halogens and free phenols afforded a diverse scope of products in the presence of $NaAuCl_4$. Based on rapid loss of deuterium at the C3 and C2 positions of labelled indoles, a catalytic cycle involving C–H auration was proposed (Scheme 46b). The transient hydroarylation product **VII** (observed by NMR) is proposed to undergo a *spiro*cyclisation at C3 to give **VIII**, followed by C3 to C2 migration based on isolation of structurally similar *spiro*-products when the reaction was performed in CH_2Cl_2.

Friedel-Crafts alkylation (vide infra, Sect. 3) has been used in tandem with alkyne hydroarylation to give C2/C3-functionalised polycyclic structures **184/185** (Scheme 47) [116]. The amido alkylation with *N*-acyliminium precursors **183** readily proceeded at room temperature in the presence of AuCl/AgOTf, whereas the hydroarylation step required two consecutive additions of PPh_3AuNTf and higher temperature. The reluctance towards hydroarylation was later addressed by replacing the all-Au-catalysed process with a dual $HNTf_2$/Au catalytic system (Scheme 47) [117]. The acid is thought to catalyse the alkylation step while also preventing the formation of inactive *gem*-diaurated species **191** by facilitating protodeauration of vinyl-Au intermediates. Indoles afforded mixtures of *6-exo-*

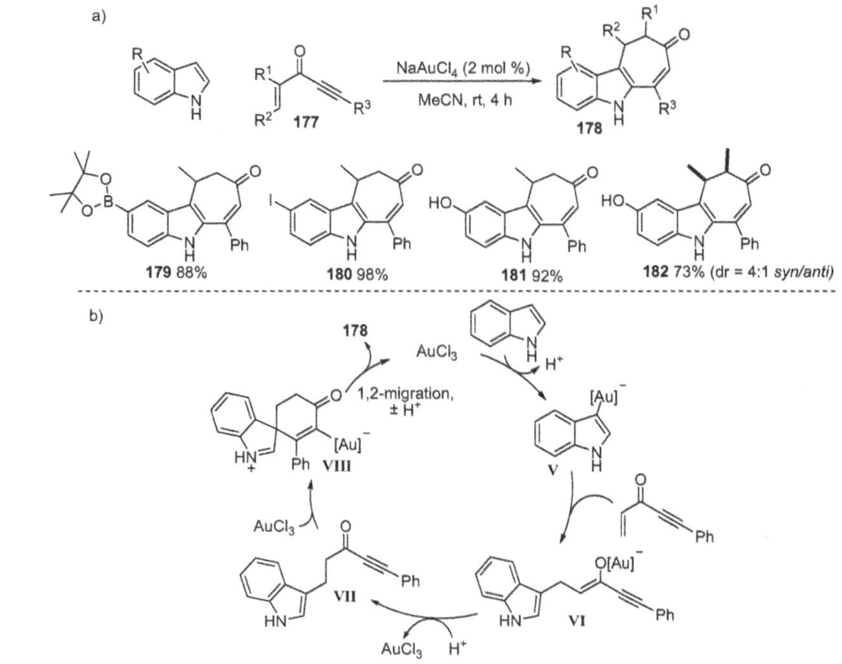

Scheme 46 Twofold hydroarylation of enynones

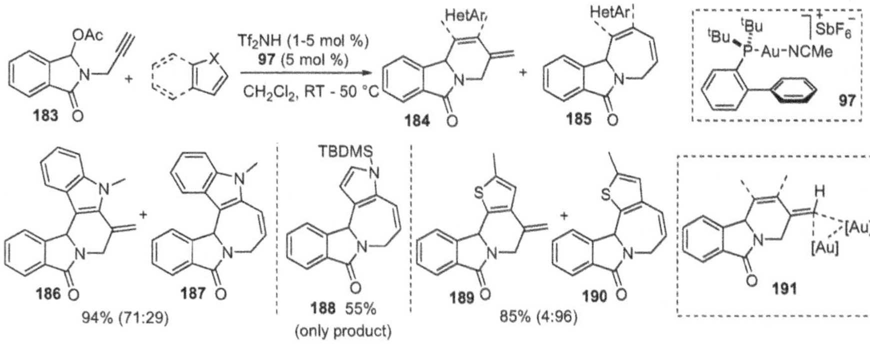

Scheme 47 Dual acid-/gold-catalysed alkylation/alkyne hydroarylation

and *7-endo*-cyclisation products **186/187**, while TBDPS pyrrole and 2-methylthiophene underwent *7-endo*-cyclisation to **188** and **190** with high selectivity but with opposite regiochemistry.

Similarly, Liu reported the preparation of indolocycloheptatrienes **193** by C3 coupling with hydroxyenynes **192** followed by C2 insertion of the alkyne (Scheme 48) [118]. This transformation was also applied to pyrroles. Use of a

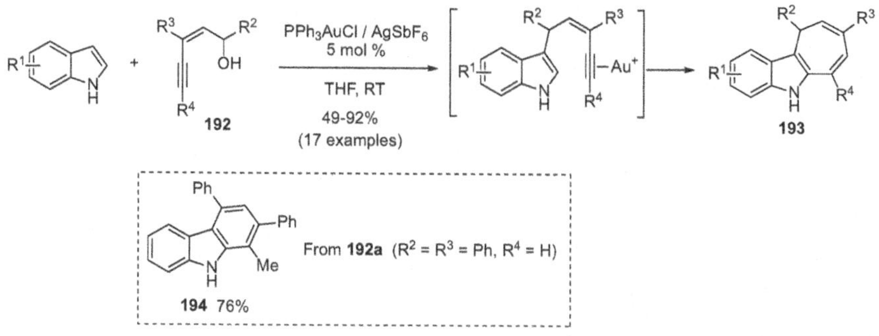

Scheme 48 Tandem Friedel-Crafts allylation/cyclisation between indoles and enynols (**192**)

terminal alkyne (**192a**, $R^2 = R^3 = Ph$, $R^4 = H$) with indole substrates gave place to the corresponding carbazole through a formal 6-*exo*-cyclisation.

3 Friedel–Crafts-Type Reactions Involving Carbon–Heteroatom Bonds

Apart from examples where gold has been used as a carbophilic Lewis acid in the activation of unsaturated bonds, Au catalysts have also been used in Lewis acid activation of oxygen- and nitrogen-based electrophiles. Coordination of the heteroatom to Au facilitates substitution reactions via a Friedel–Crafts-type reaction (S_EAr) with the arene to give alkylated products. Examples include coupling of (hetero)arenes with alcohols and acetates (Scheme 49a), ring opening of epoxides and aziridines (Scheme 49b) and addition to imines and carbonyl compounds (Scheme 49c).

Propargylic and allylic alcohols are activated by Au Lewis acids to undergo substitution reactions rather than hydroarylation (Scheme 50a–c). In a similar manner, heterocycles and arenes have been shown to afford benzylation products with benzylic acetates or alcohols (Scheme 50d). In 2005, Campagne and co-workers described a Au-catalysed nucleophilic substitution reaction of propargylic alcohols (**195**) using arenes, heteroarenes or other nucleophiles [119]. Enantiomerically enriched propargylic alcohols underwent the substitution with complete racemisation in agreement with the formation of a carbocation intermediate. Beller and co-workers used allylic acetates (**200a**), alcohols (**200b**) and carbonates (**200c**) to achieve a similar process, though at higher temperatures and employing the arene as the solvent [120]. Following Campagne's report, Chan and co-workers described the related allylic alkylation of heterocycles and other electron-rich arenes with allylic alcohols (**197**) catalysed by $AuCl_3$ [121]. The reaction, which also works to some extent with other Lewis acids such as $InCl_3$, $Cu(OTf)_2$ or Brønsted acids (TfOH), is selective for the addition of the nucleophile to the less substituted carbon of the allyl system when unsymmetrical allylic

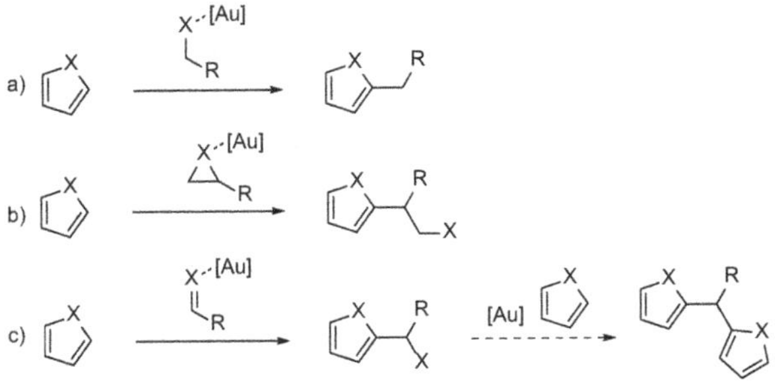

Scheme 49 Au-mediated Friedel-Crafts alkylation of heterocycles

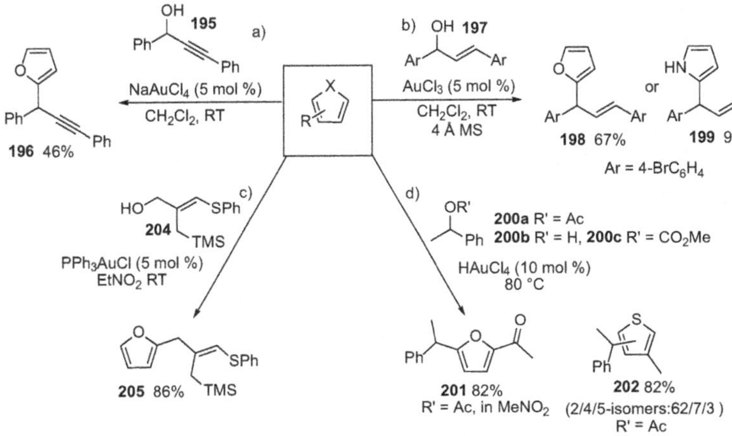

Scheme 50 Alkylation of heterocycles with (**a**) propargylic alcohols; (**b**) and (**c**) allylic alcohols; (**d**) benzylic alcohols, acetates and carbonates

alcohols were used (not shown with heterocyclic nucleophiles). A similar reaction showed the alkylation of 3-phenyl sulfanyl-substituted allylic alcohols (**204**) with furans in the presence of PPh_3AuCl [122].

Bach and co-workers also noted the superior reactivity of $AuCl_3$ over other Lewis and Brønsted acids in terms of functional group compatibility and mildness of the reaction conditions in a diastereoselective alkylation of electron-rich (hetero) arenes with benzylic acetates **206** or alcohols (Scheme 51) [123]. The *syn*- vs. *anti*-diastereoselectivity could be modulated depending on the substituent adjacent to the benzylic carbon and was independent on the relative configuration of the starting material, in line with a carbocation intermediate [124]. For example, methoxycarbonyl afforded *anti*-selective alkylation (**207**), whereas diethoxyphosphonyl group gave the corresponding *syn* product. A range of (hetero)arenes could be alkylated under mild conditions with 10 mol % $AuCl_3$, although the authors noted that the efficiency was also high with 1 mol % catalyst.

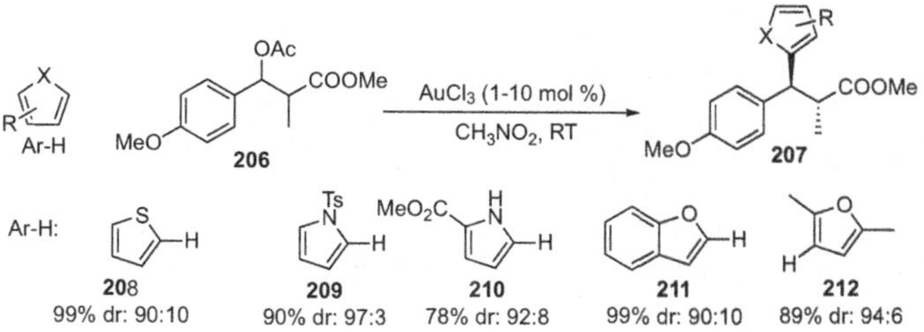

Scheme 51 Diastereoselective benzylation with benzylic acetates

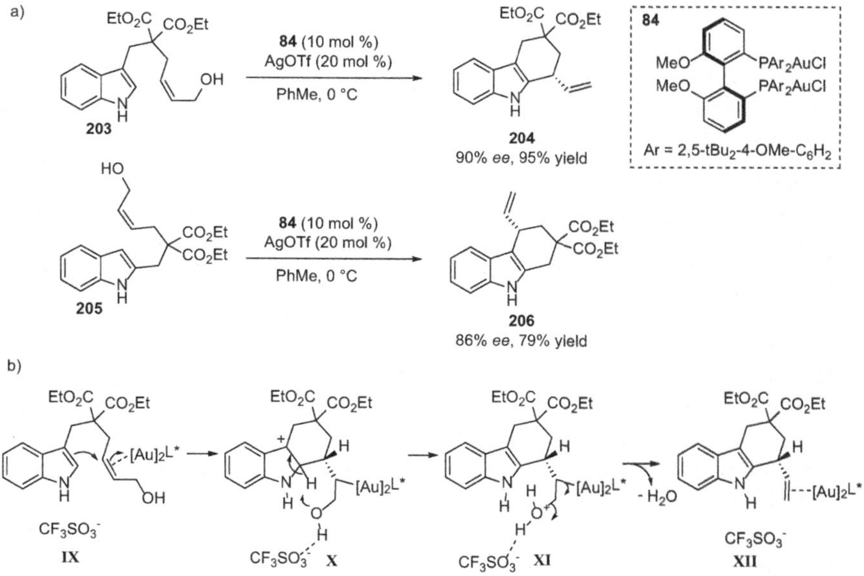

Scheme 52 Enantioselective allylation of indoles with alcohols

Intramolecular asymmetric alkylation of indoles was described by Bandini and co-workers in 2009 (Scheme 52) [125, 126]. Tetrahydrocarbazoles **204** were formed in good to excellent *ee* via C2-alkylation of indoles with *Z*-allylic alcohols **203** tethered to the C3-position of the indole using binuclear [Au(I)] complex **84**. The allylation also worked with C2-substituted indoles **205**, forming C3-allylated products **206**. Theoretical mechanistic investigations showed that the reaction proceeds via a stepwise S_N2'-type mechanism involving initial attack of the indole on the [Au(I)]-coordinated olefin **IX** leading to a σ-bonded [Au(I)] intermediate **X**. This is followed by elimination of [Au(I)]-OH_2 where the alcohol OH acts as an internal base to re-aromatise the indole, assisted by hydrogen bonding to a triflate counter ion to give place to **XII** [127].

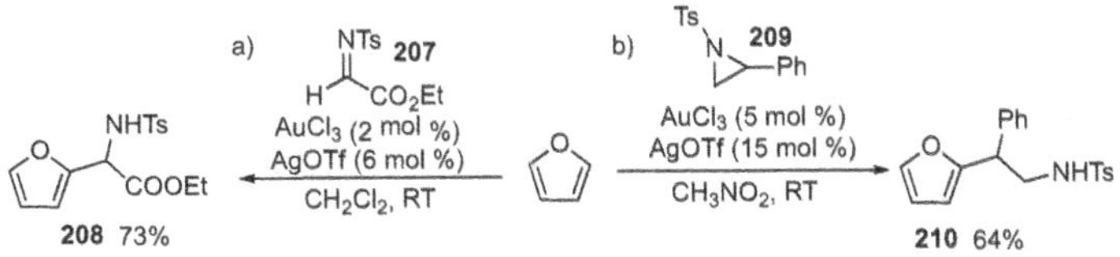

Scheme 53 Formation of amines with *N*-tosylimines and *N*-tosylaziridines

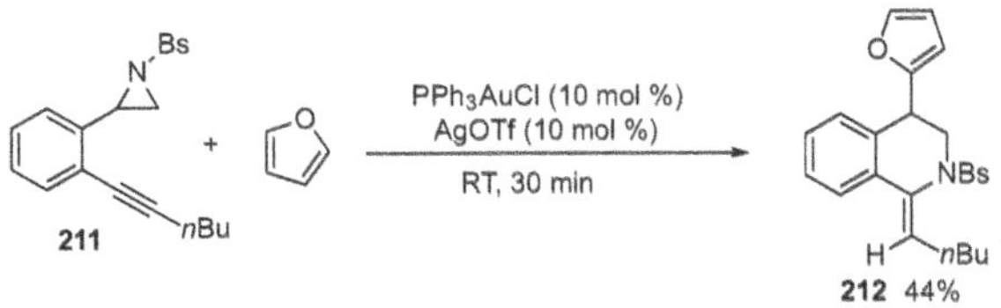

Scheme 54 Domino alkylation/hydroamination of furan

Formation of amines by alkylation of (hetero)arenes with imines and aziridines under Au or Au/Ag catalysis has also been investigated. In the reaction with *N*-tosylimines **207**, reported by Luo and Li, the $AuCl_3$/AgOTf catalytic couple was found to afford the best yields, although both metals showed catalytic activity on their own (Scheme 53a) [128]. The same catalyst combination was used by Wu and co-workers in a regioselective ring opening of aziridines **209** (Scheme 53b) [129]. Although furan was the only example of a heterocycle, the reaction was more efficient with other arenes (e.g. xylene) reaching full conversion in less than 2 min with 1 mol % $AuCl_3$. AgOTf on its own (15 mol %) or other Lewis acids in combination with AgOTf (e.g. $ZnCl_2$, $YbCl_2$, $BF_3 \cdot OEt_2$) did catalyse the reaction, but in lower yields and with longer reaction times.

Ring opening of aziridines has also been used to trigger domino cyclisations. Inspired by Wu's protocol for alkylation of arenes with aziridines, Shi and Zhang combined it with an intramolecular hydroamination to give tetrahydroisoquinolines **212** in a reaction that likely starts with the gold-promoted benzylation of the furan followed by an intramolecular alkyne hydroamination (Scheme 54) [130].

Condensation of furans, indoles and thiophenes with various aldehydes in the presence of $AuCl_3$, providing triarylmethanes **213**, has been reported by Nair and co-workers (Scheme 55a) [131]. Similarly, Hashmi and co-workers had observed the condensation of furans with acetone when the hydroarylation of α,β-unsaturated ketones was run in acetone as the solvent (Scheme 55b) [132]. Fully dehydrated products were always obtained rather than the presumed alcohol intermediate, indicating a rate limiting addition to the carbonyl followed by a fast reaction of the alcohol with a second equivalent of furan. Other Lewis and Brønsted acids also catalyse the reaction with TfOH (1 mol %) or $HClO_4$ (11 mol %), affording equally good yields.

The reaction of 1,3-dicarbonyl compounds **216** with indoles and pyrroles afford alkenylated products **217** in the presence of $NaAuCl_4$ or AuBr/AgOTf (Scheme 56)

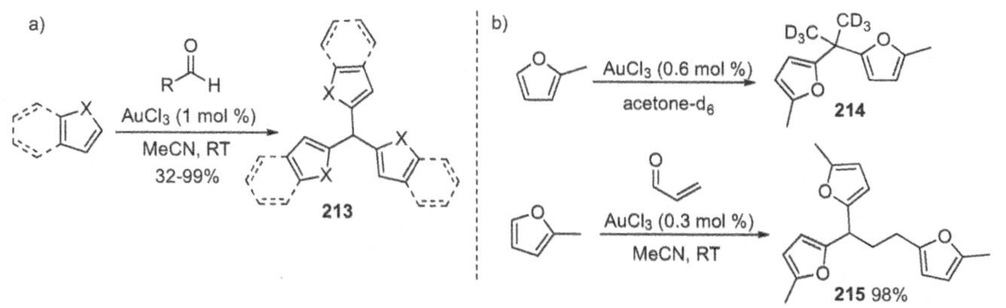

Scheme 55 Condensation of 2-methylfuran with (**a**) acetone-d_6 and (**b**) 2-methylfuran-derived aldehyde and acrolein

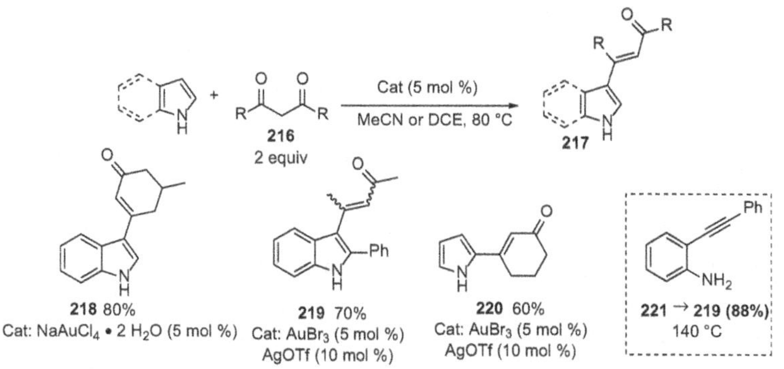

Scheme 56 Alkenylation of indoles with 1,3-dicarbonyl compounds

[133]. Indole **219** could also be accessed from 2-alkynyl imine **221** via a domino hydroamination/alkenylation sequence at 140°C.

Finally, diazo compounds have been shown to react in a similar fashion under Au catalysis. Barluenga, López and co-workers reported an allylation of heterocycles and alkenes with vinyl diazo compounds **224** (Scheme 57a) [134]. Furan underwent the allylation at the C2 position, benzo[*b*]thiophene at C3 and *N*-phenylpyrrole afforded a 1:1 mixture of C2/C3 allylation. Similarly, Lan, Shi and co-workers used α-diazoesters **225** as coupling partners with a range of heterocycles and other carbon nucleophiles to give alkylated products **223** (Scheme 57b) [135]. The selective functionalisation of C(sp^2)–H bonds [136–138] contrasts with that of Rh- or Cu-catalysts which in general promote C(sp^3)–H insertion, cyclopropanation, or N–H and O–H insertion. The ligand choice had a significant impact on the selectivity, with tris(2,4-di-*t*-butylphenyl)phosphite affording the best selectivity. This was rationalised by the better π-acceptor properties of phosphites as compared to phosphine or *N*-heterocyclic carbene ligands, resulting in a stabilisation of the carbocation resonance form **XIII** over the Au-carbene resonance form **XIV** and preferential S_EAr-type reactivity compared to typical carbene insertions (Scheme 57c). In the proposed mechanism, Au-carbenoid **XIII** reacts with the

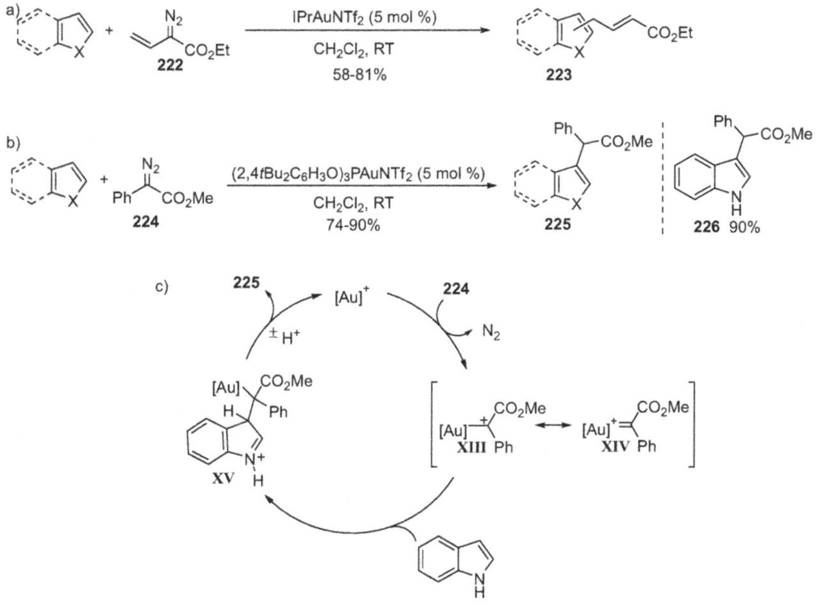

Scheme 57 Alkylation of heterocycles with diazo compounds

nucleophile to give intermediate **XV** which after deprotonation/protodeauration affords the product.

4 Coupling Reactions via C–H Auration

In this section, we will detail methods in which a direct C–H auration of heteroaromatics has been proposed during C–H functionalisation. This C–H activation may occur at either a [Au(I)] or [Au(III)] centre depending on the electronic nature of the aromatic coupling partner. Specifically, [Au(III)] complexes promote the C–H activation of electron-rich arenes **227** via an electrophilic aromatic substitution (S_EAr)-type reactivity. This type of reactivity has long been known, and refinement of the procedure has allowed for the preparation of several aryl-[Au(III)] complexes **228** (Scheme 58a) [139–143]. Similarly, though much more recently, [Au(I)] complexes have also been shown to promote C–H activation of electron-deficient arenes **229** bearing acidic C–H bonds (Scheme 58b) [144, 145]. Either (NHC)AuOH or R_3PAuCl in combination with a Ag salt can mediate this auration, though phosphine-ligated complexes show greater reactivity towards less acidic substrates. Alternatively, ^{t}Bu$_3$PAuCl in combination with NaOH can promote the C–H auration of arenes and also heteroarenes (Scheme 58c) [146]. The preparation of aryl-[Au(I/III)] compounds allows the exploration of novel transformations with gold and guides mechanistic hypotheses of already known transformations.

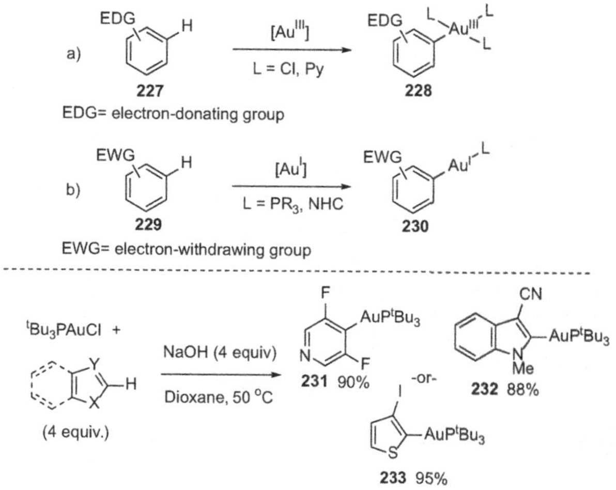

Scheme 58 Orthogonal reactivity of (**a**) [Au(III)] and (**b**) [Au(I)] complexes with electron-rich and electron-poor arenes respectively; (**c**) representative examples of direct C–H auration of electron-poor heteroarenes

4.1 Alkynylation

Although Au-catalysed hydroarylation of π-bonds represents a major success story in Au catalysis, this reactivity presents a major challenge when developing methods for the alkynylation of substrates, i.e. retaining the alkyne functionality in the product. This predicament was illustrated nicely by Reetz and Sommer who, in attempting to develop a procedure for the gold-catalysed oxidative coupling of electron-rich arenes with terminal alkynes, could only furnish the product of hydroarylation **235** (Scheme 59) [69]. Despite these challenges, a previous report on the stoichiometric coupling of aryl-[Au(III)] compounds **236**/**237** with phenyl acetylene suggested a gold-catalysed oxidative coupling was feasible (Scheme 60) [147].

In 2010, Nevado and de Haro reported the first Au-catalysed oxidative alkynylation of arenes (Scheme 61) [148] (a related study details the use of carbon-supported gold nanoparticles; however, the alkynylation of heteroaromatics was not described [149]). Interestingly, the combination of a gold catalyst in 1,2-dichloroethane with $PhI(OAc)_2$ and $NaHCO_3$ gave the desired product exclusively – no product from hydroarylation of the alkyne, or homocoupling of the arene was observed [150, 151]. The choice of an iodonium oxidant, $PhI(OAc)_2$, is a key feature of the process as other oxidants commonly used in gold redox catalysis, such as Selectfluor and tert-butylhydroperoxide, yielded no product [152]. The procedure is limited to the use of electron-rich arenes, following an S_EAr-type reactivity pattern. Benzyl-protected pyrrole and indole were also found to be reactive coupling partners; however, in the case of pyrrole, poor selectivity for the C2 product **240** vs. C3 product **241** was encountered, and with indole,

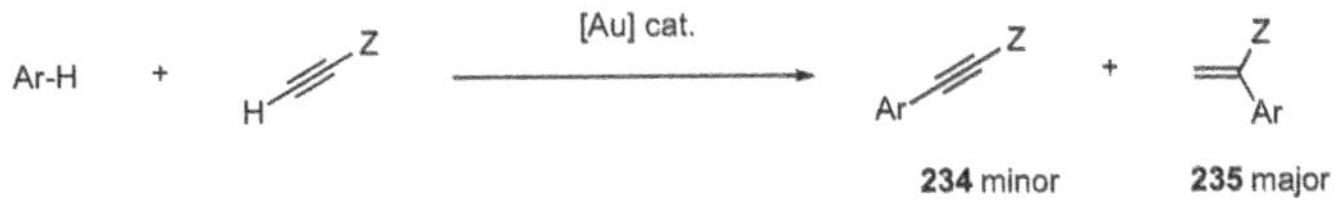

Scheme 59 Attempted Au-catalysed oxidative alkynylation of arenes prior to 2009

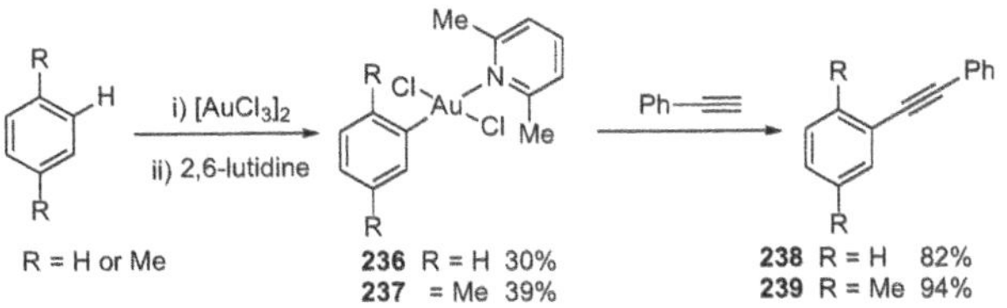

Scheme 60 Formation and stoichiometric alkynylation of aryl-[Au(III)] complexes

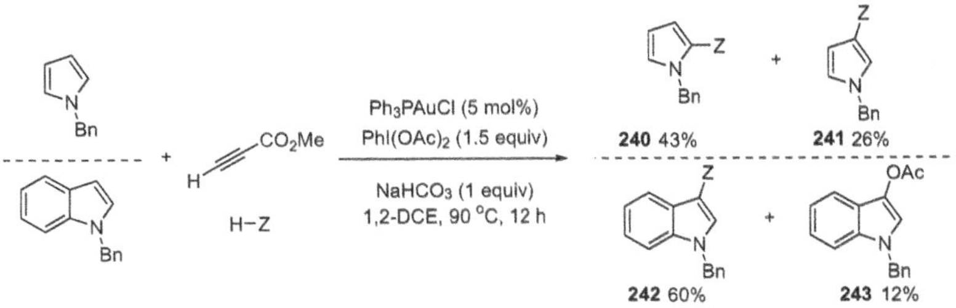

Scheme 61 Au-catalysed ethynylation of benzyl-protected pyrrole and indole

competing acylation is observed (Scheme 61). Supposedly, the formation of both indole products **242** and **243** occurs from a common [Au(III)] intermediate upon competitive reductive elimination (vide infra, Scheme 62a).

Although the oxidative coupling of arenes and terminal alkynes is a great improvement towards an ideal alkynylation procedure, the development of milder methods and their application to a wide variety of arenes, especially heteroarenes, are of high importance. In this regard, Waser and co-workers developed an alternative approach for the direct alkynylation of arenes using 1-{[tris-(1-methylethyl) silyl]ethynyl]}-1,2-benziodoxol-3(1*H*)-one (TIPS-EBX) **244** (Scheme 62) [153]. Using this privileged alkynylating agent, the authors found that a wide range of heterocycles – including indoles, pyrroles, benzothiophenes, thiophenes and furans – are suitable coupling partners and only minor changes in the reaction conditions between each class of substrate was required (Scheme 62) [155–158] (for a related alkynylation of anilines, see [154]). The requirement for prefunctionalisation of the alkyne coupling partner is a clear disadvantage in comparison to Nevado and de Haro's procedure; however, this is compensated somewhat by using lower temperatures (23 °C vs. 90 °C) and the successful use of unprotected pyrrole and indole substrates. The facile deprotection of the TIPS-acetylene unit to the

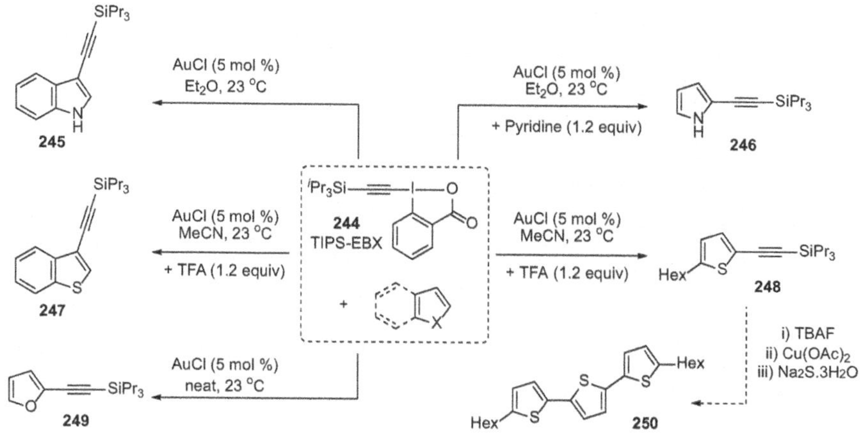

Scheme 62 Au-catalysed alkynylation of heteroarenes using TIPS-EBX reagent

terminal alkyne allows for further functionalisation, as demonstrated in the formation of an oligothiophene **250** (Scheme 62, bottom right) [155].

Several mechanistic scenarios have been proposed for the described alkynylation reactions (Scheme 63).[1] Although Nevado and de Haro's oxidative coupling (Scheme 63, part 1) and Waser's direct coupling (Scheme 63, part 2) naturally begin from different reagents, similar mechanistic scenarios have been proposed in each case (Scheme 63 paths (a), (b) and (c)). In pathway (a), a distinct C–H auration step is proposed. The catalytic cycle may begin by either oxidation of the catalyst with EBX reagent **244** directly giving place to **II** or, in the case of an oxidative coupling, the alkynyl-[Au(I)] species **I** may form prior to oxidation with $PhI(OAc)_2$. In fact, Nevado and de Haro do observe the formation of species **I** by following the reaction by NMR. From intermediate **II**, auration of the nucleophilic heteroarene would give aryl-[Au(III)] species **III** which upon reductive elimination would furnish the alkynylated product.

Although we have decided to group this reaction in the context of an aryl–Au bond forming process, we realise that other mechanisms can be proposed that do not proceed via this step. With this in mind, we shall discuss catalytic cycles that proceed via more well-known C–C triple bond activation methods also (Scheme 63b). In this case, the intermediate **IV** could form through coordination of the π-bond in EBX **244** to the [Au(I)] catalyst or alternatively by ligand exchange of **I** with $PhI(OAc)_2$. This highly electrophilic alkyne species would then react with the arene to form **V**, akin to our discussion in section 2 on hydroarylations, and the collapse of **V** through *β*-elimination would give the product. Alternatively, the arene could attack at the *β* carbon to I(III), giving **VI**. In this scenario, carbene formation via an alpha elimination, followed by a 1,2-aryl shift, could afford the

[1] A SET process is not discussed here as it has previously been deemed unlikely; see ref. [155].

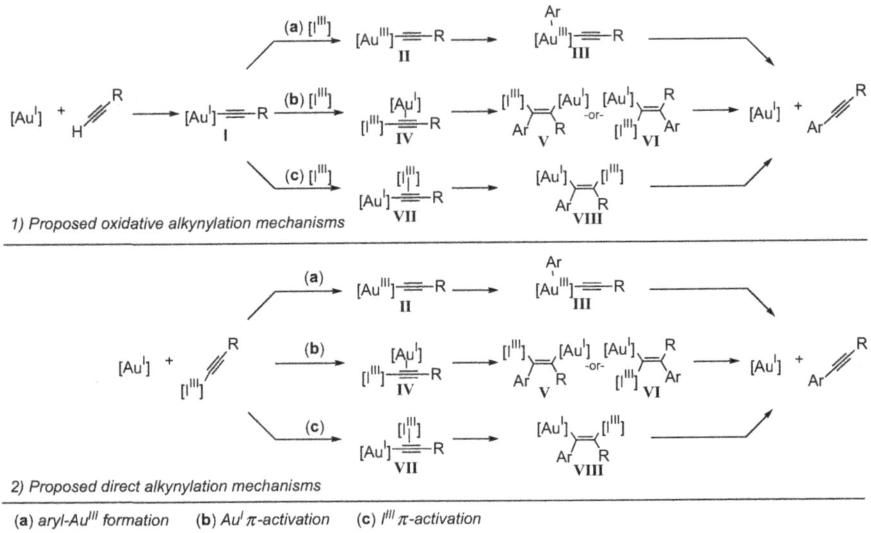

Scheme 63 Possible mechanistic pathways for the alkynylation of arenes via (a) aryl-[Au(III)] formation, (b) [Au(I)] π-activation and (c) [I(III)] π-activation

desired product.[2] In attempting to delineate the mechanism of this transformation, Waser and co-workers tested Lewis acid catalysts in place of AuCl, but no formation of product was observed, suggesting a more complex role for [Au(I)] [156].

In a similar fashion to pathway (b), species **VII** is a possible intermediate too (Scheme 63c). In this case, the role of the [Au(I)] and [I(III)] species is simply reversed in comparison to pathway (b). Although early reports had not considered this possibility, a more recent computational study found that, beginning from TIPS-EBX **244** and AuCl, the pathway via intermediate **VII** is more favourable than those via intermediates **II** or **IV**. [159].

4.2 Arylation

A seminal report by Russell, Lloyd-Jones and Ball described the ability of gold to catalyse the direct cross-coupling of aryl silanes with electron-rich arenes (Scheme 64) [160] (a more recent study has detailed the Au-catalysed direct arylation of phenylpyridines with aryl boronic acids; however, the authors do not report the C–H functionalisation of heteroarenes; see [161]). The procedure is particularly attractive as it proceeds at room temperature, requires near-equal stoichiometries of each reagent and does not require stringently dry or anaerobic conditions. As both coupling partners represent nucleophiles, an oxidant is required

[2] A 1,2-shift of the R group has been ruled out from ^{13}C labelling experiments; see ref. [155].

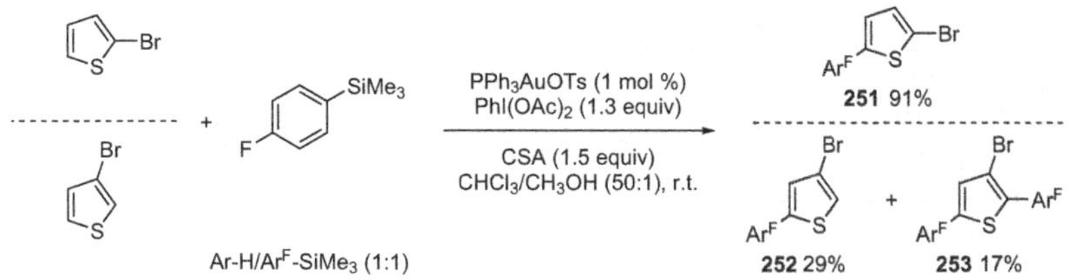

Scheme 64 Au-catalysed direct coupling of thiophenes and aryl silanes

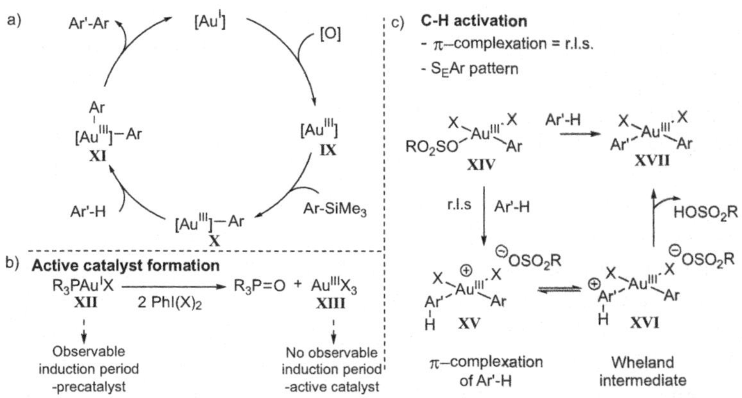

Scheme 65 (**a**) General catalytic cycle for the Au-catalysed direct coupling; (**b**) development of an improved [Au(III)] phosphineless catalyst; (**c**) investigation into the C–H auration step

in this process, which, alike with Nevado and de Haro's ethynylation, was found to be $PhI(OAc)_2$. The addition of camphorsulfonic acid (CSA) was also crucial for reactivity. The authors applied the coupling to a range of electron-rich arenes showing regioselectivity comparable to an electrophilic aromatic substitution process, suggesting that a [Au(III)] centre mediates C–H activation. Thiophenes were also suitable coupling partners, giving high yields of product in the case of **251**; however, significant quantities of bis-arylation could not be avoided with 3-bromothiophene (**252** and **253**). These examples also reveal tolerance to bromo functionality on the aromatic ring, providing clear routes for further functionalisation of the arene.

A thorough mechanistic study allowed the authors to provide strong evidence for their proposed catalytic cycle (Scheme 65a) [162]. Firstly, the observation of an induction period correlated with the consumption of 2 equivalents of oxidant, $PhI(X)_2$, and the formation of 1 equivalent of Ph_3PO, both with respect to the gold catalyst R_3PAuX **XII** (Scheme 65b). This suggested that a phosphineless Au(III) species **XIII**, formed from oxidation of both the R_3P ligand and Au(I), is the active catalyst in the cycle. Indeed, (tht)$AuBr_3$ was found to give similar yields as R_3PAuX catalysts and, importantly for kinetic measurements, proceeded with no observable induction period. Further studies established that the auration of the aryl

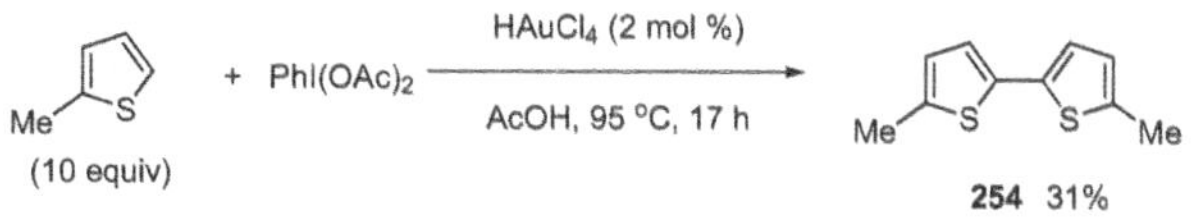

Scheme 66 Au-catalysed oxidative homocoupling of thiophene

silane occurs prior to auration of the arene (**IX** → **XI** in Scheme 65a), though both steps proceed from a [Au(III)] intermediate. The reaction exhibited first-order dependence on (tht)$AuBr_3$, arene and CSA, indicating the rate limiting step occurs at some point during the transformation of **X** → **XI**. More specifically, by plotting rate data against the relative stability of π-complexation intermediates, the authors could further speculate that coordination of the arene to [Au(III)] (**XIV** → **XV**) was the rate limiting step of the reaction (Scheme 65c).

Studies towards Au-catalysed oxidative couplings of arenes have also begun, but currently these methods are limited to homocoupling reactions [150, 151] (see also [163, 164]). Tse and co-workers reported the homocoupling of electron-rich arenes in similar conditions to the direct cross-coupling of aryl silanes with arenes, namely, $HAuCl_4$, $PhI(OAc)_2$ and AcOH (Scheme 66) [151]. Currently, this is the only description of a Au-catalysed sp^2 C–H/sp^2 C–H oxidative coupling; however, the procedure is far from ideal as large excess of the arene (10 equivalents) are required. Also, the coupling between heteroarenes gives, at present, only moderate yields of homo-coupled product **254**. Presumably the reaction proceeds through two C–H activation events at the [Au(III)] centre followed by reductive elimination and regeneration of the [Au(III)] catalyst by $PhI(OAc)_2$.

Aryl-[Au(I/III)] complexes have also been used as coupling partners in the direct arylation of nucleophilic arenes and heteroarenes. Larrosa and co-workers were the first to demonstrate this reactivity by exposing pre-isolated aryl-[Au(I)] complexes to PhI(OH)OTs and an electron-rich arene (Scheme 67) [165]. Most notable are the low levels of homocoupling products observed in all cases studied. In particular, thiophenes and furans (**255** and **256**) were successful coupling partners, reacting with **257** with high regioselectivity at the most nucleophilic position to give products **258** and **259** (Scheme 67a). The high chemoselectivity was further illustrated by performing the two C–H activation steps selectively in the presence of both arenes. Thus, an electron-deficient arene could be selectively aurated in the presence of an electron-rich arene using a [Au(I)] complex. Then, addition of PhI (OH)OTs to this mixture gave the cross-coupled biaryl in 47% yield with only traces of each homocoupling product being observed. Studies by Nevado and Hofer further demonstrated the feasibility of both aryl-[Au(III)] (**261**) and aryl-Au(I) complexes (**257**) to mediate cross-couplings with electron-rich arenes (Scheme 67b and c) [166]. In particular, *N*-methylindole was added to the list of electron-rich coupling partners for this transformation, giving place to C3-arylated product **260**.

The extension of these stoichiometric studies to a catalytic methodology that allows C–H/C–H cross couplings of arenes has yet to be realised. However, the orthogonal reactivity of [Au(I)]/[Au(III)] complexes to undergo C–H activation of

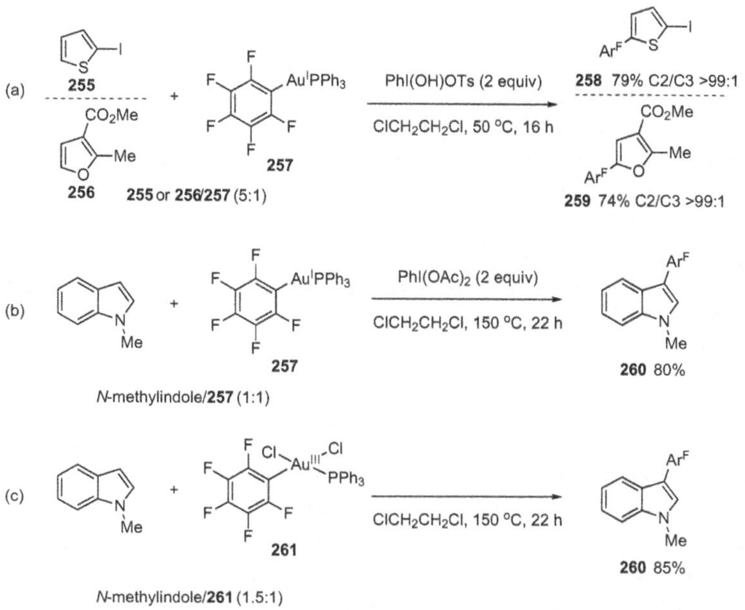

Scheme 67 Towards Au-catalysed oxidative (C–H/C–H) cross couplings by Larrosa (**a**) and Nevado with aryl-[Au(I)] (**b**) and aryl-[Au(III)]

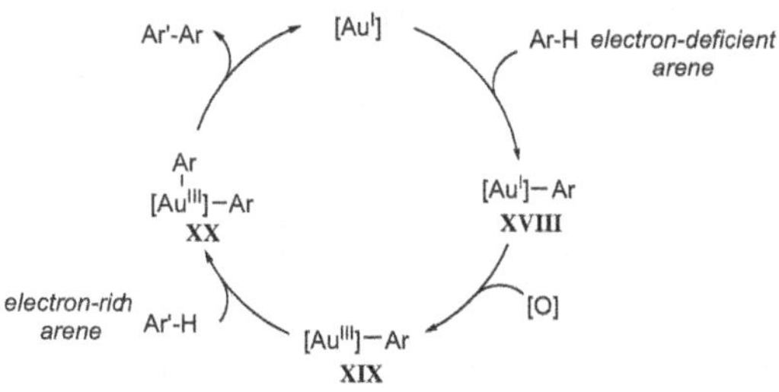

Scheme 68 Imagining Au-catalysed oxidative (C–H/C–H) cross-couplings

electron-deficient and electron-rich arenes, respectively, holds promise for the development of highly chemoselective oxidative cross couplings. A speculative catalytic cycle for this transformation may involve an initial C–H activation of an electron-deficient arene to form aryl-[Au(I)] species **XVIII** exclusively. Oxidation would then furnish the aryl-[Au(III)] species **XIX** that can then selectively react with an electron-rich arene (Scheme 68). Reductive elimination from **XX** would provide the desired cross-coupled product and close the catalytic cycle. During the preparation of this manuscript Larrosa and co-workers published a Au-catalysed oxidative cross-coupling of arenes [167].

4.3 Acyloxylation

Iodonium oxidants have been used frequently in Au catalysis. For example, in the previous section, we detailed the Au-catalysed oxidative homocoupling of arenes which required $PhI(OAc)_2$ to regenerate the active [Au(III)] catalyst (Scheme 69). Michelet, Toullec and Pradal speculated whether aryl-[Au(III)] species **XXI**, generated from the first C–H activation event, could undergo selective reductive elimination at this point to give acyloxylated arenes (Scheme 69, dotted line), rather than further reaction towards homocoupling products (Scheme 69, peripheral line) [168] (for polyacyloxylations, see [169]).

The authors found that the acyloxylation can proceed under similar conditions to those for the homocoupling reaction; however, it is the choice of substrate that dictates whether homocoupling or acyloxylation occurs; if the arene is highly sterically hindered, the arene will undergo acyloxylation rather than homocoupling. With this in mind, the reaction is limited to bulky electron-rich arenes, but several carboxylic acids were demonstrated as coupling partners, including sterically hindered *tert*-butyl carboxylic acid. On the topic of heteroarenes, indole **262** gave poor yield of the acyloxylated product **263** (Scheme 70). Although the reaction is limited to sterically hindered arenes, this shows complementary reactivity to palladium-catalysed acyloxylations which are poorly reactive with bulky substrates [170].

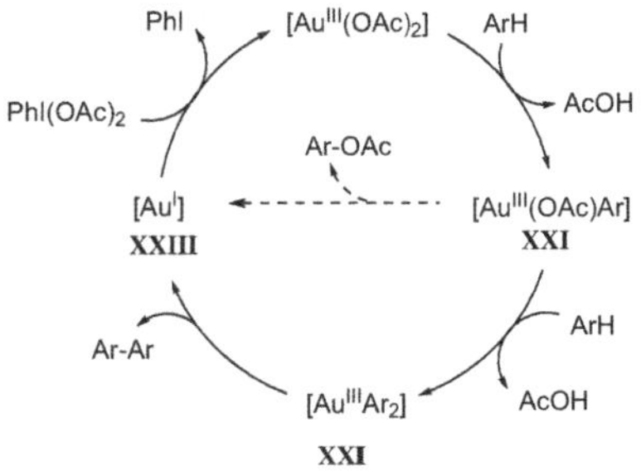

Scheme 69 Proposed cycles for the oxidative homocoupling of arenes and acyloxylation of arenes

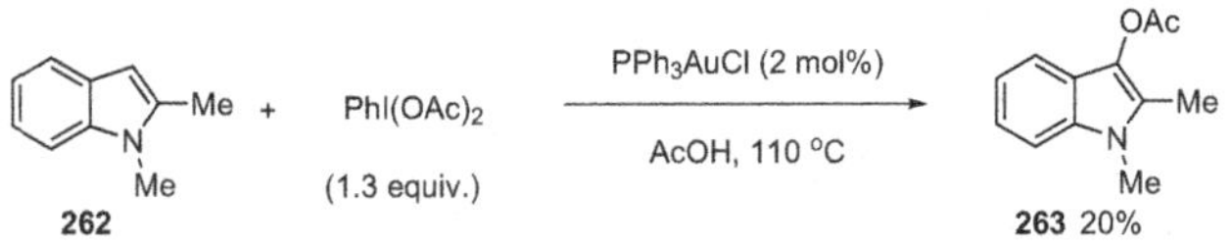

Scheme 70 Au-catalysed acyloxylation of 1,2 dimethylindole

4.4 Carboxylation

Nolan et al. developed a [Au(I)]-catalysed carboxylation of heteroaromatics bearing acidic C–H bonds with CO_2 to give carboxylic acids **264** (Scheme 71) [171]. In most cases, IPrAuOH **265** is used as catalyst, but for less acidic arenes, the more basic I^{t}BuAuOH **266** can be used. Remarkably, the transformation occurs at ambient temperature and under near atmospheric pressure of CO_2. Other methods for the C–H carboxylation of heteroarenes have been reported but require higher temperatures (>65°C) and show restricted substrate scope, for example, imidazole and benzimidazole are poorly reactive under transition metal-free and Cu-catalysed procedures [172–174]. The difference in reactivity is likely due to the more basic nature of the Au-hydroxide catalyst (IPrAuOH, $pK_a = 30.3$ and I^{t}BuAuOH, $pK_a = 32.4$ vs. IPrCuOH, $pK_a = 27.7$), allowing it to activate arenes with less acidic C–H bonds. Overall, 13 heteroarenes were successfully carboxylated including (benz)oxazoles (benzo)thiazoles and methyl-adenine.

The isolation of several possible intermediates and their exposure to the reaction conditions allowed the authors to propose a catalytic cycle according to Scheme 72. Firstly, the aryl-[Au(I)] species **XXIV** was prepared through the reaction of IPrAuOH (**265**) with oxazole, suggesting a C–H auration step occurs during the catalytic cycle. Remarkably, exposure of **XXIV** to 5 bar CO_2 at −78 °C allowed the authors to isolate the [Au(I)]-carboxylate **XXV**. Finally, the addition of KOH gave the potassium carboxylate salt of the product and regenerated IPrAuOH.

The feasibility of Nolan's proposed mechanism has been a subject of debate in the literature [175, 176]. Firstly, Geng et al. conducted a DFT study of the carboxylation of oxazole and found that the calculated barriers agreed with experimental observation; however, they propose that carboxylation occurs from a carbene intermediate **XXVIII** rather than the aryl-[Au(I)] species **XXVI** (Scheme 73a vs. b) [177]. The formation of the carbene intermediate **XXVIII** is the highest energy barrier in this case (31.7 kcal mol^{-1}), with all subsequent steps, including carboxylation, of lower energy. Conversely, a barrier of 39.9 kcal mol^{-1} was calculated for the carboxylation of oxazolyl-Au(I) **XXVI**, 8.2 kcal mol^{-1} higher in energy than formation of the carbene intermediate **XXVIII**. Ahlquist, Wendt and co-workers were also interested in delineating the mechanism as they

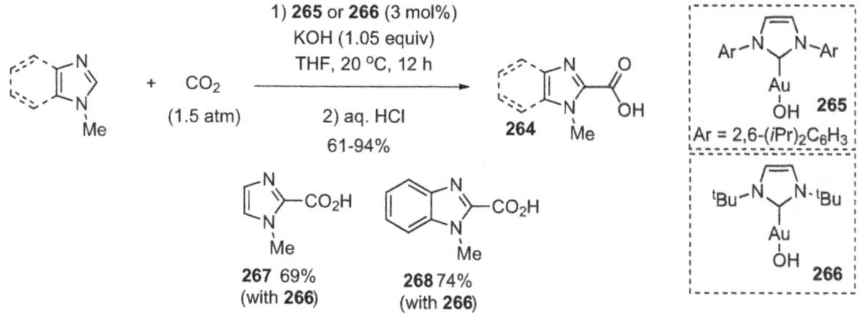

Scheme 71 Au-catalysed carboxylation of imidazole and benzimidazole

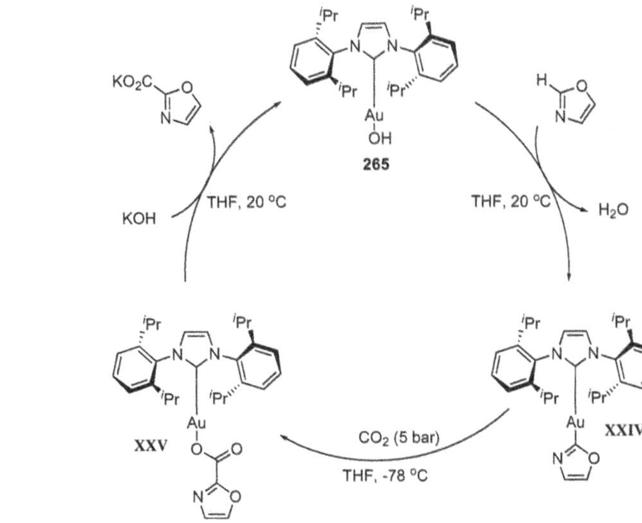

Scheme 72 Proposed mechanism for the carboxylation of oxazole

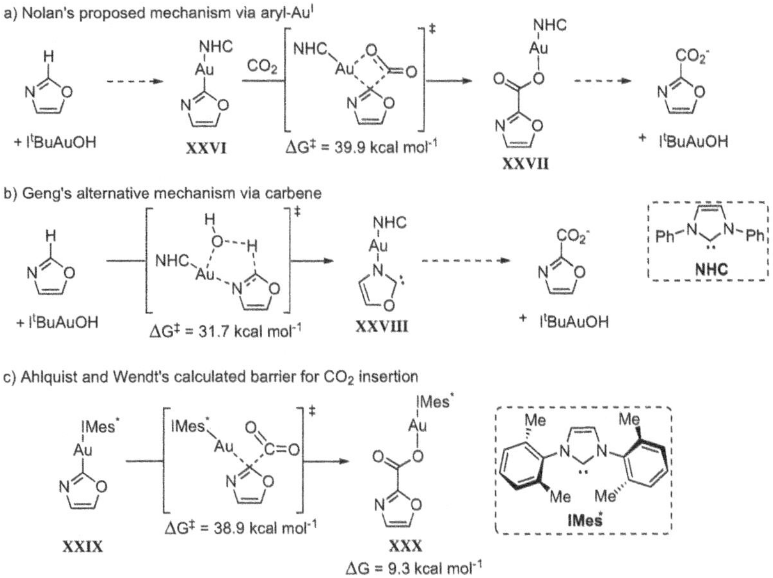

Scheme 73 Computational studies on the carboxylation of oxazole via (**a**) an aryl-Au^I intermediate and (**b**) a carbene intermediate. Only the highest calculated energy barrier for each pathway has been shown. (**c**) Carboxylation of oxazoyl-[Au(I)] **XXIX**

had found the carboxylation reported by Nolan to be irreproducible in their hands [178]. Indeed, the carboxylation of aryl-[Au(I)] species, IPrAuPh, did not undergo reaction even at 3 bar CO_2, 100°C for 5 h in C_6D_6. Related computational studies revealed a barrier for carboxylation of 38.9 kcal mol^{-1} from IMes*Au(I)-oxazoyl **XXIX**, further indicating that a different mechanism is operating (Scheme 73c).[3] Interestingly, they also calculated this transformation to be endergonic ($\Delta G = 9.3$ kcal mol^{-1}) and suggested that the formation of Au-carboxylate **XXX** should not be possible regardless of mechanism.

5 Conclusion and Future Perspectives

Gold catalysis has seen tremendous developments in the last two decades, and this chapter has outlined both early and more recent advances of its application to the C–H functionalisation of heterocycles. Although previously significantly less explored than Pd or Pt catalysts, Au catalysts have now been shown to span a diverse set of reaction types from transformations relying on Lewis acid activation of π-bonds and heteroatoms, to the use of gold catalysts in redox reactions. In addition, the specific properties of Au can provide advantages over other transition metal catalysts, for example, its reluctance to participate in oxidative addition allows for substrates containing carbon-halogen bonds to undergo C–H functionalisation with retention of the halogen functionality intact. Furthermore, the use of Au catalysts in cyclisation reactions of multiple bonds often delivers selectivity different to that obtained using other transition metal catalysts, demonstrating a unique reactivity of Au. In other examples, Au catalysts have shown superior reactivity and selectivity to other Lewis or Brønsted acids, though in some cases, a less costly alternative may be preferable.

Hydroheteroarylation of various π-bonds now provides access to complex polycyclic heteroaryl structures, and studies in the general area of hydroarylation has uncovered fascinating mechanisms and intermediates. However, the reaction outcome can be highly substrate specific, and a better understanding of the mechanisms governing these processes is expected to allow for the development of more general and predictable reactivity as well as improved enantioselective versions which so far are less common.

The ability to functionalise electron-rich arenes by C–H activation (auration) at [Au(III)] represents an exciting development in Au catalysis. It is largely due to the resistance of [Au(I)] to [Au(III)] oxidation that catalytic examples of coupling reactions proceeding via such pathways have only recently been developed. Nevertheless, the use of suitable oxidants, in particular iodonium-based oxidants, has allowed for the first catalytic examples of this type, including alkynylation, biaryl formation and heteroatom C–H bond functionalisation. The C–H auration of electron-deficient arenes at [Au(I)] has also been demonstrated, opening up for

[3] IMes*- For the computational studies, the mesityl groups in the IMes ligand was replaced with 2,6-xylyl groups. IMes = (1,3-bis(2,4,6-trimethylphenyl)imidazol-2-ylidene); IMes* = (1,3-bis (2,4-dimethylphenyl)imidazol-2-ylidene)

interesting substrate selectivity between the [Au(I)] and [Au(III)] oxidation states. The vast potential in this relatively young area and the growing number of emerging transformations will undoubtedly deliver new, unique and selective Au-catalysed C–H bond functionalisations in the future.

References

1. Trost BM (1995) Angew Chem Int Ed 34:259–281
2. Gimeno MC, Laguna A (1997) Chem Rev 97:511–522
3. Schmidbaur H, Cronje S, Djordjevic B, Schuster O (2005) Chem Phys 311:151–161
4. Boorman TC, Larrosa I (2011) Chem Soc Rev 40:1910–1925
5. de Haro T, Nevado C (2011) Synthesis 2530–2539
6. Hopkinson MN, Gee AD, Gouverneur V (2011) Chemistry 17:8248–8262
7. Skouta R, Li CJ (2008) Tetrahedron 64:4917–4938
8. Jiménez-Núñez E, Echavarren AM (2007) Chem Commun 43:333–346
9. Sohel SMA, Liu RS (2009) Chem Soc Rev 38:2269–2281
10. Muratore ME, Homs A, Obradors C, Echavarren AM (2014) Chem Asian J 9:3066–3082
11. Arcadi A (2008) Chem Rev 108:3266–3325
12. Li Z, Brouwer C, He C (2008) Chem Rev 108:3239–3265
13. Hashmi ASK (2007) Chem Rev 107:3180–3211
14. Fürstner A (2009) Chem Soc Rev 38:3208–3221
15. Shapiro ND, Toste FD (2010) Synlett 2010:675–691
16. Dorel R, Echavarren AM (2015) Chem Rev 115:9028–9072
17. Obradors C, Echavarren AM (2014) Acc Chem Res 47:902–912
18. Obradors C, Echavarren AM (2014) Chem Commun 50:16–28
19. Jiménez-Núñez E, Echavarren AM (2008) Chem Rev 108:3326–3350
20. Michellet V, Toullec PY, Genêt JP (2008) Angew Chem Int Ed 47:4268–4315
21. Ma S, Yu S, Gu Z (2006) Angew Chem Int Ed 45:200–203
22. Michelet V, Toullec PY, Genêt JP (2008) Angew Chem Int Ed 47:4268–4315
23. Chiarucci M, Bandini M (2013) Beilstein J Org Chem 9:2586–2614
24. Yang W, Hashmi ASK (2014) Chem Soc Rev 43:2941–2955
25. Krause N, Winter C (2011) Chem Rev 111:1994–2009
26. Shen HC (2008) Tetrahedron 64:3885–3903
27. Bandini M (2011) Chem Soc Rev 40:1358–67
28. Fürstner A, Davies PW (2007) Angew Chem Int Ed 46:3410–3449
29. de Mendoza P, Echavarren AM (2010) Pure Appl Chem 82:801–820
30. Hashmi ASK (2007) Catal Today 122:211–214
31. Kanno O, Kuriyama W, Wang ZJ, Toste FD (2011) Angew Chem 123:10093–10096
32. Brooner REM, Robertson BD, Widenhoefer RA (2014) Organometallics 33:6466–6473
33. Dyker G, Muth E, Hashmi ASK, Ding L (2003) Adv Synth Catal 345:1247–1252
34. Homs A, Escofet I, Echavarren AM (2013) Org Lett 15:5782–5785
35. Mézailles N, Ricard L, Gagosz F (2005) Org Lett 7:4133–4136
36. Hashmi ASK, Schwarz L, Choi J-H, Frost TM (2000) Angew Chem Int Ed 122:2285–2288
37. Aguilar D, Navarro R, Urriolabeitia EP (2007) Organometallics 26:4604–4611
38. Aguilar D, Contel M, Navarro R, Soler T, Urriolabeitia EP (2009) J Organomet Chem 694:486–493
39. Arcadi A, Bianchi G, Chiarini M, D'Anniballe G, Marinelli F (2004) Synlett 6:944–950
40. Alfonsi M, Arcadi A, Aschi M, Bianchi G, Marinelli FJ (2005) J Org Chem 70:2265–2273
41. Hashmi ASK, Salathé R, Frey W (2007) Eur J Org Chem 10:1648–1652
42. Li Z, Shi Z, He C (2005) J Organomet Chem 690:5049–5054

43. Alfonsi M, Arcadi A, Bianchi G, Marinelli F, Nardini A (2006) Eur J Org Chem 10:2393–2402
44. Pirovano V, Dell'Acqua M, Facoetti D, Nava D, Rizzato S, Abbiati G, Rossi E (2013) Eur J Org Chem 28:6267–6279
45. Pirovano V, Facoetti D, Dell'Acqua M, Della Fontana E, Abbiati G, Rossi E (2013) Org Lett 15:3812–3815
46. Wang M-Z, Wong M-K, Che C-M (2008) Chem Eur J 14:8353–8364
47. Xiao Y-P, Liu X-Y, Che C-M (2009) J Organomet Chem 694:494–501
48. Gandon V, Lemière G, Hours A, Fensterbank L, Malacria M (2008) Angew Chem Int Ed 47:7534–7538
49. Zhang Z, Liu C, Kinder RE, Han X, Qian H, Widenhoefer RA, Carolina N (2006) J Am Chem Soc 128:9066–9073
50. Liu Z, Wasmuth AS, Nelson SG (2006) J Am Chem Soc 128:10352–10353
51. Qiu Y, Zhou J, Li J, Fu C, Guo Y, Wang H, Ma S (2015) Chem Eur J 21:15939–15943
52. Tarselli MA, Gagné MR (2008) J Org Chem 73:2439–2441
53. Weber D, Tarselli MA, Gagné MR (2009) Angew Chem Int Ed 48:5733–5736
54. Weber D, Gagné MR (2013) Chem Sci 4:335–338
55. Brown TJ, Weber D, Gagné MR, Widenhoefer RA (2012) J Am Chem Soc 134:9134–9137
56. Barluenga J, Piedrafita M, Ballesteros A, Suárez-Sobrino ÁL, González JM (2010) Chem Eur J 16:11827–11831
57. Chen B, Fan W, Chai G, Ma S (2012) Org Lett 14:3616–3619
58. Qiu Y, Fu C, Zhang X, Ma S (2014) Chem Eur J 20:10314–10322
59. Alcaide B, Almendros P, Alonso JM, Quirós MT, Gadziński P (2011) Adv Synth Catal 353:1871–1876
60. Alvarez E, García-García P, Fernández-Rodríguez MA, Sanz RJ (2013) J Org Chem 78:9758–9771
61. Rudolph M, Hashmi ASK (2012) Chem Soc Rev 41:2448–2462
62. Zeldin RM, Toste FD (2011) Chem Sci 2:1706–1709
63. Zhang Z, Widenhoefer RA (2008) Org Lett 10:2079–2081
64. Toups KL, Liu GT, Widenhoefer RA (2009) J Organomet Chem 694:571–575
65. Kona CN, Shinde MH, Ramana CV (2015) Org Biomol Chem 13:5358–5362
66. Pirovano V, Decataldo L, Rossi E, Vicente R (2013) Chem Commun 49:3594–3596
67. Liu C, Widenhoefer RA (2007) Org Lett 9:1935–1938
68. Wang M-Z, Zhou C-Y, Guo Z, Wong EL-M, Wong M-K, Che C-M (2011) Chem Asian J 6:812–824
69. Reetz MT, Sommer K (2003) Eur J Org Chem 2003:3485–3496
70. Ferrer C, Amijs CHM, Echavarren AM (2007) Chem Eur J 13:1358–1373
71. Hashmi ASK, Blanco MC (2006) Eur J Org Chem 2006:4340–4342
72. Hashmi ASK, Frost TM, Bats JW (2000) J Am Chem Soc 122:11553–11554
73. Hashmi ASK, Frost TM, Bats JW (2001) Org Lett 3:3769–3771
74. Hashmi ASK, Blanco MC, Kurpejović E, Frey W, Bats JW (2006) Adv Synth Catal 348:709–713
75. Huguet N, Leboeuf D, Echavarren AM (2013) Chem Eur J 19:6581–6585
76. Hashmi ASK, Weyrauch JP, Kurpejović E, Frost TM, Miehlich B, Frey W, Bats JW (2006) Chem Eur J 12:5806–5814
77. Hashmi ASK, Kurpejović E, Frey W, Bats JW (2007) Tetrahedron 63:5879–5885
78. Rudolph M, McCreery MQ, Frey W, Hashmi ASK (2011) Beilstein J Org Chem 7:794–801
79. Hashmi ASK, Haufe P, Schmid C, Rivas Nass A, Frey W (2006) Chem Eur J 12:5376–5382
80. Hashmi ASK, Rudolph M, Siehl HU, Tanaka M, Bats JW, Frey W (2008) Chem Eur J 14:3703–3708
81. Hashmi ASK, Häffner T, Rudolph M, Rominger F (2011) Chem Eur J 17:8195–8201
82. Chen Y, Li G, Liu Y (2011) Adv Synth Catal 353:392–400
83. Chen Y, Liu Y (2011) J Org Chem 76:5274–5282

84. Chen Y, Lu Y, Li G, Liu Y (2011) Org Lett 11:3838–3841
85. Barluenga J, Fernández A, Rodríguez F, Fañanás FJ (2009) J Organomet Chem 694:546–550
86. Samala S, Mandadapu AK, Saifuddin M, Kundu B (2013) J Org Chem 78:6769–6774
87. Black DSC, Craig DC, Kumar N (1991) Tetrahedron Lett 32:1587–1590
88. Praveen C, Perumal PT (2011) Synlett 2011:521–524
89. Qiu Y, Kong W, Fu C, Ma S (2012) Org Lett 14:6198–6201
90. Qiu Y, Zhou J, Fu C, Ma S (2014) Chem Eur J 20:14589–14593
91. Qiu Y, Ma D, Fu C, Ma S (2013) Org Biomol Chem 11:1666–1671
92. Kong W, Fu C, Ma S (2011) Chem Eur J 17:13134–13137
93. Hashmi ASK, Yang W, Rominger F (2011) Angew Chem Int Ed 50:5762–5765
94. Sun N, Xie X, Liu Y (2014) Chem Eur J 20:7514–7519
95. Matsude Y, Naoe S, Oishi S, Fujii N, Ohno H (2015) Chem Eur J 21:1463–1467
96. Hirano K, Inaba Y, Watanabe T, Oishi S, Fujii N, Ohno H (2010) Adv Synth Catal 352:368–372
97. Hirano K, Inaba Y, Takahashi N, Shimano M, Oishi S, Fujii N, Ohno H (2011) J Org Chem 76:1212–1227
98. Ferrara G, Jin T, Oniwa K, Zhao J, Asiri AM, Yamamoto Y (2012) Tetrahedron Lett 53:914–918
99. Wang L, Li G, Liu Y (2011) Org Lett 13:3786–3789
100. Ferrer C, Echavarren AM (2006) Angew Chem Int Ed 45:1105–1109
101. Liu Z, Zhang L (2012) Angew Chem Int Ed 51:7301–7304
102. Hashmi ASK, Yang W, Rominger F (2012) Adv Synth Catal 354:1273–1279
103. Pan B, Lu X, Wang C, Hu Y, Wu F, Wan B (2012) Org Lett 16:2244–2247
104. Loh CCJ, Badorrek J, Raabe G, Enders D (2011) Chem Eur J 17:13409–13414
105. Hack D, Loh CCJ, Hartmann JM, Raabe G, Enders D (2014) Chem Eur J 20:3917–3921
106. Peshkov VA, Pereshivko OP, Van der Eycken EV (2012) Adv Synth Catal 354:2841–2848
107. Modha SG, Kumar A, Vachhani DD, Jacobs J, Sharma SK, Parmar VS, Van Meervelt L, Van der Eycken EV (2012) Angew Chem Int Ed 51:9572–9575
108. Kimar A, Vachhani DD, Modha SG, Sharma SK, Parmar VS, Van der Eycken EV (2013) Beilstein J Org Chem 9:2097–2102
109. Kumar A, Li Z, Sharma SK, Parmar VS, Van der Eycken EV (2013) Chem Commun 49:6803–6805
110. Modha SG, Kumar A, Vachhani DD, Sharma SK, Parmar VS, Van der Eycken EV (2012) Chem Commun 48:10916–10918
111. Kumar A, Vachhani DD, Modha SG, Sharma SK, Parmar VS, Van der Eycken EV (2013) Synthesis 45:2571–2582
112. Kumar A, Vachani DD, Modha SG, Sharma SK, Parmar VS, Van der Eycken EV (2013) Eur J Org Chem 2013:2288–2292
113. Li Z, Kumar A, Vachhani DD, Sharma SK, Parmar VS, Van der Eycken EV (2014) Eur J Org Chem 2014:2084–2091
114. Xie X, Du X, Chen Y, Liu Y (2011) J Org Chem 76:9175–9181
115. Heffernan SJ, Tellam JP, Queru ME, Silvanus AC, Benito D, Mahon MF, Hennessy AJ, Andrews BI, Carbery DR (2013) Adv Synth Catal 355:1149–1159
116. Boiaryna L, El Mkaddem MK, Taillier C, Dalla V, Othman M (2012) Chem Eur J 18:14192–14200
117. Boiaryna L, Azizi MS, El Bouakher A, Picard B, Taillier C, Othman M, Trabelsi-Ayadi M, Dalla V (2015) Org Lett 17:2130–2133
118. Lu Y, Du X, Jia X, Liu Y (2009) Adv Synth Catal 351:1517–1522
119. Georgy M, Boucard V, Campagne J-M (2005) J Am Chem Soc 127:14180–14181
120. Mertins K, Iovel I, Kischel J, Zapf A, Beller M (2006) Adv Synth Catal 348:691–695
121. Rao W, Chan PWH (2008) Org Biomol Chem 6:2426–2433
122. Topinka M, Tata RR, Harmata M (2014) Org Lett 16:4476–4479
123. Rubenbauer P, Bach T (2008) Adv Synth Catal 350:1125–1130

124. Mühlthau F, Stadler D, Goeppert A, Olah GA, Prakash GKS, Bach T (2006) J Am Chem Soc 128:9668–9675
125. Bandini M, Eichholzer A (2009) Angew Chem Int Ed 48:9533–9537
126. Bandini M, Gualandi A, Monari M, Romaniello A, Savoia D, Tragni M (2011) J Organomet Chem 696:338–347
127. Bandini M, Bottoni A, Chiarucci M, Cera G, Miscione GP (2012) J Am Chem Soc 134:20690–20700
128. Luo Y, Li C-J (2004) Chem Commun 17:1930–1931
129. Sun X, Sun W, Fan R, Wu J (2007) Adv Synth Catal 349:2151–2155
130. Zhang Z, Shi M (2010) Chem Eur J 16:7725–7729
131. Nair V, Abhilash KG, Vidya N (2005) Org Lett 26:5857–5859
132. Hashmi ASK, Schwarz L, Rubenbauer P, Blanco MC (2006) Adv Synth Catal 348:705–708
133. Arcadi A, Alfonsi M, Bianchi G, D'Anniballe G, Marinelli F (2006) Adv Synth Catal 348:331–338
134. Barluenga J, Lonzi G, Tomás M, López L (2013) Chem Eur J 19:1573–1576
135. Xi Y, Su Y, Yu Z, Dong B, McClain EJ, Lan Y, Shi X (2014) Angew Chem Int Ed 53:9817–9821
136. Fructos MR, Belderrain TR, de Frémont P, Scott NM, Nolan SP, Díaz-Requejo MM, Pérez PJ (2005) Angew Chem Int Ed 44:5284–5288
137. Rivilla I, Gómez-Emeterio BP, Fructos MR, Díaz-Requejo MM, Pérez PJ (2011) Organometallics 30:2855–2860
138. Yu Z, Ma B, Chen M, Wu H, Liu L, Zhang J (2014) J Am Chem Soc 136:6904–6907
139. Kharasch MS, Isbell HS (1931) J Am Chem Soc 53:3053–3059
140. Kharasch MS, Beck TM (1934) J Am Chem Soc 56:2057–2060
141. Liddle KS, Parkin C (1972) J Chem Soc Chem Commun 26a
142. Calderazzo F, Dell'Amico DB (1974) J Organomet Chem 76:C59–C60
143. de Graaf PWJ, Boersma J, van der Kerk GJM (1976) J Organomet Chem 105:399–406
144. Lu P, Boorman TC, Slawin AMZ, Larrosa I (2010) J Am Chem Soc 132:5580–5581
145. Gaillard S, Slawin AMZ, Nolan SP (2010) Chem Commun 46:2742–2744
146. Ahlsten N, Perry GJP, Cambeiro XC, Boorman TC, Larrosa I (2013) Catal Sci Technol 3:2892–2897
147. Fuchita Y, Utsunomiya Y, Yasutake M (2001) J Chem Soc Dalton Trans 2330–2334
148. De Haro T, Nevado C (2010) J Am Chem Soc 132:1512–1513
149. Leyva-Pérez A, Oliver-Meseguer J, Cabrero-Antonino JR, Rubio-Marqués P, Serna P, Al-Resayes SI, Corma A (2013) ACS Catal 3:1865–1873
150. Kar A, Mangu N, Kaiser HM, Beller M, Tse MK (2008) Chem Commun 386–388
151. Kar A, Mangu N, Kaiser HM, Tse MK (2009) J Organomet Chem 694:524–537
152. Hopkinson MN, Gee AD, Gouverneur V (2011) Chem Eur J 17:8248–8262
153. Brand JP, Li Y, Waser J (2013) Isr J Chem 53:901–910
154. Brand JP, Charpentier J, Waser J (2009) Angew Chem Int Ed 48:9346–9349
155. Brand JP, Waser J (2010) Angew Chem Int Ed 49:7304–7307
156. Brand JP, Chevalley C, Scopelliti R, Waser J (2012) Chem Eur J 18:5655–5666
157. Li Y, Brand JP, Waser J (2013) Angew Chem Int Ed 52:6743–6747
158. Brand JP, Waser J (2012) Org Lett 14:744–747
159. Ariafard A (2014) ACS Catal 4:2896–2907
160. Ball LT, Lloyd-Jones GC, Russell CA (2012) Science 337:1644–1648
161. Wu Q, Du C, Huang Y, Liu X, Long Z, Song F, You J (2015) Chem Sci 6:288–293
162. Ball LT, Lloyd-Jones GC, Russell CA (2014) J Am Chem Soc 136:254–264
163. Ishida T, Aikawa S, Mise Y, Akebi R, Hamasaki A, Honma T, Ohashi H, Tsuji T, Yamamoto Y, Miyasaka M, Yokoyama T, Tokunaga M (2015) ChemSusChem 8:695–701
164. Perea-Buceta JE, Wirtanen T, Laukkanen O-V, Mäkelä MK, Nieger M, Melchionna M, Huittinen N, Lopez-Sanchez JA, Helaja J (2013) Angew Chem Int Ed 52:11835–11839
165. Cambeiro XC, Boorman TC, Lu P, Larrosa I (2013) Angew Chem Int Ed 52:1781–1784

166. Hofer M, Nevado C (2013) Tetrahedron 69:5751–5757
167. Cambeiro XC, Ahlsten N, Larrosa I (2015) J Am Chem Soc 137:15636–15639
168. Pradal A, Toullec PY, Michelet V (2011) Org Lett 13:6086–6089
169. Pradal A, Faudot dit Bel P, Toullec PY, Michelet V (2012) Synthesis 44:2463–2468
170. Yoneyama T, Crabtree RH (1996) J Mol Catal A Chem 108:35–40
171. Boogaerts IIF, Nolan SP (2010) J Am Chem Soc 132:8858–8859
172. Zhang L, Cheng J, Ohishi T, Hou Z (2010) Angew Chem Int Ed 49:8670–8673
173. Boogaerts IIF, Fortman GC, Furst MRL, Cazin CSJ, Nolan SP (2010) Angew Chem Int Ed 49:8674–8677
174. Vechorkin O, Hirt N, Hu X (2010) Org Lett 12:3567–3569
175. Yeung CS, Dong VM (2014) Top Catal 57:1342–1350
176. Yang L, Wang H (2014) ChemSusChem 7:962–998
177. Zhang X, Geng Z, Wang Y, Hou X, Wang D (2012) J Mol Catal A Chem 363–364:31–40
178. Johnson MT, van Rensburg JMJ, Axelsson M, Ahlquist MSG, Wendt OF (2011) Chem Sci 2:2373–277

Top Heterocycl Chem (2016) 46: 227–248
DOI: 10.1007/7081_2015_5003

Published online: 12 December 2015

Enantioselective Gold-Catalyzed Synthesis of Heterocyclic Compounds

Dillon H. Miles and F. Dean Toste

Abstract Asymmetric gold-catalyzed syntheses of a diverse range of heterocycles are described herein. This chapter outlines efforts toward the construction of these molecules via intermolecular, intramolecular, and tandem cyclization strategies. Additionally, asymmetric hydrogenation and epoxidation approaches are detailed. These methodologies generally utilize substrates containing soft, unsaturated carbon-carbon bonds, which can easily interact with the carbophilic gold atom(s). Mostly through the use of bis- or monophosphine ligands, modest to excellent yields and enantioselectivities are obtained for a variety of partially and fully saturated heterocycles. Although steric variation of the phosphine catalyst is the most commonly used strategy for enantioinduction, the use of chiral counteranions has proven useful for more challenging transformations. This chapter includes only examples of syntheses where a chiral gold catalyst is used in the direct formation of the heterocycle; examples of pendant functionalization of an existing heterocycle or chirality transfer from enantioenriched substrate to product are not included.

Keywords Alkenes • Alkynes • Allenes • Asymmetric catalysis • Gold • Hydroalkoxylation • Hydroamination • Hydroarylation • Oxazolines • Pyrrolidines • Tetrahydrofurans

Contents

D.H. Miles and F.D. Toste (✉)
Department of Chemistry, University of California, Berkeley, CA 94720, USA
e-mail: fdtoste@berkeley.edu

Abbreviations

BINAP	(1,1′-Binaphthalene-2,2′-diyl)bis(diphenylphosphine)
BINOL	1,1′-Bi-2-naphthol
C3-TunePhos	1,13-Bis(diphenylphosphino)-7,8-dihydro-6H-dibenzo[f,h][1,5]dioxonin
Cl-MeO-BIPHEP	(5,5′-Dichloro-6,6′-dimethoxy-1,1'-biphenyl)-2,2′-diyl-bis(diphenylphosphine)
DCE	1,2-Dichloroethane
DCM	Dichloromethane
DIPAMP	1,2-Bis[(2-methoxyphenyl)(phenylphosphino)]ethane
DM-BINAP	1,1′-Binaphthalene-2,2′-diyl)bis[bis(3,5-dimethylphenyl)phosphine]
DM-SEGPHOS	5,5′-Bis[di(3,5-xylyl)phosphino]-4,4′-bi-1,3-benzodioxole
dppm	1,1-Bis(diphenylphosphino)methane
DTBM-MeO-BIPHEP	2,2′-Bis[di(3,5-di-t-butyl-4-methoxyphenyl)phosphino]-6,6′-dimethoxy-1,1′-biphenyl
DTBM-SEGPHOS	5,5′-Bis[di(3,5-di-tert-butyl-4-methoxyphenyl)phosphino]-4,4′-bi-1,3-benzodioxole
ee	Enantiomeric excess
IMes	1,3-Bis(2,4,6-trimethylphenyl)imidazol-2-ylidene
NAC	Nitrogen acyclic carbene
NHC	*N*-Heterocyclic carbene
PNB	4-Nitrobenzoate
SEGPHOS	5,5′-Bis(diphenylphosphino)-4,4′-bi-1,3-benzodioxole
SPINOL	1,1′-Spirobiindane-7,7′-diol
TADDOL	α,α,α,α-Tetraaryl-1,3-dioxolane-4,5- dimethanol
TFA	Trifluoroacetate
THF	Tetrahydrofuran
TRIP	3,3′-Bis(2,4,6-triisopropylphenyl)-1,1′-binaphthyl-2,2′-diyl phosphate

1 Introduction

In the past few decades, the construction of enantioenriched heterocyclic molecules has remained an area of intensive research within the synthetic community. Their ubiquitous presence in both medicinal and naturally occurring compounds justifies the discovery of new and improvement of existing catalytic methodologies in this field [1, 2]. Many functionalized heterocycles can be produced using homogeneous organo- or transition metal catalysis. More specifically, asymmetric gold catalysis, first applied in the 1986 and again popularized in the early 2000s, has become a powerful tool for the modern organic chemist with a variety of reaction manifolds [3–7].

Utilizing the inherent carbophilicity and Lewis acidity of cationic gold, several strategies for heterocycle formation have surfaced during the growth of this field: intermolecular cycloadditions mainly utilizing 1,3-dipolar molecules, intramolecular cyclizations utilizing nucleophilic heteroatoms, and intramolecular cyclizations with heteroatom linkers. These three strategies have proven uniquely valuable methods to produce diverse heterocyclic space.

As will be demonstrated throughout this survey, enantioinduction strategies primarily rely on the use of enantiopure bimetallic bisphosphine gold (I) complexes to impart high levels of selectivity, generally relying on unfavorable steric interactions in the transition state for enantiodifferentiation. Due to their additional modularity, BINOL-based phosphoramidite ligands have also emerged as competent scaffolds, frequently surpassing their bisphosphine counterparts in terms of enantioselectivity and reactivity.

In many scenarios, the linear environment of the gold(I) complex places the reactive center relatively far away from the source of chiral information on the phosphorous ligand, resulting in nearly racemic products depending on the substrate used. One final strategy, introduced by Toste and coworkers in 2007, seeks to solve this problem by changing the counteranion, typically achiral and non-coordinating, to a BINOL-based phosphate, relocating the source of chiral information to a potentially more productive location in the transition state.

Owing to the diversity of gold-catalyzed transformations, this summary is organized by related cyclization strategies published before July 2015. Examples where a preexisting heterocycle is further functionalized resulting in the formation of a pendant acyclic or carbocyclic stereocenter have been omitted to sharpen the focus of the chapter. We have also excluded examples where another catalytic component sets the stereocenter, but an achiral gold catalyst performs the cyclization step. In the case of chiral phosphoric acid-catalyzed reactions, this is sometimes ambiguous; only examples including evidence for gold chiral phosphate catalyst competency are included (i.e., the ee is comparable). These examples are presented in Sect. 4. Finally, examples demonstrating transfer of chirality from an enantioenriched substrate to the desired heterocycle have been omitted, since an achiral gold catalyst is utilized. A review covering this subfield has been published [8].

2 Intermolecular Cycloadditions

The first example of asymmetric gold-catalyzed formation of heterocycles was reported in 1986 by Hayashi et al., utilizing a chiral ferrocenylphosphine ligand for the synthesis of enantioenriched oxazolines. Depending on aldehyde identity, the authors usually obtain the *trans* product with enantioselectivities up to 97% (Scheme 1) [9]. The presence and length of the diamine side chain was critical, as other bisphosphine ligands give nearly racemic products. In a follow-up study, Hayashi et al. showed the terminal dialkylamino group could be changed to

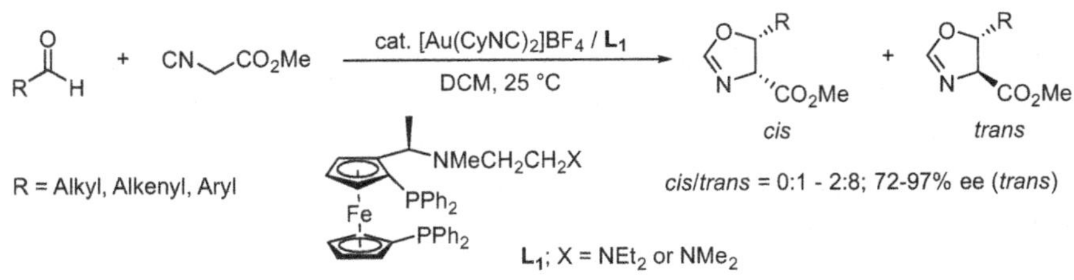

Scheme 1 Initial synthesis of enantioenriched oxazolines

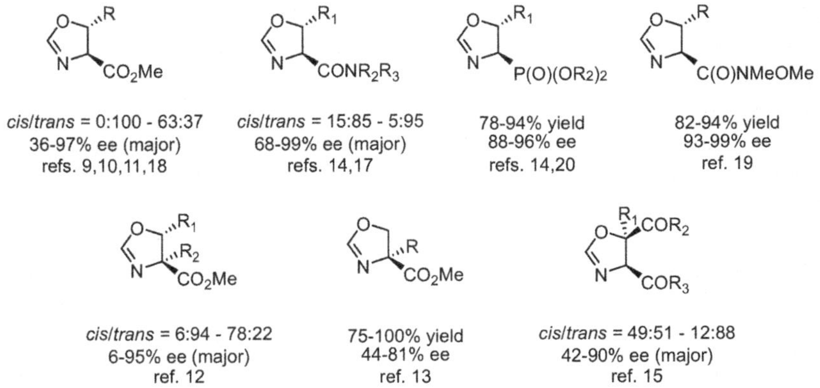

Scheme 2 Scope of enantioenriched oxazolines

morpholine, resulting in improvements to both stereo- and enantioselectivity [10]. Over the course of a few years Hayashi et al. extended this methodology to the synthesis of a variety of enantioenriched substituted oxazolines using various aldehyde and ketone electrophiles, which are shown in Scheme 2 [11–20]. An application of this methodology to the total synthesis of (−)-α-kainic acid was accomplished by Bachi et al. in 1997 [21].

Beginning in 1989, but publishing with Hayashi concurrently, Pastor et al. published detailed investigations into the structure-activity relationship of various substrates and ferrocenylphosphine ligands [22–27]. Although the central chirality of the stereogenic carbon atom was dominant in terms of the absolute configuration of the product, the authors described how matched planar chirality of the ligand was crucial for high levels of enantioinduction. Interestingly, this was presented as "the first example in a chiral transition metal ligand containing both central and planar chirality of internal cooperativity of chirality [25]." Additionally Togni and Pastor showed similar reactivity was realized utilizing a neutral complex composed of a ferrocenylphosphine ligand and gold(I) chloride [28].

Besides aldehydes, other electrophiles have been used in conjunction with isocyanoacetates. Lin et al. showed *N*-sulfonylimines can be substituted to obtain enantioenriched imidazoline products [29]. Carretero and coworkers utilized *N*-phenylmaleimide to obtain pyrroline products with high ee [30].

The first example of a non-isocyanide intermolecular cycloaddition to form heterocyclic products was published in 2007. Toste et al. used azlactones (münchnones) in conjunction with various electron-deficient alkenes to form a variety of enantioenriched pyrroline products (Scheme 3) [31]. In contrast to most other gold-catalyzed cyclizations, the reactions proceed through activation of the pro-nucleophilic component as opposed to typical gold/electrophile coordination. These mechanistic studies and additional electrophile scope was published in a follow-up article in 2011 [32]. The same transformation utilizing cationic gold trifluoroacetate complexes has also been published [33, 34].

In 2010, Nájera, Sansano, et al. utilized a similar strategy for the synthesis of enantioenriched chiral proline derivatives (Scheme 4) [35]. With the use of azomethine ylides, the authors obtained the desired heterocycles with good to excellent yields and ees. Comparative studies of this reaction with expanded scope and comparisons to other coinage metal catalysts in addition to DFT studies can be found in subsequent publications [34, 36]. This methodology was applied to the synthesis of a second-generation GSK-hepatitis C virus inhibitor [37].

A different 1,3-dipolar cycloaddition utilizing nitrones and *N*-allenamides resulting in the asymmetric formation of substituted isoxazolidines was published in 2013. Although chiral bisphosphine ligands gave no reaction, use of 3,3′-BINOL-derived phosphoramidite ligands furnished the desired products with 45–99% yield and 63–99% ee [38]. In a similar vein, Mascareñas, López, et al. used allenamides and carbonyl-tethered alkenes to form enantioenriched oxa-bridged carbocycles [39].

A study by Gong et al. focused on the hetero-Diels-Alder reaction of simple dienes with diazine nucleophiles [40]. Excellent yields and ees of the tetrahydropyridazine products are usually obtained, along with good regioselectivity.

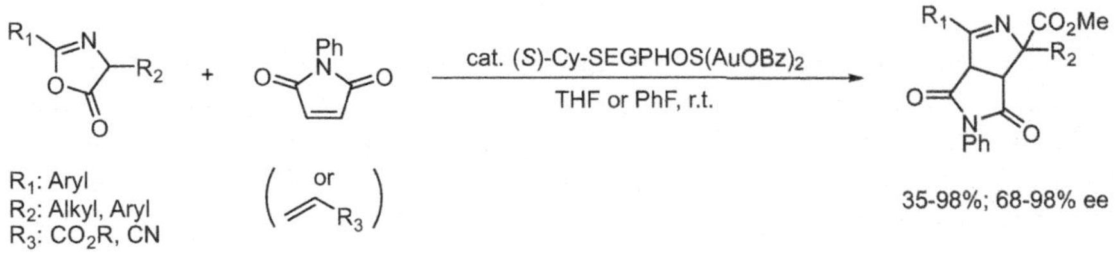

Scheme 3 Synthesis of enantioenriched pyrrolines

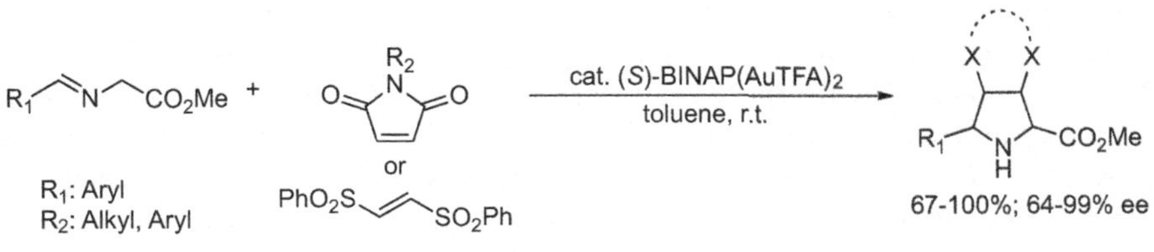

Scheme 4 Synthesis of enantioenriched proline derivatives

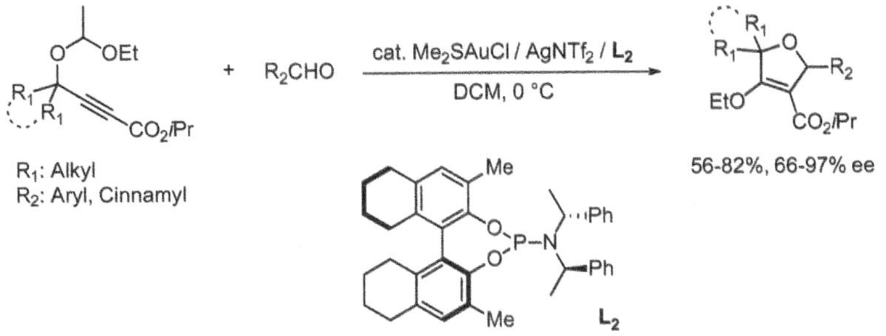

Scheme 5 Synthesis of enantioenriched dihydrofurans

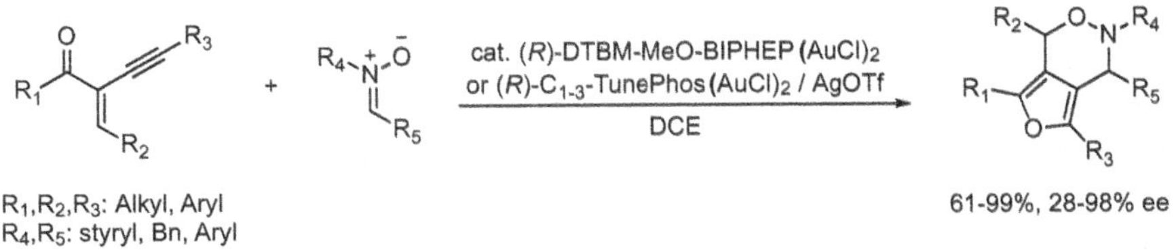

Scheme 6 Synthesis of enantioenriched tetrahydro-1,4-oxazines

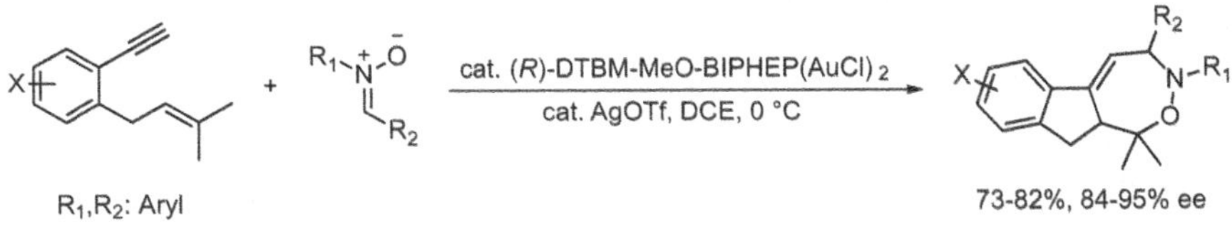

Scheme 7 Synthesis of enantioenriched tetrahydrooxazepines

A synthesis of enantioenriched dihydrofurans was outlined by Zhang and coworkers in 2015 [41]. The authors employed propargyl acetals that are proposed to initially form cationic vinylgold 1,3-dipolar intermediates (Scheme 5). Attack by the external aldehyde and subsequent cyclization gives the desired product. The scope of the electrophile was expanded to nitrones by Toste et al. shortly thereafter [42].

In the field of cascade reactions, an enantioselective example of a formal [4 + 3] cycloaddition was published by Zhang and coworkers in 2010, utilizing 2-(1-alkynyl)-2-alken-1-ones [43]. This proof of concept was applied to a cycloaddition with nitrones in the same year [44]. As shown in Scheme 6, good to excellent yields and ees were obtained for a variety of tetrahydro-1,4-oxazine products. In 2014, a significant improvement in this methodology came about with introduction of chiral sulfinamide monophosphine (Ming-Phos) ligands. In addition to expanded substrate scope, the authors were able to select for a new product diastereomer [45]. A related reaction via the kinetic resolution of 1-(1-alkynyl)cyclopropyl ketones with nitrones was also demonstrated by the group of Zhang [46].

Another example of a cycloaddition cascade was reported in 2012 by Liu et al. [47]. As seen in Scheme 7, a variety of fused tetrahydro-azepine products were obtained with excellent enantioselectivities. The same group reported a similar nitrone cycloaddition with allenyl acetals to afford tricyclic oxazolidine products, although only one modestly enantioselective example was reported [48]. Liu and coworkers also outlined a single example where a nitrone-like intermediate reacts with an alkene to form a bridged isoxazolidine product in 73% ee [49]. For a review concerning gold(I)-catalyzed enantioselective cycloaddition reactions in general, see [50].

3 Intramolecular Cyclizations

The first example of intramolecular enantioselective cyclization of aminoalkene substrates was reported by Mikami et al. in 2011, utilizing *N*-alkenyl ureas as substrates to form pyrrolidines with good yields and modest enantioselectivities (Scheme 8) [51]. Enantioselectivity was slightly improved (48% ee) by the addition of a chiral phosphate and triflic acid. Shi et al. published a similar transformation with similarly modest ees utilizing novel types of axially chiral gold catalysts [52].

In 2014, Michon et al. utilized chiral phosphoramidite ligands for a similar transformation, obtaining enantioselectivities up to 40% [53]. In subsequent publications, both Michon and Widenhoefer reported significantly improved enantioselectivity using dinuclear gold(I) phosphine complexes as catalysts (Scheme 9) [54, 55]. Both groups demonstrated that inclusion of water as an additive (Michon) or use of a protic solvent (Widenhoefer) improved the level of enantioselectivity, with Widenhoefer proposing that protodeauration occurs though a concerted stereodetermining process involving a hydrogen-bonded methanol bridge.

One example of an asymmetric diene hydroamination was reported by Toste et al. in 2011 [56]. The authors showed that an alcohol solvent or additive

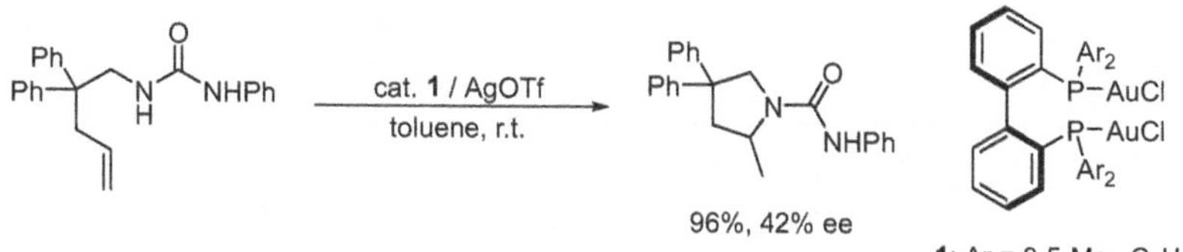

1; Ar = 3,5-Me_2-C_6H_3-

Scheme 8 Intramolecular enantioselective cyclization of aminoalkenes

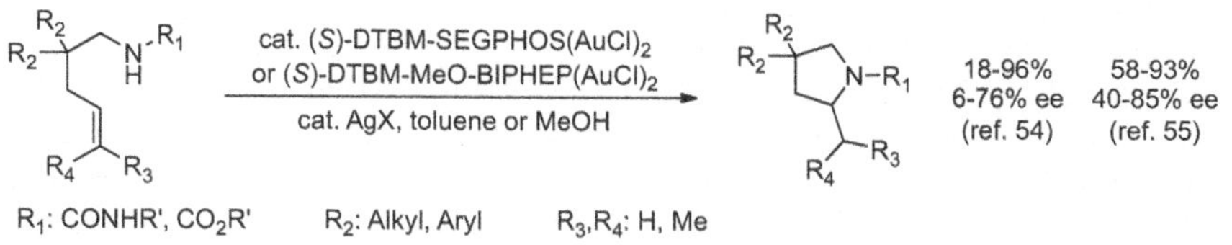

Scheme 9 Improved intramolecular enantioselective cyclization of aminoalkenes

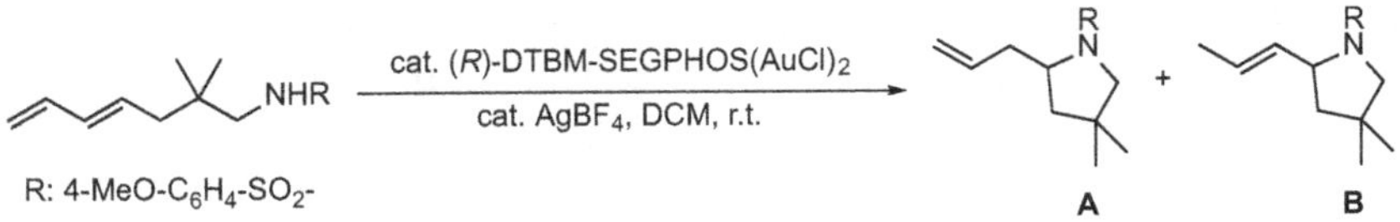

entry	additive	% conv.	A:B	% ee **A**	% ee **B**
1		18	1:0	35	--
2	10 eq CyOH	97	1:1.5	21	97
3	10 eq (+)-menthol	94	1:1.9	19	94
4	10 eq (-)-menthol	>99	1:9.0	55	95

Scheme 10 Intramolecular enantioselective cyclization of aminodienes

significantly accelerated the rate of substrate conversion, resulting in formation of two products (Scheme 10); enantiomer match/mismatch of the alcohol and ligand was also important. Mechanistic studies suggested product **A** formed through a traditional hydroamination pathway similar to other examples, while product **B** proceeded through a menthol/gold Brønsted-acidic species, which activated the diene and induced cyclization.

During 2007, several seminal papers by Toste and Widenhoefer were published on the intramolecular hydrofunctionalization of allenes, producing a variety of enantioenriched pyrrolidine, piperidine, tetrahydrofuran, tetrahydropyran, and lactone products. The report by Widenhoefer focused on the hydroalkoxylation of mono- and disubstituted allenes (Scheme 11) [57]. In the case of chiral 1,3-disubstituted allene substrates, the authors noted that the starting enantiopurity determined the corresponding *E*:*Z* ratio of the alkene products, although the sp^3 stereocenter configuration was determined by the catalyst in all cases.

Toste et al. published a related methodology on the hydroamination of allenes to afford chiral 2-substituted pyrrolidine and piperidine products in high yields and enantioselectivities (Scheme 11) [58]. Interestingly, use of the 4-nitrobenzoate counterion was crucial for transmission of ligand chiral information to the coordinated substrate. Although high levels of enantioinduction were obtained, substrates were limited to symmetrical 1,1-disubstituted allenes. A mechanistic study of this system suggested that the catalyst contained an aurophilic Au-Au interaction, in addition to a strong interaction between the sulfonamide and gold, crucial for enantioselectivity [59]. To corroborate these observations, Gade et al. showed that in an alternative catalyst system capable of incorporating up to three gold atoms, higher selectivities were exclusively observed as the number of gold centers increased to three, suggesting aurophilic interactions are at play [60]. Gade also utilized enantiopure ditopic cyclophosphazane ligands to accomplish this transformation with good to modest ees [61].

Using DTBM-MeO-BIPHEP$(AuCl)_2$ and $AgClO_4$ as an alternative catalytic system, Widenhoefer et al. successfully formed enantioenriched pyrrolidines from both terminal and substituted allene substrates, albeit with slightly lower selectivity [62]. Later, a follow-up paper demonstrated racemic unsymmetrical 1,1-disubstituted allenes were able to undergo dynamic kinetic asymmetric

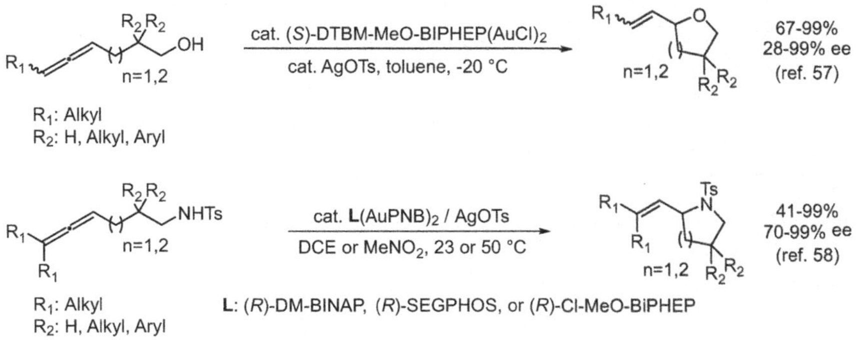

Scheme 11 Synthesis of enantioenriched pyrrolidines and piperidines

transformation (DYKAT) to afford reasonably *E*/*Z* selective mixtures with varying levels of enantioselectivity [63]. Widenhoefer also utilized ureas as nucleophiles to obtain similar chiral pyrrolidine products [64].

A significant advance in intermolecular hydroamination methodology arose with the introduction of a chiral counterion strategy by Toste et al. in 2007 [65]. Although hydroamination of allenes with sulfonamides was shown to be previously feasible with chiral bisphosphine ligands and achiral counterions, efforts into expansion of the asymmetric hydroalkoxylation and carboxylation of allenes were unsuccessful. The authors proposed successful enantioinduction was limited by the linear environment of the gold(I) catalyst, projecting the catalytically active gold atom away from the chiral environment of the ligand. To solve this problem, a chiral silver phosphate co-precatalyst in combination with either an achiral or chiral bisphosphine gold complex was utilized. This pair forms a transition state where the chiral anion is closely associated with the cationic gold allene complex, resulting in high levels of enantioselectivity for substituted tetrahydrofurans, tetrahydropyrans, lactones, and pyrrolidines (Scheme 12). In a later study, Mikami et al. showed formation of the monocationic complex through reduction of silver phosphate loading allowed for slightly improved ees for selected hydroxyallene substrates [66]. Espinet et al. also showed hydrogen-bonded heterocyclic carbene (HBHC) dinuclear gold(I) catalysts are also competent at facilitating this transformation [67]. This methodology was applied to the synthesis of pyrazolidines, isoxazolidines, and tetrahydrooxazines by Toste et al. in 2010 [68].

Mononuclear catalysts have also been employed for the hydrofunctionalization of allenes (Scheme 13). One example is the use of a TADDOL-based phosphoramidite gold complex for the synthesis of indolines, pyrrolidines, and tetrahydrofurans by Fürstner et al. [69]. In addition to the modularity provided by a convergent catalyst synthesis, this system was attractive as only one gold atom is required to accomplish previously published transformations with similar levels of conversion and enantioselectivity. Michon et al. have also shown BINOL-based phosphoramidite catalysts are relatively competent at forming enantioenriched pyrrolidines [70].

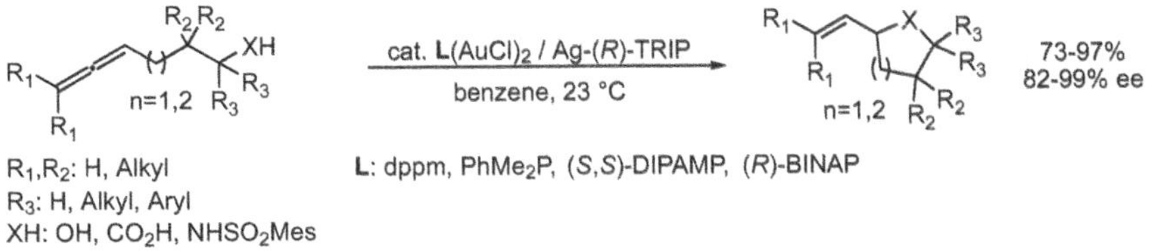

Scheme 12 Chiral counteranion-assisted enantioselective hydrofunctionalization

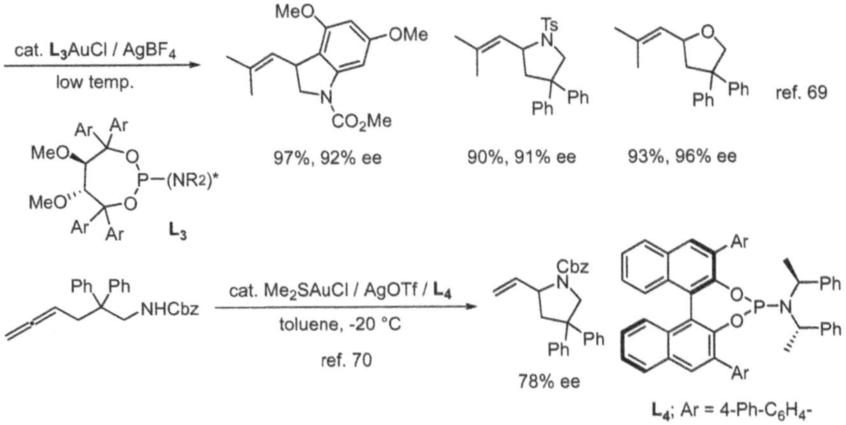

Scheme 13 Enantioselective hydrofunctionalization of allenes

In addition to these conventional approaches to hydrofunctionalization of allenes, Lipshutz et al. have demonstrated the ability to synthesize enantioenriched lactones inside a PEGylated micelle using water as the bulk solution [71]. Partial recyclability of the catalyst was demonstrated, a potentially attractive feature if this methodology is to be applied in the areas of green chemistry. In 2015, utilizing a bifunctional silica-supported gold phosphine heterogeneous catalyst, Toste et al. were able to demonstrate production of significantly enantioenriched substituted lactones [72]. Significantly faster reaction times than analogous homogenous conditions were attributed to the acidity of the silica surface, increasing the rate of the turnover-limiting protodeauration step. Enantioselectivity was preserved after 11 cycles, making this an attractive method for future developments in heterogeneous gold catalysis.

In 2010, Bandini et al. published the first enantioselective synthesis of heterocycles utilizing a gold-catalyzed activation of allylic alcohols [73]. The substituted morpholine and 1,4-oxazepane products were obtained with good yields and enantioselectivities producing water as the sole by-product (Scheme 14).

Several subsequent publications utilized this dehydrative strategy for a variety of enantioenriched heterocyclic compounds (Scheme 15). The group of Bandini published an additional related study where indole substrates with a nitrogen-

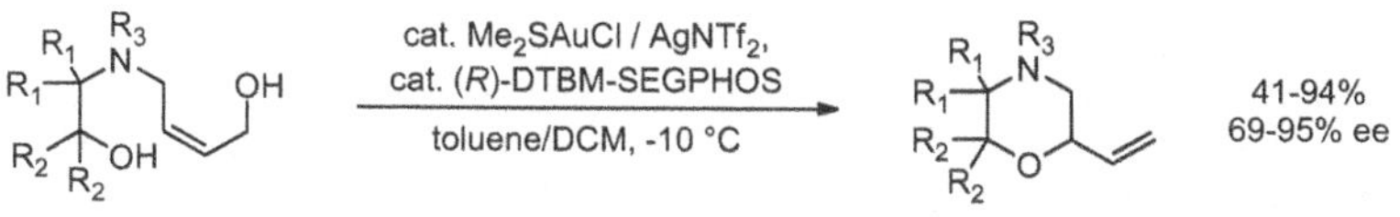

Scheme 14 Enantioselective allylic alcohol activation

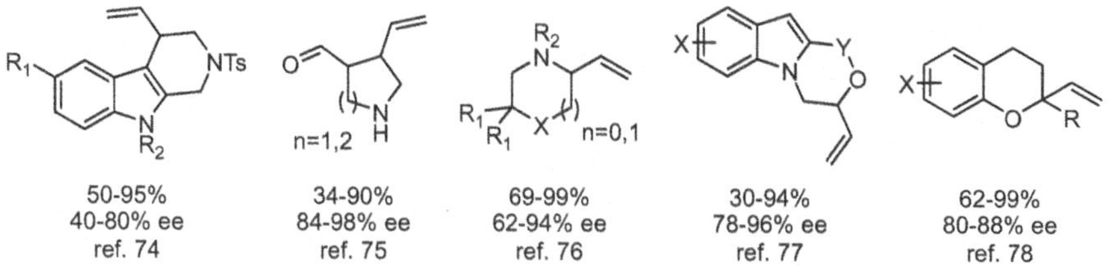

Scheme 15 Scope of enantioselective allylic alcohol activation

containing tether were cyclized to form tetrahydro-β-carbolines with good enantioselectivity [74]. A similar work published by Bandini et al. utilized a synergistic enamine/gold catalyst system to form substituted pyrrolidines with good to excellent selectivities [75]. In 2012, Widenhoefer et al. published a similar study forming enantioenriched substituted pyrrolidines, piperidines, and piperazines [76]. The authors also showed that in the case of chiral substrates, the absolute configuration of the product was determined only by the catalyst, with *E*:*Z* selectivity determined by the enantiomeric excess of the starting material. Bandini et al. demonstrated the feasibility of a cascade reaction, with gold-catalyzed alkyne hydroamination followed by the typical allylic alcohol activation [77]. Finally, the enantioselective total synthesis of α- and γ-tocopherol (vitamin E family members) was realized by Rueping et al. in 2014, utilizing this strategy [78]. In many of these cases, the starting olefin geometry of the substrate is important, with the other isomer generally giving reduced yield and ee.

In related work by Bandini et al., extensive experimental and computational studies were performed to elucidate the mechanism of the reaction [79]. The authors conclude the reaction proceeds via a stepwise S_N2'-like mechanism, with carboauration of the alkene, rearomatization, and subsequent elimination of water (Scheme 16). The triflate anion was also shown to play an important role by both assisting in proton transfer and bringing the reactive sites of the substrate closer together via a "folding effect."

In 2009, Toste et al. published a strategy for the enantioselective synthesis of benzopyrans via rearrangement of allylic oxonium intermediates [80]. The authors proposed the mechanism of this reaction proceeds via the formation of an allylic cation as control experiments rule out a 1,4- or a 2,3-/3,3-rearrangement pathway (Scheme 17). In a similar substrate/product vein, Toste's group demonstrated a dynamic kinetic resolution of propargyl esters facilitated by a chiral acyclic

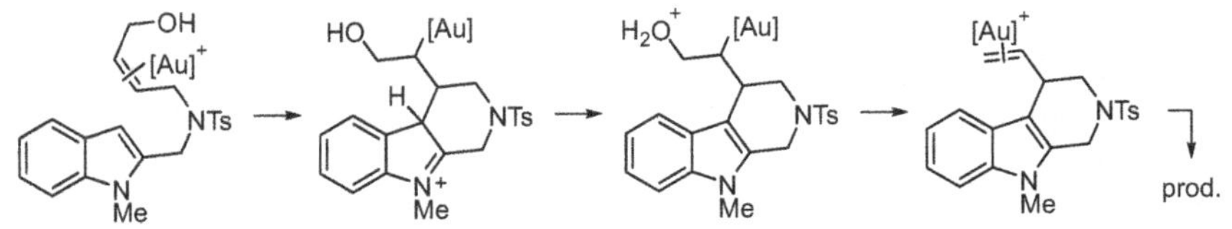

Scheme 16 Mechanism of enantioselective allylic alcohol activation

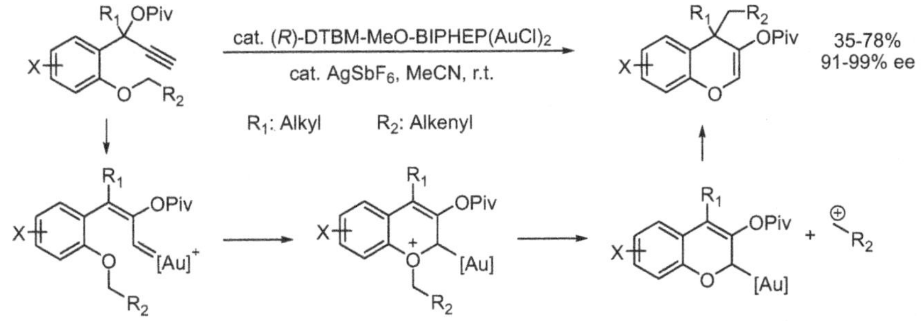

Scheme 17 Enantioselective synthesis of benzopyrans

diaminocarbene gold(I) complex, moving away from the typical chiral phosphine gold(I) regime [81].

Also in 2009, Czekelius et al. reported the enantioselective desymmetrization of diynamides with up to 60% ee enabled by a bisphosphine gold(I) catalyst [82]. Several other chiral bis(tetrahydroisoquinoline)-derived carbene ligands were also reported to give modest enantioselectivity [83]. A breakthrough came about in 2012 through the use of a chiral phosphate counteranion-based catalyst system, allowing for high levels of enantioinduction to be achieved (Scheme 18) [84]. A summary of these and related studies has been published [85].

Other desymmetrization reactions have been reported, including the conversion of *meso*-alkynylbenzenediols into enantioenriched isochromenes [86]. This methodology was also applicable to the kinetic resolution of enantiomers, where *S* factors range from 20 to 201. In a related vein, Uemura et al. successfully synthesized planar chiral isochromene chromium complexes stating from a similar *meso*-diol substrate [87]. This methodology was subsequently expanded, a summary of which is shown in Scheme 19 [88]. Also utilizing alkynyl diols, Brimble et al. produced benzannulated spiroacetals using a chiral counterion strategy [89].

In 2012, Slaughter et al. reported the use of mononuclear gold(I) acyclic diaminocarbene complexes to access enantioenriched isochromenes through alkynylbenzaldehyde cyclization in the presence of an external alcohol nucleophile [90]. Approaching the problem through the comparison of different catalyst structures in the solid state and DFT calculations, the authors suggested the presence of a Au-arene interaction on the catalyst enhances its enantioselectivity and stability. In

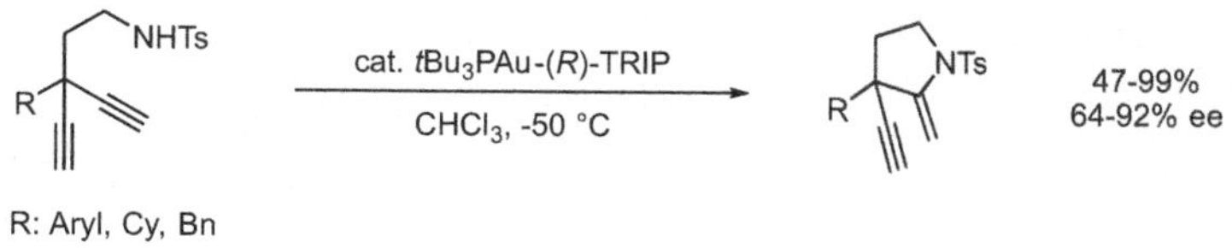

Scheme 18 Enantioselective desymmetrization of diynamides

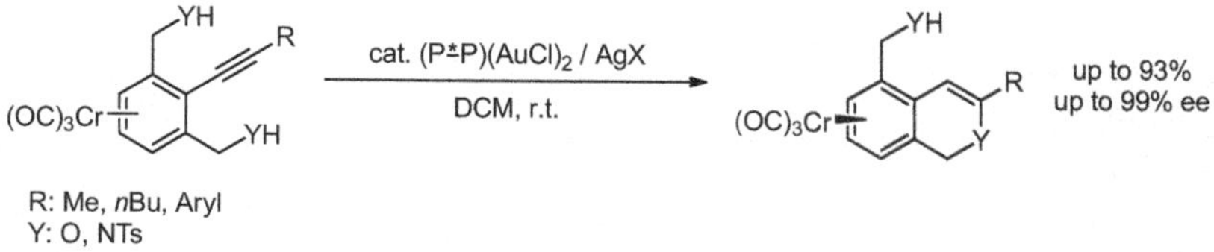

Scheme 19 Enantioselective desymmetrization of *meso*-alkynylbenzenediols

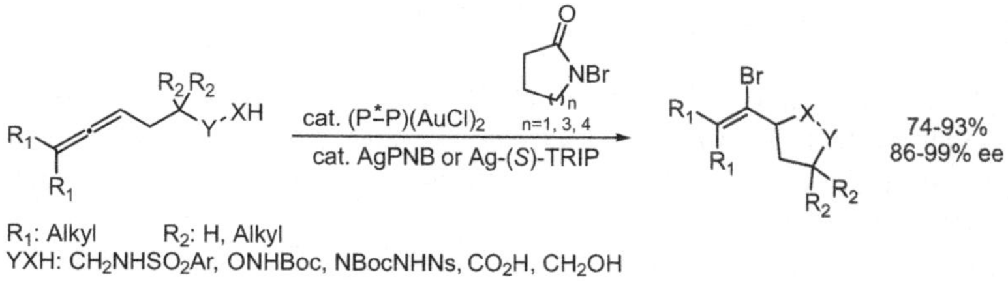

Scheme 20 Enantioselective allene bromofunctionalization

a 2010 work, Toste and coworkers were able to carry out an enantioselective alkyne polycyclization producing a variety of bi-, tri-, and tetracyclic heterocycles [91].

Recently two examples of gold-catalyzed heterocycle synthesis have been published which avoid the traditional protodeauration step, instead utilizing an external electrophile. Toste et al. published the first example of gold-catalyzed allene bromofunctionalization, utilizing *N*-bromolactams as an electrophilic halogen sources (Scheme 20) [92]. The more commonly used *N*-bromosuccinimide was shown to give exclusively racemic product, demonstrating the necessity for a fine balance between gold catalyst and halogen electrophilicity. Utilizing a similar methodology, Gong and coworkers showed that atropisomeric oxaborinines can be generated through the use of a diazene-1,2-dicarboxylate as an electrophilic nitrogen source [93].

Bandini et al. reported the synthesis of enantioenriched polycyclic indolines in 2012 [94]. Through initial alkyne activation and subsequent carboauration, a vinylgold iminium intermediate is formed, which undergoes further attack by the vinyl alcohol moiety to afford dihydropyranylindoline products (Scheme 21). A similar mechanism of iminium formation and subsequent attack was reported by Chen and coworkers in 2015 [95]. The use of an electrophilic allene and chiral phosphoramidite catalyst afforded various pyrroloindoline derivatives (Scheme 22).

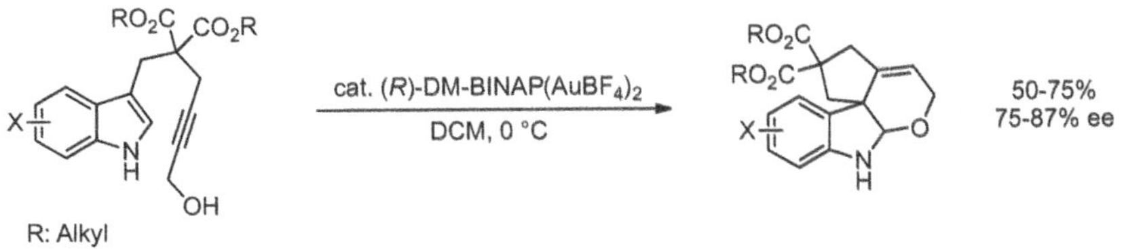

Scheme 21 Synthesis of enantioenriched polycyclic indolines

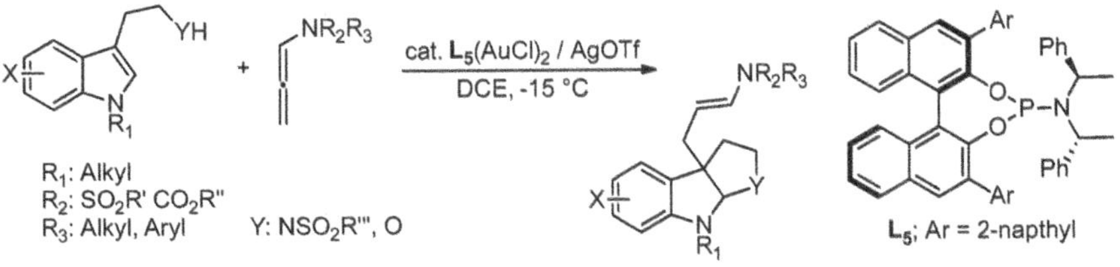

Scheme 22 Synthesis of enantioenriched pyrroloindolines

An additional example of iminium intermediate attack was shown by Zhang et al. via cyclization of an unsaturated 2-alkynyl enone and subsequent 1,5-hydride shift. The resultant azepine products are formed in high yield and enantioselectivity [96].

In the field of enyne cyclization, Michelet and coworkers reported in 2009 the enantioselective synthesis of functionalized cyclopropanated dihydrofurans and tetrahydropyridines utilizing a chiral biaryl bisphosphine gold catalyst (Scheme 23) [97]. Although yields were generally modest, excellent enantioselectivities were obtained in all cases. Additional substrates were published in a follow-up work [98].

Later in 2011, Shi et al. reported the cyclization of 1,6 enynes and interception of the proposed gold carbenoid intermediate [99]. Utilizing either an acetic acid or diphenylsulfoxide in the presence of a binaphthyl-based NHC-gold complex, enantioenriched functionalized pyrrolidines were obtained in excellent yields and with modest selectivities (Scheme 24). An extension of this work using substituted indoles as nucleophiles was also reported in 2011 [100]. Michelet and coworkers applied this reaction to the synthesis of tetrahydrofurans with modest enantioselectivities using indoles, 1,3,5-trimethoxybenzene and water as nucleophiles [101]. Although containing only one example of an enantioenriched heterocycle, 8-methylquinoline *N*-oxide has also been used to form a ketone-functionalized lactam using a similar strategy [102].

This methodology has been applied to the synthesis of the triple reuptake inhibitor GSK1360707 first by Elitzin et al. and later by Fürstner and coworkers with (*R*)-DM-BINAP and TADDOL-based phosphoramidite ligands, respectively [103, 104]

A significant improvement in synthesis of these heterocycles came through the use of TADDOL-based phosphoramidite ligands by Fürstner et al. [69]. Substituted tetrahydropyridines and dihydropyrans were obtained with up to 98% ee.

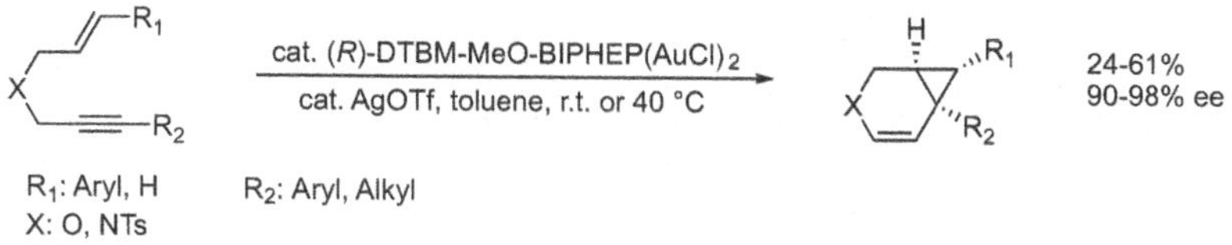

Scheme 23 Enantioselective synthesis of functionalized cyclopropanated dihydrofurans

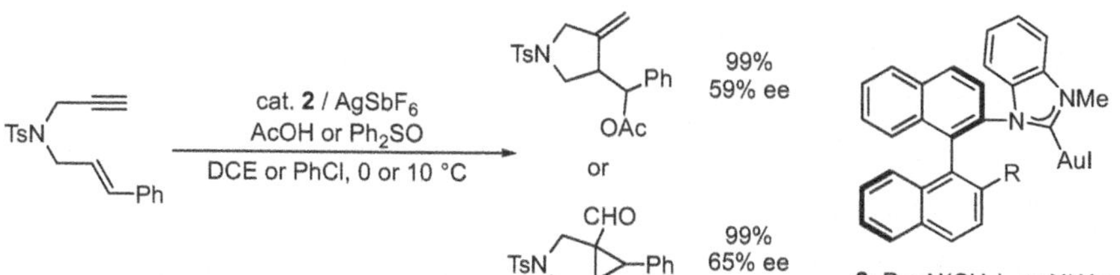

2; R = N(CH$_2$)$_4$ or NHAc

Scheme 24 Enantioselective synthesis of functionalized pyrrolidines

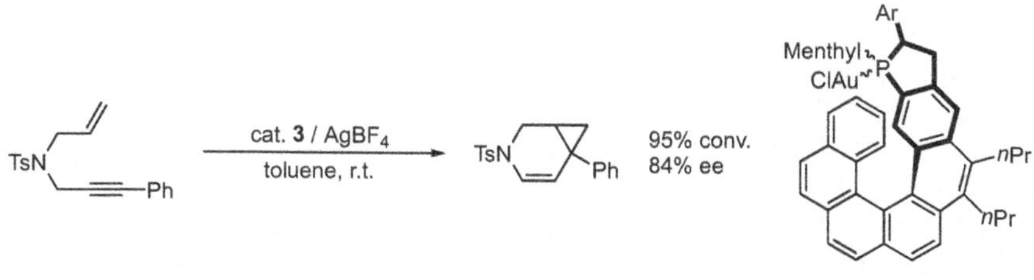

3; Ar = 3,5-Me$_2$-C$_6$H$_3$-

Scheme 25 Enantioselective synthesis of tetrahydropyridines

A creative strategy for enantioenriched tetrahydropyridines via enyne cycloisomerization was outlined by Voituriez et al., through the use of a helically chiral gold phosphine catalyst as shown in Scheme 25 [105]. A related class of ligands, phosphathiahelicenes, was also used for this transformation, allowing for even higher enantioselectivities with some substrates [106].

In 2009, Mascareñas et al. reported the synthesis of pyrrolidine-containing bicycles starting from allenediene substrates using mononuclear chiral gold phosphoramidite catalysts (Scheme 26) [107]. Superior levels of enantio- and regioselectivity were observed compared with bisphosphine ligands. Toste and coworkers published a similar study using 3,3′-pyrenyl substituents to further increase the ee [108]. Additional pyrrolidine products with fused cycloheptadiene rings were synthesized in a follow-up work by Mascareñas et al. using the same catalyst system [109]. In 2012, Fürstner and coworkers showed TADDOL-based phosphoramidite gold complexes were also competent catalysts for this reaction, achieving greater than 90% ee [69, 110].

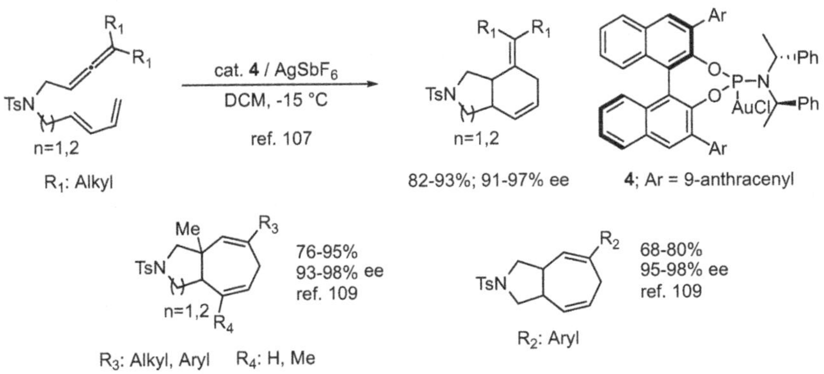

Scheme 26 Enantioselective synthesis of pyrrolidine-containing bicycles

In 2007, Toste et al. reported the first asymmetric cycloisomerization of an allenene substrate to form a cyclobutane-fused pyrrolidine product in 70% yield and 54% ee [111]. The authors demonstrate evidence of a reversible stepwise mechanism, initially proceeding through alkene attack of the gold-coordinated allene to form a benzylic carbocation. Vinyl gold attack of the benzylic cation then results in formation of the cyclobutane ring. Fürstner and coworkers succeeded in improving the enantioselectivity to 95% using a TADDOL-based phosphoramidite ligand [69, 110]. A more detailed study by Toste et al. in 2011 showed that a variety of enantioenriched products can be synthesized using a SPINOL-based phosphoramidite ligand (Scheme 27) [112]. Additionally, the authors found the benzylic carbocation intermediate could be trapped by addition of alcohol or water to furnish different pyrrolidines with good to excellent yields and enantioselectivities.

In 2014, Tanaka and coworkers reported the asymmetric synthesis of azahelicenes using chiral dinuclear bisphosphine complexes (Scheme 28) [113]. Formation of the helically chiral product proceeds through initial formation of an axially chiral species, which will only undergo further cyclization in the presence of excess silver triflate relative to gold. Synthesis of similar enantioenriched axially chiral molecules was published in a preliminary work by the same group [114]. Tanaka et al. also published a synthesis of highly substituted pyrrolinones employing a dearomatization strategy on similar substrates [115].

Xu, Shi, et al. reported the synthesis of enantioenriched bis(indoyl)tetrahydropyridines in a 2013 work shown in Scheme 29 [116]. The reaction proceeds through a pseudo-1,5-indole shift, yielding products which resemble several naturally occurring bis(indole) alkaloids which are pharmacologically active.

Recently, Hashmi and coworkers published a novel synthesis of functionalized chromanes using chiral NAC- and NHC-gold(I) complexes. Reaction enantioselectivities were modest for all catalyst scaffolds explored [117]. An enantioselective Conia-ene-type reaction was recently published by Bezzenine-Lafollée et al., resulting in the synthesis of enantioenriched pyrrolidinones with good to excellent selectivity [118].

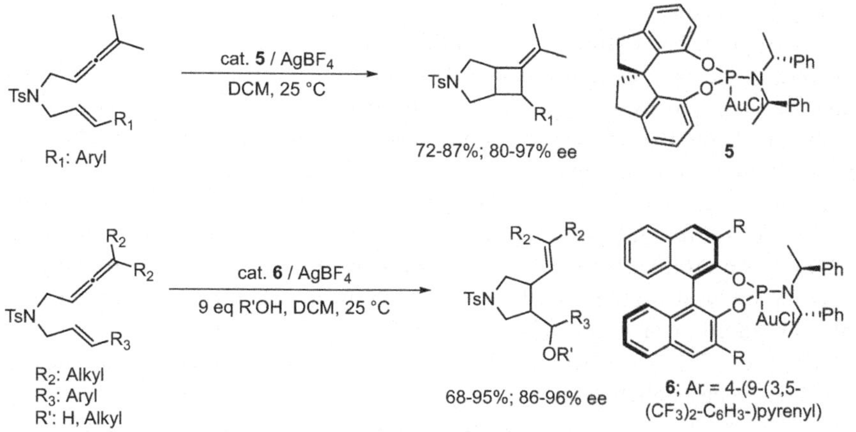

Scheme 27 Enantioselective cycloisomerization of an allenenes

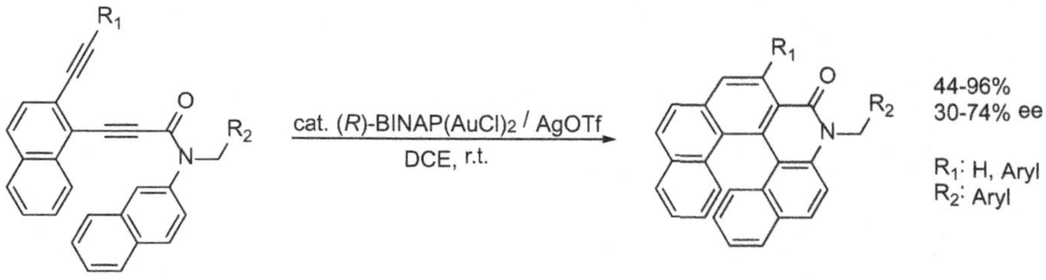

Scheme 28 Asymmetric synthesis of azahelicenes

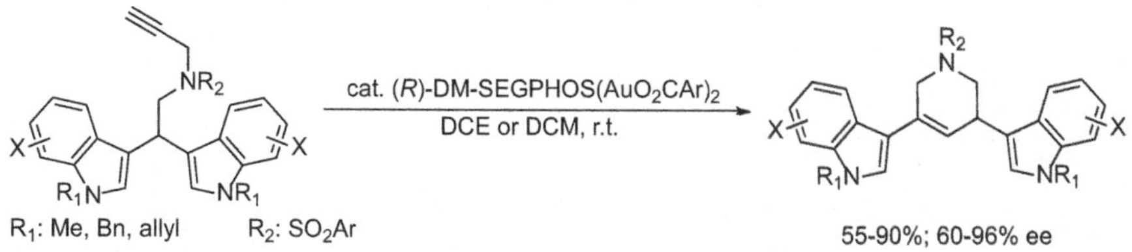

Scheme 29 Synthesis of enantioenriched bis(indoyl)tetrahydropyridines

4 Tandem Reactions

In 2011, the first instance of a purely gold phosphate-catalyzed tandem reaction was reported. Although similar strategies had been published since 2009, these instances required the addition of excess chiral phosphoric acid to achieve high enantioselectivity. Generally when excess phosphoric acid is excluded, these reactions were slower and give lower enantioselectivity, indicating the necessity of a protonated phosphate cocatalyst. These studies will not be covered, but reviews have been published outlining most of this work [119, 120].

Gong et al. have demonstrated the coupling of an azlactone, and an alkynol could be achieved by using an achiral phosphine gold methyl complex with an equimolar amount of chiral phosphoric acid (Scheme 30). Although addition of excess phosphoric sped up the reaction, ee values were nearly identical [121]. Also using an in situ-generated gold phosphate, an enantioselective three-component synthesis of spiroacetals was reported in 2013. Attack of an iminium phosphate ion pair by the preformed enol ether was proposed to be the stereodetermining step [122]. A study also published by Gong and coworkers shows similar reactivity, although addition of excess phosphoric acid resulted in slight improvements to yield and ee [123]. A synthesis of a bezoxasilole with preliminary enantioselectivity (28%) has been published [124].

A three-component enantioselective synthesis of cyclic carbamimidates was reported in 2011 by Toste et al. [125] (Scheme 31). The proposed mechanism proceeds through gold acetylide addition to the imine, followed by reaction of the isocyanate and subsequent cyclization.

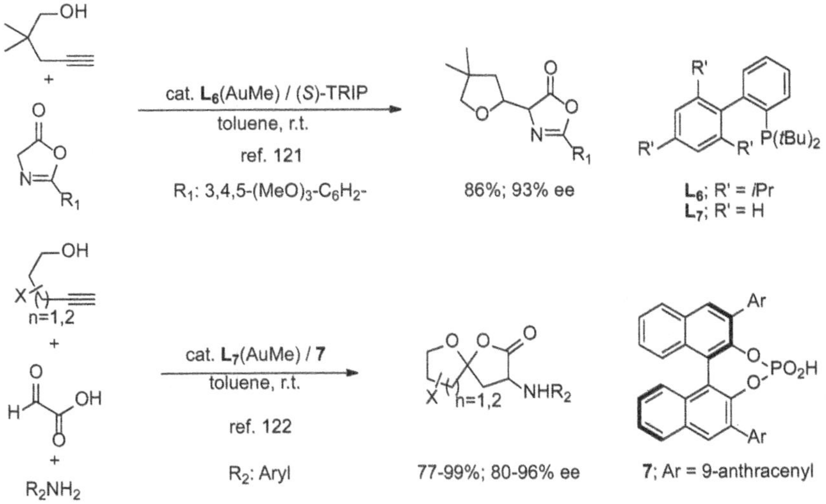

Scheme 30 Enantioselective tandem reactions proceeding via alkynol cyclization

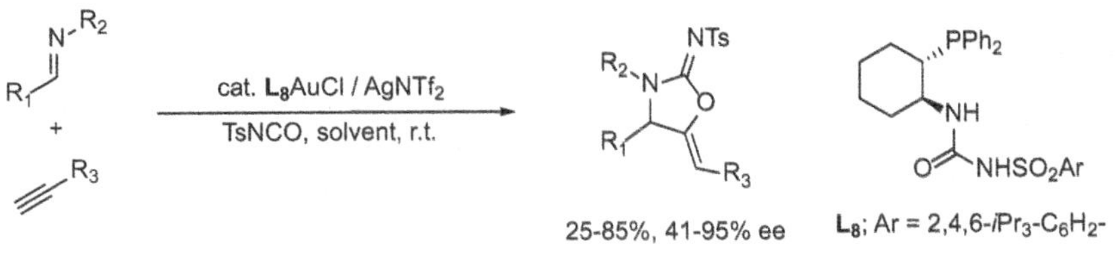

Scheme 31 Three-component enantioselective synthesis of cyclic carbamimidates

5 Miscellaneous Reactions

In 2012, Gong and coworkers reported a highly enantioselective transfer hydrogenation of quinolines to tetrahydroquinolines utilizing a chiral gold phosphate catalyst generated in situ from a phosphoric acid and a methylgold precatalyst mixture (Scheme 32) [126]. Catalyst loadings down to 0.01 mol% were still able to impart sufficient selectivity. Although the phosphoric acid was capable of catalyzing the reaction, the use of a carbene ligand increased the rate of reaction. Patil et al. also showed a hindered gold(I) phosphine complex was capable of performing the same reduction in a control experiment, although the reaction time was much slower than the phosphoric acid alone [127]. A similar study was published where an initial hydroamination forms a dihydroquinoline that undergoes subsequent tautomerization and reduction to the same products with similar selectivity (Scheme 32) [128].

A unique methodology by Sabater et al. uses a proposed chiral gold(III) precatalyst to generate enantioenriched epoxides with modest selectivity utilizing iodosobenzene or sodium hypochlorite as an oxidant under an oxygen atmosphere (Scheme 33) [129]. The authors showed that the reaction does not proceed through a radical chain process; a gold oxo or peroxo species was proposed as the active catalyst.

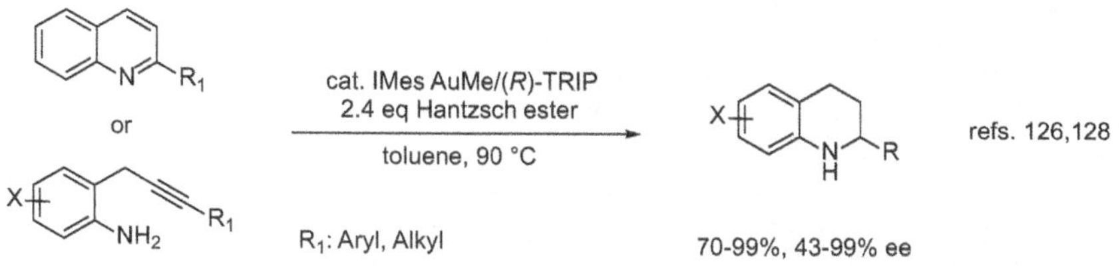

Scheme 32 Enantioselective transfer hydrogenation of quinolines

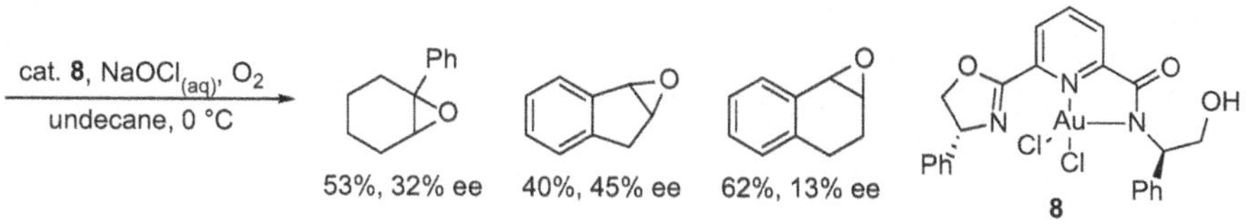

Scheme 33 Synthesis of enantioenriched epoxides

References

1. Yu J, Shi F, Gong L-Z (2011) Acc Chem Res 44:1156
2. Huang Y-Y, Yang X, Chen Z, Verpoort F, Shibata N (2015) Chemistry 21:8664
3. Widenhoefer RA (2008) Chemistry 14:5382
4. Cera G, Bandini M (2013) Isr J Chem 53:848
5. Yang W, Hashmi ASK (2014) Chem Soc Rev 43:2941
6. Qian D, Zhang J (2014) Chem Rec 14:280
7. Wang Y-M, Lackner AD, Toste FD (2014) Acc Chem Res 47:889
8. Patil NT (2012) Chem Asian J 7:2186
9. Ito Y, Sawamura M, Hayashi T (1986) J Am Chem Soc 108:6405
10. Ito Y, Sawamura M, Hayashi T (1987) Tetrahedron Lett 28:6215
11. Ito Y, Sawamura M, Hayashi T (1988) Tetrahedron Lett 29:239
12. Ito Y, Sawamura M, Shirakawa E, Hayashizaki K, Hayashi T (1988) Tetrahedron 44:5253
13. Ito Y, Sawamura M, Shirakawa E, Hayashizaki K, Hayashi T (1988) Tetrahedron Lett 29:235
14. Ito Y, Sawamura M, Kobayashi M, Hayashi T (1988) Tetrahedron Lett 29:6321
15. Ito Y, Sawamura M, Hamashima H, Emura T, Hayashi T (1989) Tetrahedron Lett 30:4681
16. Sawamura M, Ito Y, Hayashi T (1989) Tetrahedron Lett 30:2247
17. Soloshonok VA, Hayashi T (1994) Tetrahedron Asymmetry 5:1091
18. Soloshonok VA, Hayashi T (1994) Tetrahedron Lett 35:2713
19. Sawamura M, Nakayama Y, Kato T, Ito Y (1995) J Org Chem 60:1727
20. Togni A, Pastor SD (1989) Tetrahedron Lett 30:1071
21. Bachi MD, Melman A (1997) J Org Chem 62:1896
22. Togni A, Pastor SD (1989) Helv Chim Acta 72:1038
23. Pastor SD, Togni A (1989) J Am Chem Soc 111:2333
24. Pastor SD, Togni A (1990) Tetrahedron Lett 31:839
25. Togni A, Pastor SD (1990) J Org Chem 55:1649
26. Pastor SD, Togni A (1991) Helv Chim Acta 74:905
27. Pastor SD, Kesselring R, Togni A (1992) J Organomet Chem 429:415
28. Togni A, Pastor SD, Rihs G (1990) J Organomet Chem 381:C21
29. Zhou X-T, Lin Y-R, Dai L-X, Sun J, Xia L-J, Tang M-H (1999) J Org Chem 64:1331
30. Padilla S, Adrio J, Carretero JC (2012) J Org Chem 77:4161
31. Melhado AD, Luparia M, Toste FD (2007) J Am Chem Soc 129:12638
32. Melhado AD, Amarante GW, Wang ZJ, Luparia M, Toste FD (2011) J Am Chem Soc 133:3517
33. Martín-Rodríguez M, Nájera C, Sansano J (2011) Synlett 2012:62
34. Nájera C, Sansano JM (2014) J Organomet Chem 771:78
35. Martín-Rodríguez M, Nájera C, Sansano JM, Wu F-L (2010) Tetrahedron Asymmetry 21:1184
36. Martín-Rodríguez M, Nájera C, Sansano JM, de Cózar A, Cossío FP (2011) Chemistry 17:14224
37. Martín-Rodríguez M, Nájera C, Sansano JM, de Cózar A, Cossío FP (2011) Beilstein J Org Chem 7:988
38. Li G-H, Zhou W, Li X-X, Bi Q-W, Wang Z, Zhao Z-G, Hu W-X, Chen Z (2013) Chem Commun (Camb) 49:4770
39. Faustino H, Alonso I, Mascareñas JL, López F (2013) Angew Chem Int Ed Engl 52:6526
40. Liu B, Li K-N, Luo S-W, Huang J-Z, Pang H, Gong L-Z (2013) J Am Chem Soc 135:3323
41. Liu F, Wang Y, Ye W, Zhang J (2015) Org Chem Front 2:221
42. Navarro C, Shapiro ND, Bernasconi M, Horibe T, Toste FD (2015) Tetrahedron 71:5800
43. Gao H, Wu X, Zhang J (2010) Chem Commun (Camb) 46:8764
44. Liu F, Qian D, Li L, Zhao X, Zhang J (2010) Angew Chem Int Ed Engl 49:6669
45. Zhang Z-M, Chen P, Li W, Niu Y, Zhao X-L, Zhang J (2014) Angew Chem Int Ed Engl 53:4350

46. Zhang Y, Zhang J (2012) Chem Commun (Camb) 48:4710
47. Gawade SA, Bhunia S, Liu R-S (2012) Angew Chem Int Ed Engl 51:7835
48. Vasu D, Liu R-S (2012) Chemistry 18:13638
49. Jadhav AM, Bhunia S, Liao H-Y, Liu R-S (2011) J Am Chem Soc 133:1769
50. López F, Mascareñas JL (2013) Beilstein J Org Chem 9:2250
51. Kojima M, Mikami K (2011) Synlett 2012:57
52. Sun Y-W, Xu Q, Shi M (2013) Beilstein J Org Chem 9:2224
53. Michon C, Abadie M-A, Medina F, Agbossou-Niedercorn F (2014) Catal Today 235:2
54. Abadie M-A, Trivelli X, Medina F, Capet F, Roussel P, Agbossou-Niedercorn F, Michon C (2014) ChemCatChem 6:2235
55. Du Lee S, Timmerman JC, Widenhoefer RA (2014) Adv Synth Catal 356:3187
56. Kanno O, Kuriyama W, Wang ZJ, Toste FD (2011) Angew Chem Int Ed Engl 50:9919
57. Zhang Z, Widenhoefer RA (2007) Angew Chem Int Ed Engl 46:283
58. LaLonde RL, Sherry BD, Kang EJ, Toste FD (2007) J Am Chem Soc 129:2452
59. Kim JH, Park S-W, Park SR, Lee S, Kang EJ (2011) Chem Asian J 6:1982
60. Rodríguez L-I, Roth T, Lloret Fillol J, Wadepohl H, Gade LH (2012) Chemistry 18:3721
61. Roth T, Wadepohl H, Wright DS, Gade LH (2013) Chemistry 19:13823
62. Zhang Z, Bender CF, Widenhoefer RA (2007) Org Lett 9:2887
63. Zhang Z, Bender CF, Widenhoefer RA (2007) J Am Chem Soc 129:14148
64. Li H, Du Lee S, Widenhoefer RA (2011) J Organomet Chem 696:316
65. Hamilton GL, Kang EJ, Mba M, Toste FD (2007) Science 317:496
66. Aikawa K, Kojima M, Mikami K (2010) Adv Synth Catal 352:3131
67. Bartolomé C, García-Cuadrado D, Ramiro Z, Espinet P (2010) Inorg Chem 49:9758
68. Lalonde RL, Wang ZJ, Mba M, Lackner AD, Toste FD (2010) Angew Chem Int Ed Engl 49:598
69. Teller H, Corbet M, Mantilli L, Gopakumar G, Goddard R, Thiel W, Fürstner A (2012) J Am Chem Soc 134:15331
70. Michon C, Medina F, Abadie M-A, Agbossou-Niedercorn F (2013) Organometallics 32:5589
71. Handa S, Lippincott DJ, Aue DH, Lipshutz BH (2014) Angew Chem Int Ed Engl 53:10658
72. Shu X-Z, Nguyen SC, He Y, Oba F, Zhang Q, Canlas C, Somorjai GA, Alivisatos AP, Toste FD (2015) J Am Chem Soc 137:7083
73. Bandini M, Monari M, Romaniello A, Tragni M (2010) Chemistry 16:14272
74. Bandini M, Gualandi A, Monari M, Romaniello A, Savoia D, Tragni M (2011) J Organomet Chem 696:338
75. Chiarucci M, di Lillo M, Romaniello A, Cozzi PG, Cera G, Bandini M (2012) Chem Sci 3:2859
76. Mukherjee P, Widenhoefer RA (2012) Angew Chem Int Ed Engl 51:1405
77. Chiarucci M, Mocci R, Syntrivanis L-D, Cera G, Mazzanti A, Bandini M (2013) Angew Chem Int Ed Engl 52:10850
78. Uria U, Vila C, Lin M-Y, Rueping M (2014) Chemistry 20:13913
79. Bandini M, Bottoni A, Chiarucci M, Cera G, Miscione GP (2012) J Am Chem Soc 134:20690
80. Uemura M, Watson IDG, Katsukawa M, Toste FD (2009) J Am Chem Soc 131:3464
81. Wang Y-M, Kuzniewski CN, Rauniyar V, Hoong C, Toste FD (2011) J Am Chem Soc 133:12972
82. Wilckens K, Uhlemann M, Czekelius C (2009) Chemistry 15:13323
83. Wilckens K, Lentz D, Czekelius C (2011) Organometallics 30:1287
84. Mourad AK, Leutzow J, Czekelius C (2012) Angew Chem Int Ed Engl 51:11149
85. Mourad A, Czekelius C (2013) Synlett 24:1459
86. Sota Y, Yamamoto M, Murai M, Uenishi J, Uemura M (2015) Chemistry 21:4398
87. Murai M, Uenishi J, Uemura M (2010) Org Lett 12:4788
88. Murai M, Sota Y, Onohara Y, Uenishi J, Uemura M (2013) J Org Chem 78:10986
89. Quach R, Furkert DP, Brimble MA (2013) Tetrahedron Lett 54:5865
90. Handa S, Slaughter LM (2012) Angew Chem Int Ed Engl 51:2912

91. Sethofer SG, Mayer T, Toste FD (2010) J Am Chem Soc 132:8276
92. Miles DH, Veguillas M, Toste FD (2013) Chem Sci 4:3427
93. Guo R, Li K-N, Liu B, Zhu H-J, Fan Y-M, Gong L-Z (2014) Chem Commun (Camb) 50:5451
94. Cera G, Chiarucci M, Mazzanti A, Mancinelli M, Bandini M (2012) Org Lett 14:135
95. Shen Z-Q, Li X-X, Shi J-W, Chen B-L, Chen Z (2015) Tetrahedron Lett 56:4080
96. Zhou G, Liu F, Zhang J (2011) Chemistry 17:3101
97. Chao C-M, Beltrami D, Toullec PY, Michelet V (2009) Chem Commun (Camb) 6988
98. Pradal A, Chao C-M, Toullec PY, Michelet V (2011) Beilstein J Org Chem 7:1021
99. Wang W, Yang J, Wang F, Shi M (2011) Organometallics 30:3859
100. Yang J, Zhang R, Wang W, Zhang Z, Shi M (2011) Tetrahedron Asymmetry 22:2029
101. Pradal A, Chao C-M, Vitale MR, Toullec PY, Michelet V (2011) Tetrahedron 67:4371
102. Qian D, Hu H, Liu F, Tang B, Ye W, Wang Y, Zhang J (2014) Angew Chem Int Ed Engl 53:13751
103. Deschamps NM, Elitzin VI, Liu B, Mitchell MB, Sharp MJ, Tabet EA (2011) J Org Chem 76:712
104. Teller H, Fürstner A (2011) Chemistry 17:7764
105. Yavari K, Aillard P, Zhang Y, Nuter F, Retailleau P, Voituriez A, Marinetti A (2014) Angew Chem Int Ed Engl 53:861
106. Aillard P, Voituriez A, Dova D, Cauteruccio S, Licandro E, Marinetti A (2014) Chemistry 20:12373
107. Alonso I, Trillo B, López F, Montserrat S, Ujaque G, Castedo L, Lledós A, Mascareñas JL (2009) J Am Chem Soc 131:13020
108. González AZ, Toste FD (2010) Org Lett 12:200
109. Alonso I, Faustino H, López F, Mascareñas JL (2011) Angew Chem Int Ed 50:11496
110. Teller H, Flügge S, Goddard R, Fürstner A (2010) Angew Chem Int Ed Engl 49:1949
111. Luzung MR, Mauleón P, Toste FD (2007) J Am Chem Soc 129:12402
112. González AZ, Benitez D, Tkatchouk E, Goddard WA, Toste FD (2011) J Am Chem Soc 133:5500
113. Nakamura K, Furumi S, Takeuchi M, Shibuya T, Tanaka K (2014) J Am Chem Soc 136:5555
114. Shibuya T, Nakamura K, Tanaka K (2011) Beilstein J Org Chem 7:944
115. Oka J, Okamoto R, Noguchi K, Tanaka K (2015) Org Lett 17:676
116. Huang L, Yang H-B, Zhang D-H, Zhang Z, Tang X-Y, Xu Q, Shi M (2013) Angew Chem Int Ed Engl 52:6767
117. Göker V, Kohl SR, Rominger F, Meyer-Eppler G, Volbach L, Schnakenburg G, Lützen A, Hashmi ASK (2015) J Organomet Chem 795:45
118. Fang W, Presset M, Guérinot A, Bour C, Bezzenine-Lafollée S, Gandon V (2014) Org Chem Front 1:608
119. Inamdar SM, Konala A, Patil NT (2014) Chem Commun (Camb) 50:15124
120. Zhang S, Wei F, Song C, Jia J, Xu Z (2014) Chin J Chem 32:937
121. Han Z-Y, Guo R, Wang P-S, Chen D-F, Xiao H, Gong L-Z (2011) Tetrahedron Lett 52:5963
122. Cala L, Mendoza A, Fañanás FJ, Rodríguez F (2013) Chem Commun (Camb) 49:2715
123. Wu H, He Y-P, Gong L-Z (2013) Org Lett 15:460
124. Horino Y, Takahashi Y, Nakashima Y, Abe H (2014) RSC Adv 4:6215
125. Campbell MJ, Toste FD (2011) Chem Sci 2:1369
126. Tu X-F, Gong L-Z (2012) Angew Chem Int Ed Engl 51:11346
127. Patil NT, Raut VS, Tella RB (2013) Chem Commun (Camb) 49:570
128. Du Y-L, Hu Y, Zhu Y-F, Tu X-F, Han Z-Y, Gong L-Z (2015) J Org Chem 80:4754
129. Corma A, Domínguez I, Doménech A, Fornés V, Gómez-García CJ, Ródenas T, Sabater MJ (2009) J Catal 265:238

Top Heterocycl Chem (2016) 46: 249–284
DOI: 10.1007/7081_2015_5007

Published online: 3 February 2016

Gold Catalysis in the Synthesis of Natural Products: Heterocycle Construction via Direct C–X-Bond-Forming Reactions

Yu-Hui Wang, Zhong-Yan Cao, and Jian Zhou

Abstract In this chapter, a summarization of the application of gold-catalyzed heterocycle synthesis involving direct C-X bond formation in natural product synthesis is given, with particular emphasis on the advantage of gold catalysis in tackling the chemo- and regioselectivity problem.

Keywords Chemoselectivity • Gold catalysis • Heterocycle construction • Natural product synthesis • Regioselectivity

Contents

1 Introduction

While transition-metal catalysis has been well established as a powerful strategy in the total synthesis of natural products, the potential of gold catalysis in organic synthesis is just gradually recognized in this century with the golden rush. Possibly,

Y.-H. Wang, Z.-Y. Cao, and J. Zhou (✉)
Shanghai Key Laboratory of Green Chemistry and Chemical Process, School of Chemistry and Molecular Engineering, East China Normal University, Shanghai 200062, P. R. China
e-mail: jzhou@chem.ecnu.edu.cn

two major prejudices retard the exploration of gold catalysis: one is that gold catalysts are expensive, and the other is that gold is inert. However, recent comprehensive studies have unveiled interesting properties of gold catalysts, including the superior π-acidity, the ability to stabilize adjacent carbocation, and the reluctance to undergo oxidative additions and reductive eliminations, which are strongly affected by the relativistic effects [1, 2]. In addition, as a type of excellent soft acids, gold(I) catalysts preferentially bind to "soft" bases such as alkynes and allenes, rather than "hard" bases such as oxygen, which contributes to the high air and moisture compatibility of gold catalysts, very helpful to develop operationally friendly synthetic methods [3–12]. Another important merit is that the catalytic properties of gold catalysis can be readily tuned by varying the structure of ligands or counteranions [13], which provides a flexible way to maximize the reaction outcome. In light of these attractive features, the exploration of gold catalysis in new methodology development is at the frontier of catalysis research during the past decade, which further fuels the application of gold catalysis to the total synthesis of natural products.

In 2000, Hashmi et al. first demonstrated the capacity of homogeneous gold catalysis in the construction of complex ring systems, which represents a very important task in natural product synthesis [14, 15]. This seminar work, along with several elegant protocols developed in 2004, including the Conia-ene reaction of β-ketoesters and 3,3-rearrangement reaction developed by Toste et al. [16, 17], the enyne cyclization reactions disclosed by Echavarren et al. [18], and the enyne cycloisomerization process reported by Zhang and Kozmin [19], greatly arouse the enthusiasm in exploring the potential of gold catalysis in the synthesis of complex natural products. Today, it is no exaggeration to say that gold catalysis has proved to be a valuable tool in the total synthesis of natural products, owing to the good functional group tolerance, mild reaction condition, and the powerfulness in building up molecular complexity from readily available starting materials. In 2008, Hashmi et al. first reviewed this application of gold catalysis in natural product synthesis and then updated it in 2012 [20, 21], which aimed at introducing basic reaction types of gold catalysis and was organized by the substrate types. Most recently, Yang et al. reviewed gold-catalyzed chemo- and regioselective transformations in complex natural product synthesis [22]. In addition to these excellent contributes, we present herein a comprehensive survey of synthesis of natural product involving direct C–O and C–N bond formation events. The synthesis of oxygen- or nitrogen-containing heterocycles without C–X bond-forming reactions is not the scope of this review (for examples of natural product synthesis using Au-catalyzed oxygen or nitrogen-containing heterocycle formation but without direct C–X bonding forming reactions [23–30]). Furthermore, the scope of the chapter deals also with highlighting the advantage of gold catalysis in tackling the challenges associated with the control of chemo- and regioselectivity.

2 Applications in the Synthesis of Natural Products

2.1 *Oxygen-Containing Heterocycle Formation*

2.1.1 Hydrofunctionalization of Alkynes

A remarkable example of [Au(I)]-catalyzed hydroalkoxylation to construct pyran ring system was reported by Trost and Dong in the first asymmetric total synthesis of bryostatin 16 [31, 32] that exhibits remarkable anticancer activity [33]. To furnish the acid and/or base-sensitive C ring of this macrocycle, a 6-*endo-dig* cyclization was designed at the end of the synthesis. The use of cationic [Au(I)] catalysts enabled highly chemo- and regioselective intramolecular hydroalkoxylation of compound **4** to construct the six-membered pyran ring under mild condition. As shown in Scheme 1, when multifunctional macrocycle **4** was subjected to 20 mol% PPh_3AuSbF_6, prepared from PPh_3AuCl and $AgSbF_6$, the electron-deficient alkynyl moiety was effectively activated for the intramolecular attack by adjacent hydroxyl group in a 6-*endo-dig* manner, forming the C ring of bryostatin 16 successfully. Since the resultant **5** was acid sensitive, an excess $NaHCO_3$ was employed. Noteworthy, the use of Pd catalyst in this step resulted in poor results in terms of 5-*exo* vs. 6-*endo* selectivity [34], which highlighted the advantage of [Au(I)] catalysis.

A second example of [Au(I)]-mediated intramolecular oxo-Michael addition of a hydroxy group to electron-deficient alkyne was reported in 2012 by Bihelovic and Saicic [35, 36], during their effort in the enantioselective total synthesis of (−)-atrop-abyssomicin C, the first known substance in bacteria that inhibits the biosynthesis of *p*-aminobenzoate (*p*ABA) [37]. It was found that under the catalysis of 10 mol% Gagosz's gold catalyst ($Ph_3PAuNTf_2$), the cyclization carried out in CH_2Cl_2 at 120°C failed to afford the desired tricyclic ether **8** and only led to the

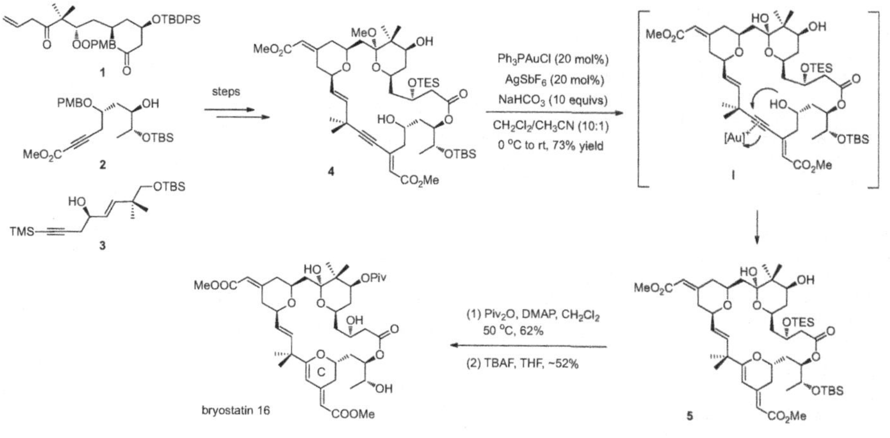

Scheme 1 The total synthesis of bryostatin 16

undesired bicyclic ketone **7** (Scheme 2). Nevertheless, this result was encouraging, as it suggested the formation of (*Z*)-**II**, which further underwent a rearrangement to ketone **7**. After screening other reaction parameters, it was found that when using ^{i}PrOH as the solvent, intermediate (*Z*)-**II** readily underwent an isomerization to the desired *E* intermediate upon irradiation with UV light (254 nm), followed by a base-mediated lactonization to the desired tricyclic ether **8**, paving the way for the total synthesis of naturally occurring (−)-atrop-abyssomicin C. Noticeably, no reaction took place when using $AuCl_3$ as catalyst, no matter AgOTf was used or not, which suggested the activation of alkyne by [Au(I)] was important for reaction course.

The [Au(I)]-catalyzed hydroalkoxylation of electron-neutral alkyne was employed in Fürstner's second-generation total synthesis of spirastrellolide F methyl ester [38], to furnish fused spirocyclic BC ring system (Scheme 3). Initial investigation disclosed that under AuCl·SMe_2 catalysis (10 mol%), only regiospecific 5-*exo-dig* adduct **12** was produced from macrocycle **9**, while the use of Echavarren's catalyst **10** enabled the selective 6-*endo-dig* addition reaction. Although the *endo*/*exo* ratio was not high (ca. 5:1), the desired compound **11** was achieved in 62% yield, which contributed to the total synthesis of the target product.

Besides six-membered pyran ring systems, [Au(I)]-catalyzed cyclization of homopropargylic alcohols also provided a facile way to access five-membered oxoheterocycles, as exemplified by the total synthesis of (+)-cephalostatin 1, isolated from the marine worm *Cephalodiscus gilchristi* in only $8.36 \times 10^{-4}\%$ yield. As it proved to be a potent cell growth inhibitor for the *p16* tumor suppressor gene [39, 40], Shair et al. attempted to develop an enantioselective synthetic route to prepare this compound in sufficient amount for medicinal studies (Scheme 4) [41]. Commenced with commercial available steroid trans-androsterone **13**, diol intermediate **14** could be obtained. Noticeably, the [Au(I)]-catalyzed gram-scale 5-*endo* cyclization took place on the highly hindered internal alkyne of **14** chemoselectively to afford dihydrofuran **15** in 88% yield. The possible 4-*exo* cyclization of diol **14** was inhibited due to very constrained four-membered-ring intermediate.

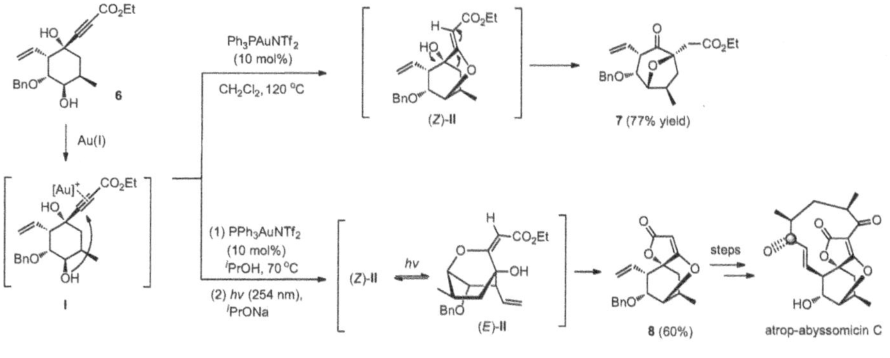

Scheme 2 The total synthesis of (−)-atrop-abyssomicin C

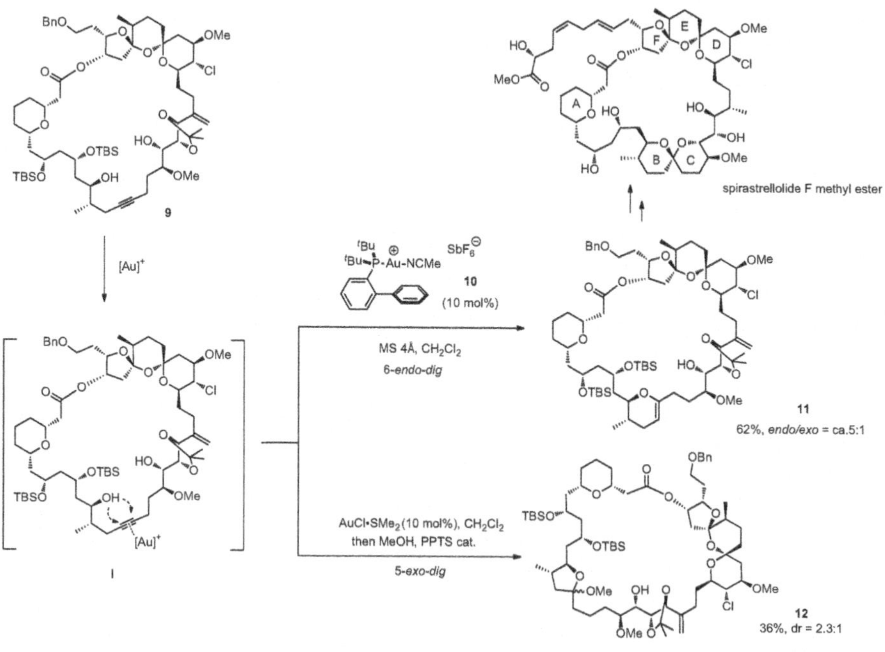

Scheme 3 The total synthesis of spirastrellolide F methyl ester

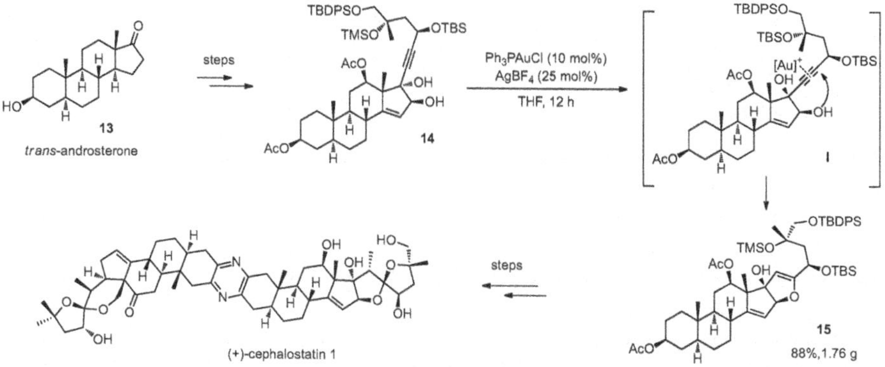

Scheme 4 The total synthesis of (+)-cephalostatin 1

In 2013, the Ley group developed a facile gold-catalyzed alkoxycyclization as the key reaction to construct the tetrahydropyran subunit of natural product (−)-hennoxazole A [42], which displays potency against herpes simplex virus type 1, as well as peripheral analgesic activity [43]. The functionalized diol **16** underwent a ring closure under the catalysis of 5 mol% of (IMes)$AuNTf_2$ **17** in MeOH at 55°C, giving tetrahydropyran **18** as a single diastereomer in 91% yield (Scheme 5). Then

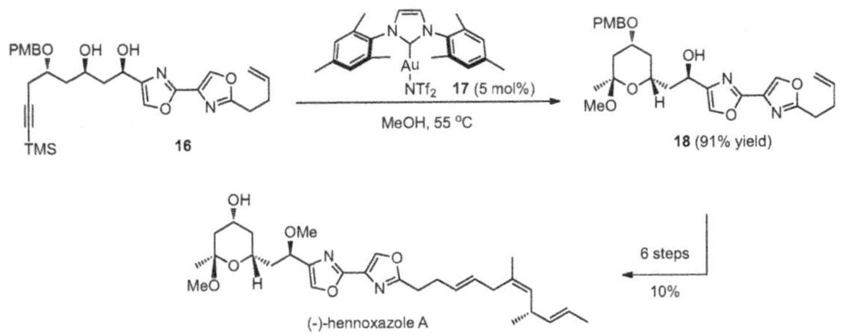

Scheme 5 The alkoxycyclization in the total synthesis of (−)-hennoxazole A

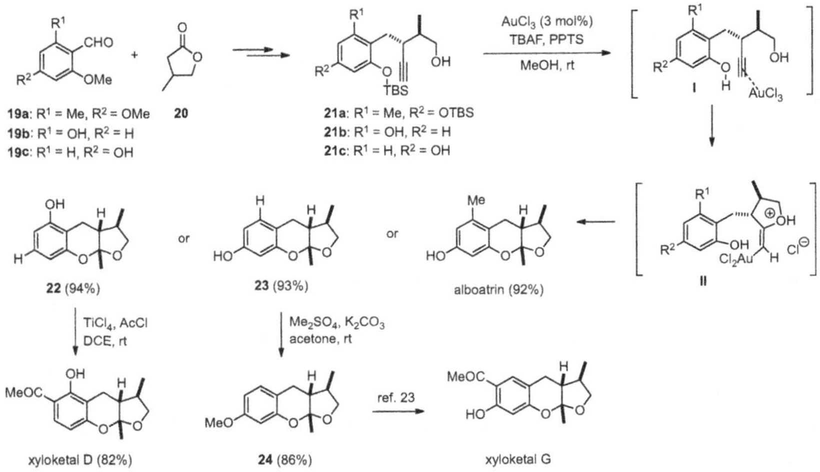

Scheme 6 One-pot desilylation-gold-catalyzed cycloisomerization in the total synthesis of alboatrin, xyloketal D and xyloketal G

18 was taken forward to complete the total synthesis of (−)-hennoxazole A over six steps.

Bioactive natural products, alboatrin [44, 45] and xyloketal D and G, share a common structural motif, tetrahydrofuranobenzopyran ring system. In 2013, Panda and Sarkar developed a novel one-pot desilylation–gold-catalyzed cycloisomerization to furnish this core tricyclic ketal unit (Scheme 6) [46]. The precursor alkynes **21** contained a TBS-protected phenolic OH and a terminal alcoholic OH. The TBS ether was first removed by tetrabutylammonium fluoride (TBAF) to release intermediate **I** for ring closure. An $AuCl_3$-catalyzed 5-*exo-dig* annulation afforded enol ether **II**, which further cyclized to finalize the thermodynamically more stable product having a cis ring junction. All three natural products could be accessible in excellent yields using different substituted aromatic aldehydes

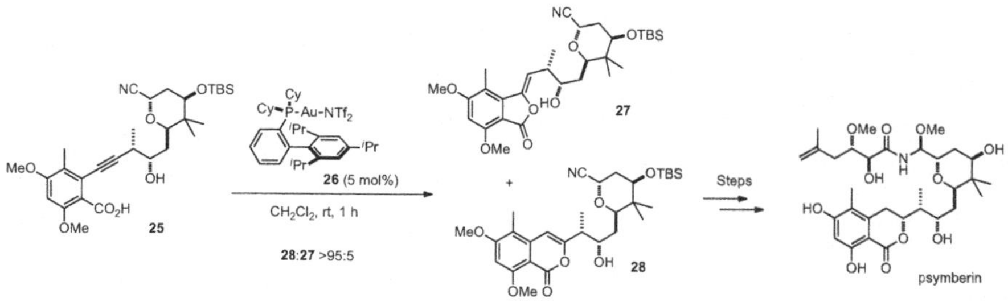

Scheme 7 The total synthesis of psymberin

[47]. This process was also adaptable to chiral synthesis with the enantiopure lactone (*R*)-**20** in the initial condensation step.

Apart from hydroxyl group, carboxylic acid has also been used as viable nucleophile for hydrofunctionalization of alkyne enticed by gold catalyst. For example, in Brabander's total synthesis of psymberin [48], the key isocoumarin formation from *o*-alkynyl benzoic acid **25** was achieved by using [Au(I)] catalysis (Scheme 7). Two challenges were on the way to the desired transformation: one is the coexistence of an extra hydroxy group which might result in competitively chemoselective addition problem, and the other was the regioselective issue arising from the 5-*exo* or 6-*endo* cyclization. When using *p*-TsOH, TFA, or $AuCl_3$ as catalyst, neither the desired 6-*endo* product **28** nor 5-*exo* product **27** was detected. When $AgSbF_6$, $[Pt(C_2H_4)Cl_2]_2$, JohnPhos-ligated AuCl, $Ph_3PAuSbF_6$, or $Ph_3PAuNTf_2$ was used, the ratio of **28** to **27** was poor. Helpfully, $XPhosAuNTf_2$ **26** turned out to be potent, allowing the reaction to work efficiently at room temperature, affording dihydroisocoumarin **28** in excellent regioselectivity (>95:5).

2.1.2 Spiroketalization

An important extension of hydroalkoxylation of alkynes is the formation of spiroketals via [Au(I)]-catalyzed spiroketalization of functionalized alkynes bearing two tethered hydroxy groups or mixed hydroxy/alkoxy moieties. This approach has been successfully utilized in the total synthesis of a broad range of spiroketal-bearing natural products that display various important biological activities. The use of an alkyne as a dehydrated surrogate for a ketone brings about some attractive features. First, it is possible to use the multifunctional ethyne-based building blocks free from protecting or activating groups, which are often indispensable when multifunctional ketone substrates are involved. Second, it offers the promise to achieve the desired selective reactions under mild conditions, even with the coexistence of multiple functionalities, by carefully tuning the structure of substrates, catalytic properties of gold catalysts, and other reaction parameters.

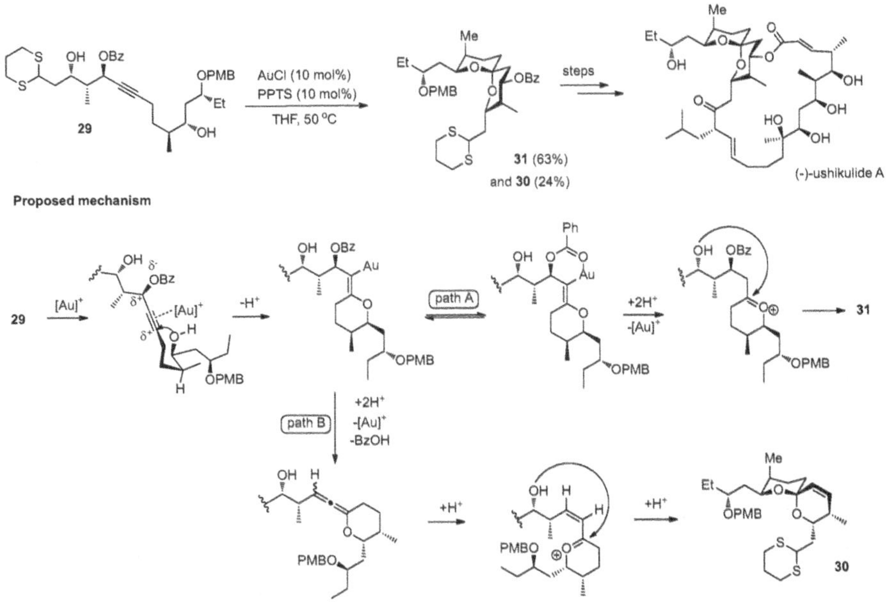

Scheme 8 The total synthesis of (−)-ushikulide A

For example, in the first total synthesis of (−)-ushikulide A [49], a natural product isolated from a culture broth of *Streptomyces* sp. IUK-102 with immunosuppressant activity [50], Trost et al. found that [Au(I)] catalysis played an important role in converting diol-alkyne **29** to the desired spiroketal **31** in 63% yield, although unwanted elimination product **30** was still obtained in 24% yield via the hydrolysis of benzoate (Scheme 8). In the model reaction using an analogue of **29**, where the thioacetal moiety of **29** was replaced by a CH_2OBn group, it was found that no desired product was observed when using 10 mol% of $PdCl_2(MeCN)_2$ with pyridinium *p*-toluenesulfonate (PPTS) at 50°C in MeOH, or 10 mol% of Pt (COD) Cl_2 with PPTS at 60°C in *i*-Pr_2O, but the merging of AuCl and PPTS worked even at room temperature. The benzoyl group mattered to secure high yield of the desired spiroketal **31** (path A); otherwise, the corresponding elimination side product 30 dominated. During condition optimization, it turned out that the product distribution was very sensitive to the catalyst system, and the merger of AuCl and PPTS (10 mol%, each) and the use of THF as the solvent were the choice for the desired cyclization while suppressing the unwanted elimination.

Okadaic acid is a polyether natural product isolated from the marine sponges *Halichondria okadai* and *H. melanodocia* in 1981 [51]. Forsyth and co-workers nicely utilized the Au-catalyzed spiroketalization to construct both C19 and C34 spiroketal moieties of this complex molecule (Scheme 9) [52]. First, it was found that a catalytic amount of AuCl enabled the spiroketalization of intermediate **32** to synthesize the C19 spiroketal of okadaic acid with high diastereoselectivity. On the

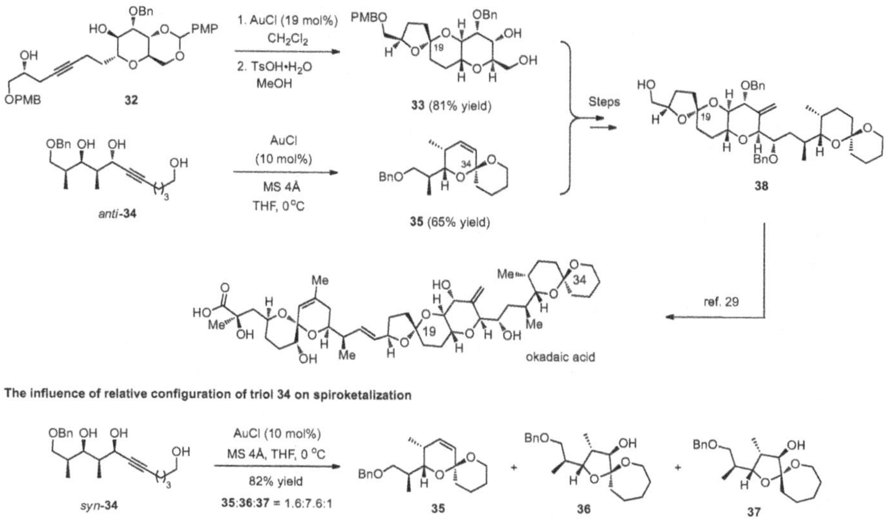

Scheme 9 The formal total synthesis of okadaic acid

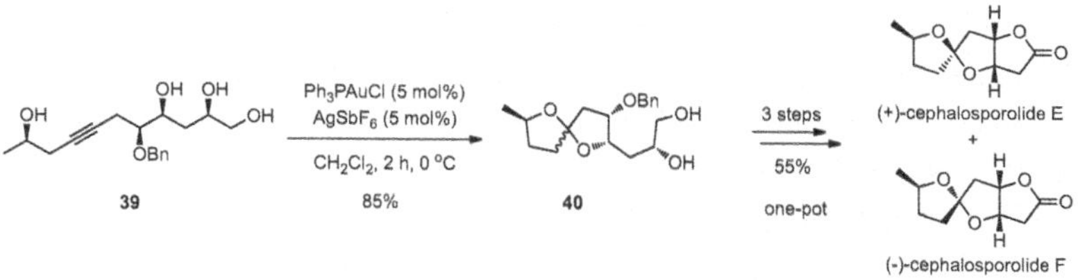

Scheme 10 The total synthesis of (+)-cephalosporolide E and (−)-cephalosporolide F

other hand, for the spiroketalization to **35**, the relative configuration of triol **34** greatly affected the selectivity. While anti-**34** delivered [6.6]-spiroketal **35** in 65% yield, three different resultants (**35**:**36**:**37** = 1.6:7.6:1) were observed in the case of syn-**34**, due to poor regioselectivity. To avoid material loss, a sequential oxidation and diastereoselective reduction were used to convert syn-**34** to anti-**34**. Based on the two ketal fragments **33** and **35**, compound **38** was then obtained, a known intermediate in Isobe's or Ley's route for the total synthesis of okadaic acid [53–55].

[Au(I)]-catalyzed alkynediol spiroketalization was also coupled into a modular total synthesis of cephalosporolides E and F, by Ramana et al., to forge the central bicyclic spiroketal core [56]. The use of 5 mol% $PPh_3AuCl/AgSbF_6$ allowed alkynetetrol **39** to afford the desired spiroketal **40** as a mixture of diastereomers in 85% yield under mild condition, whereas the Pd catalysts tested gave product **40** in poor yield with side products (Scheme 10). Based on intermediate **40**, subsequent one-pot transformations provided both cephalosporolides E and F in 55% combined yield.

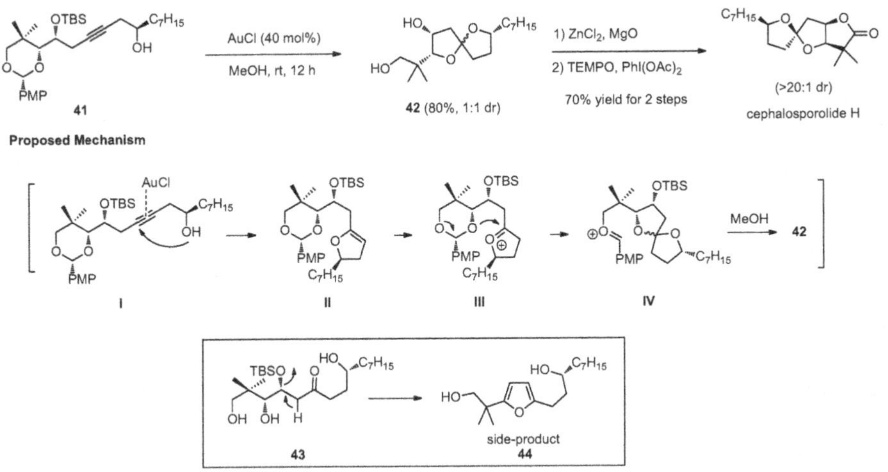

Scheme 11 The total synthesis of cephalosporolide H

As shown above, the control of regioselectivity of the spiroketalization is difficult in some cases, especially when multiple hydroxy groups coexist. To meet this challenge, acetal or ether groups are attempted as masked hydroxy groups for spiroketalization; nevertheless, for a better reactivity, an unprotected hydroxyl is usually used to initiate the nucleophilic addition, followed by the participation of another masked oxygen-containing group.

For instance, Dudley et al. employed a gold-catalyzed spiroketalization involving an acetal moiety in the total synthesis of cephalosporolide H [57, 58], which was isolated from the culture broth of the marine fungus *Penicillium* sp. and has a structure similar to cephalosporolides E and F (Scheme 11) [59]. Initial attempts based on unprotected triols ended in the formation of complex mixtures, due to poor regioselectivity. Therefore, homopropargyl **41** with an acetal moiety was used for reaction development, as the gold-catalyzed 5-*endo-dig* cyclization should proceed in high regioselectivity, and the following protonation of enol ether **II** induced the opening of acetal and furnished spiroketal **42** via alcoholysis. In the presence of 40 mol% of AuCl, the cycloisomerization of **41**, carried out in MeOH, afforded the desired spiroketal **42** in 80% yield as a 1:1 mixture of diastereomers. The two spiroketal epimers of **42** could be converted into a single diastereomer upon coordination with $ZnCl_2$. The subsequent TEMPO oxidation completed the synthesis of cephalosporolide H. During condition optimization, furan **44** was detected as a by-product. Possibly, a hydration of alkyne to ketone **43** took place, and a further elimination reaction led to this furan side product. This result implied the advantage of gold-catalyzed alkynediol spiroketalization over traditional ketone-diol condensation method to spiroketals.

Forsyth et al. also utilized the ether as the decorated hydroxyl group, in the synthesis of the A–D rings of azaspiracid via [Au(I)]-catalyzed bis-spiroketalization (Scheme 12) [60]. Under acidic condition, 8 mol% of AuCl catalyst

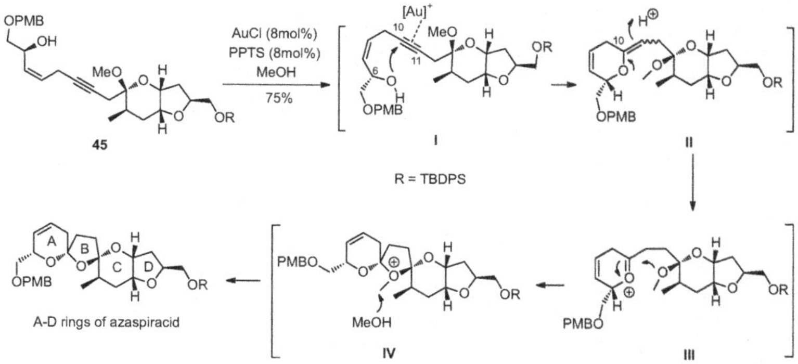

Scheme 12 The synthesis of A–D rings of azaspiracid

allowed the formation of bis-spirocycle as a single isomer in 75% yield. This reaction was proposed to be initiated by the addition of the hydroxy group to the activated alkyne to form enol ether **II**, the protonation of which enticed the methoxy group to attack C10 in a 5-*exo-trig* fashion to close the B ring. A final transfer of the methyl group to solvent MeOH quenched the oxonium species **IV** and generated the desired target.

2.1.3 Addition of Carbonyl Group to Alkynes

The oxygen atom of carbonyl group is also capable of nucleophilic addition to π-coordinated alkyne-Au complexes as well. In this context, acetylenic β-ketoester **46** is a type of popular substrates for reaction design. In addition to readily accessibility, they can undergo cyclization reaction under the catalysis of gold catalysts under mild condition to construct pyrone skeleton, widely present in natural products. Furthermore, by varying substitute R in acetylenic β-ketoesters, two different types of products may be obtained, as shown in Scheme 13. Once the triple bond of **46** coordinates to [Au(I)], the ester carbonyl attacks the electron-deficient site of the π-system to form a six-membered-ring intermediate **II** or **II′**. If R is not *t*-Bu group, it will traverse tautomerization and protodeauration, leading to the corresponding 2-alkoxy-4-pyrone **47** (path A). On the other hand, when R is a *t*-Bu group, the mechanism follows path B that the drop of *tert*-butyl group in **II′** may release intermediate **III′**, leading to the formation of 2-pyrone **48**.

The viability of path A was recently demonstrated by Fürstner et al. in the concise synthesis of an exceptional brominated 4-pyrone derivative of algal origin [61], isolated from red algae of *Phacelocarpus labillardieri* in southern Australian waters [62]. As shown in Scheme 14, the coupling of two fragments **49** and **50** via esterification gave acetylenic β-ketoesters **51** in 70% yield. Upon treating with 3 mol% of SPhosAuNTf$_2$ **52** in a mixed solvent of MeCN/HOAc, compound **51** afforded 2-alkoxy-4-pyrone **53** in 97% yield, through path A in Scheme 13. Specifically, the presence of another two electron-rich C–C triple bonds did not affect

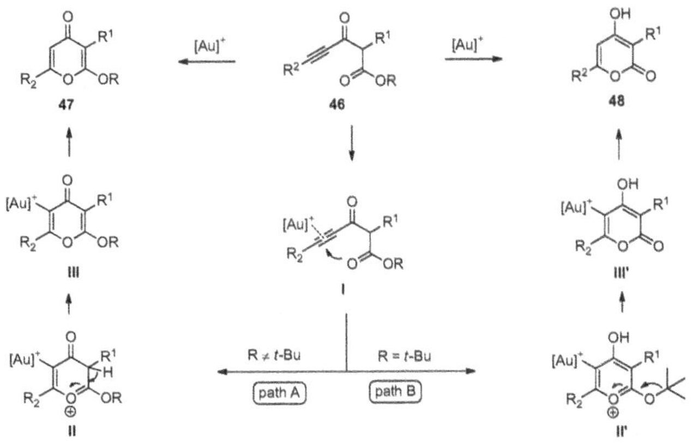

Scheme 13 General mechanism for transformation from acetylenic β-ketoesters to 2-pyrone or 4-pyrone via gold catalysis

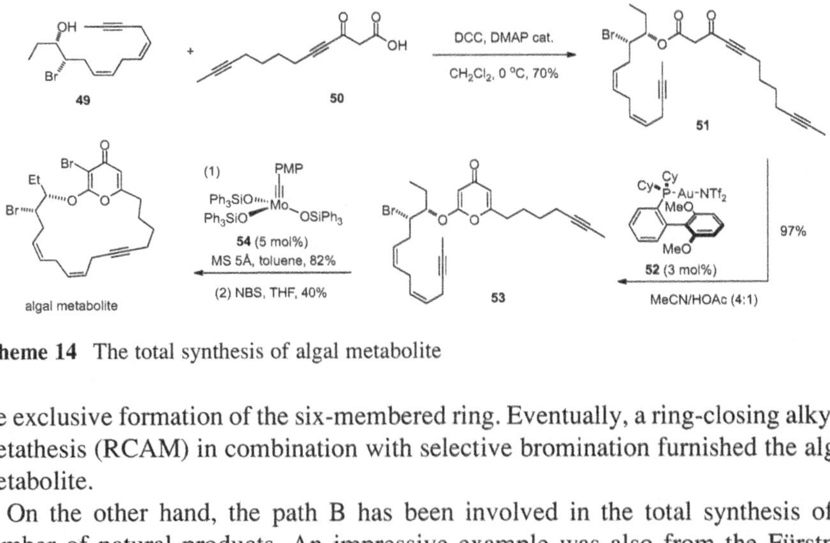

Scheme 14 The total synthesis of algal metabolite

the exclusive formation of the six-membered ring. Eventually, a ring-closing alkyne metathesis (RCAM) in combination with selective bromination furnished the algal metabolite.

On the other hand, the path B has been involved in the total synthesis of a number of natural products. An impressive example was also from the Fürstner group, who utilized this protocol for the efficient construction of the 4-hydroxy-2-pyrone motif in neurymenolide A [63], which was isolated from the Fijian red alga *Neurymenia fraxinifolia* and shows activity against methicillin-resistant *Staphylococcus aureus* and vancomycin-resistant *Enterococcus faecium* [64]. As shown in Scheme 15, the critical phase of the total synthesis of neurymenolide A was how to selectively convert highly sensitive and functionalized precursor **55** to product **56** featuring the desired 4-AcO-2-pyrone moiety, as an appreciable challenge was that the embedded multiple unsaturated bonds might cause side reactions [16]. As expected, initial trials in carrying out the ring closure in aprotic condition using SPhosAuNTf$_2$ **52** only resulted in the decomposition, but the use of gold complex

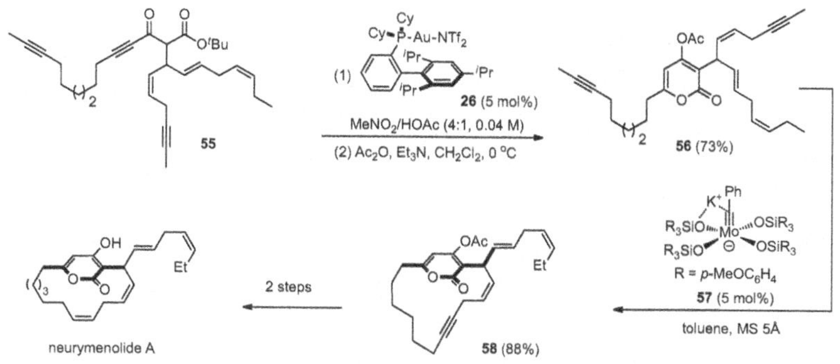

Scheme 15 The total synthesis of neurymenolide A

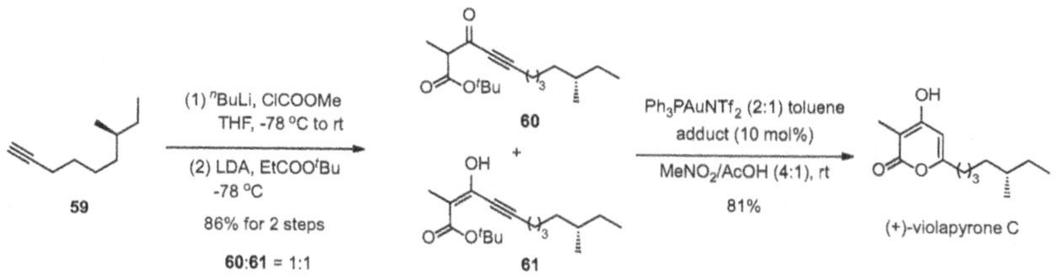

Scheme 16 The total synthesis of (+)-violapyrone C

26 derived from a bulkier ligand XPhos gave the desired product **56** in 26% yield, accompanied by other side products and 37% of starting material **55** recovered. Fortunately, the use of HOAc as cosolvent greatly improved the yield of pyrone **56** to 73%, possibly via facilitating the protodeauration step within the catalytic cycle. With pyrone **56** in hand, the total synthesis of neurymenolide A was finally achieved.

Likewise, the [Au(I)]-catalyzed 2-pyrone synthesis was also employed by Lee et al. in the total synthesis of (+)-violapyrone C (Scheme 16) [65]. Appending an ester group to alkyne **59** followed by a Claisen condensation afforded **60** along with its tautomer **61** in 1:1 ratio. The two inseparable intermediates were then subjected to 10 mol% of $Ph_3PAuNTf_2$ in a mixed solvent of $MeNO_2$/AcOH, and (+)-violapyrone C was readily obtained in 81% yield at ambient temperature. Various metal catalysts, including $PdCl_2$, Ag_2CO_3, $InCl_3$, $Sc(OTf)_3$, and $AuCl_3$, were examined in this step, but all exhibited lower efficiency than that of [Au(I)] catalysis.

In the first total synthesis of wailupemycin reported by Li and Liu in 2014, [Au(I)] catalysis was implemented in two critical steps, including the pyrone formation (Scheme 17) [66, 67]. The synthesis started from the cyclization of furanyne **62** with a propargylic alcohol moiety, under the catalysis of 2 mol% of Echavarren's gold complex **10** in DCE at 80°C. The *endo* attacking of C2 carbon atom of furan ring to the triple bond, followed by ring opening and aromatization terminated by a 1,2-migration of proton to the vicinal carbon finalized the (*E*)-**63** as major isomer in

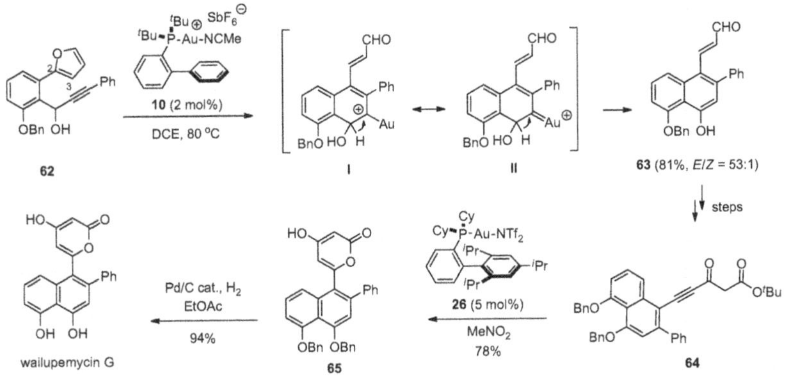

Scheme 17 The total synthesis of wailupemycin G

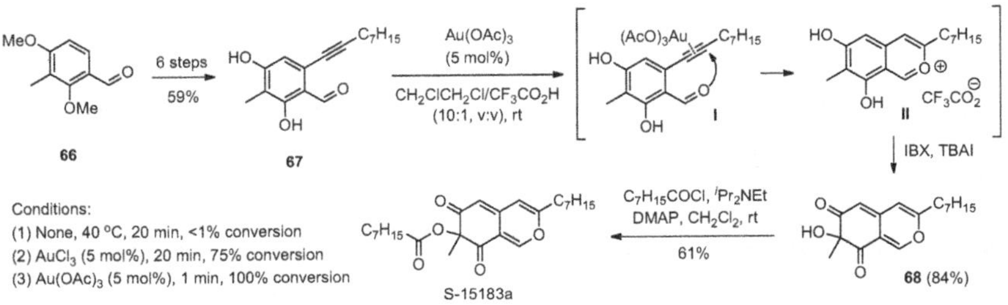

Scheme 18 Gold-catalyzed azaphilone formation in the total synthesis of S-15183a

81% yield. This process offered an efficient access to multisubstituted 1-naphthols with an enal moiety at the C4 position. In late stage, the gold-catalyzed 2-pyrone synthesis emerged again to complete the recurring 2-pyrone substructure, and 78% yield for product **65** was achieved when using XPhosAuNTf$_2$ **26** in MeNO$_2$. The natural wailupemycin G was then attained in 94% yield after Pd/C hydrogenation.

Apart from the addition of ester carbonyl group to alkynes for ring closure, a gold-catalyzed hydrofunctionalization of alkyne using aldehyde carbonyl group to 4*H*-pyran system was also developed by the Porco, Jr group in the total synthesis of S-15183a [68] that has a sphingosine kinase inhibitory activity (Scheme 18) [69]. With **67** as the model substrate, condition optimization for the cycloisomerization step indicated that almost no conversion was observed without catalyst. In contrast, the use of 5 mol% of AuCl$_3$ allowed the promising result in **68** in 75% conversion within 20 min. When switching to Au(OAc)$_3$, full conversion was achieved within 1 min. The synthesis of azaphilone **68** from alkyne-aldehyde **67** possibly proceeded via the attack of aldehyde carbonyl group to triple bond activated by [Au(III)], which afforded the benzopyrylium salt **II** that was oxidized to generate the azaphilone **68** in 84% yield by using *o*-iodoxybenzoic acid (IBX) in a mixed solvent of 1,2-dichloroethane/trifluoroacetic acid.

2.1.4 Hydrofunctionalization of Allenes

Owing to the easy availability and rich chemistry of allenes [70–73], gold-catalyzed functionalization of allene derivatives also emerges as a fruitful strategy in construction of heterocycles for the total synthesis of natural products. However, compared with alkynes, the presence of more reactive sites in allenes [74] often makes it difficult to realize good selectivity. A common solution is to develop reactions generating 5- or 6-membered ring to simplify the problem. From this aspect, α-hydroxyallenes such as **69** are widely applied for the total synthesis of dihydrofuran-containing natural products. Generally, the gold catalyst first interacts with the π-system of the allenyl group in **69** to form the intermediate **I**, followed by the nucleophilic oxygen atom attack the π-system in a 5-*endo-trig* fashion. The subsequent protodeauration of the vinyl gold intermediate **II** sets free the product 2,5-dihydrofuran **70** and the gold catalyst (Scheme 19).

Krause et al. attempted an early example in the synthesis of citreoviral [75, 76], which could be derived from the 2,5-dihydrofuran framework. The α-hydroxyallene **72** was easily prepared from enyne **71** as a mixture of diastereomers, which was then converted to polysubstituted dihydrofuran **73** in 80% yield, under the catalysis of 5 mol% $AuCl_3$ (Scheme 20). In this case, gold(III) catalysis proved to be superior to Ag catalysis, since the same reaction reported by Marshall [77] involved the use of 0.8 equivs of $AgNO_3$ catalyst, and no TON of silver catalyst was observed.

Meanwhile, the Krause group also worked on the total synthesis of (+)-linalool oxide, another tetrahydrofuran-containing natural product that widely exists in the many essential oils as well as oolong, green and black tea [78–80]. Since the (2*R*)-isomer of linalool oxide processes a leafy earthy smell, while the (2*S*)-isomer has a sweet floral creamy flavor [81–83], stereoselective synthesis of this natural product would benefit its potential application in perfumery production. Their synthetic route [84] embarked on the conversion of ester **74** to chiral dihydroxyallene **75**. Upon treatment with 0.1 mol% of $AuCl_3$, the stereochemical information of **75** was successfully transferred into dihydrofuran **76** (Eq. 1, Scheme 21), contributing to the synthesis of the (2*S*,5*R*)-linalool oxide in >99% ds and 97% ee. This procedure

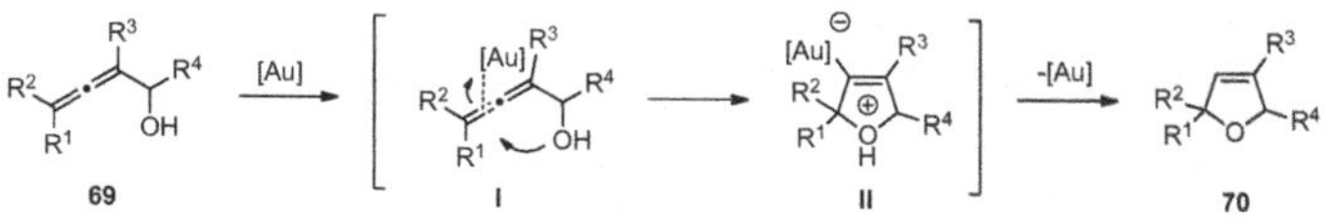

Scheme 19 General mechanism for intramolecular hydroalkoxylation of allenes

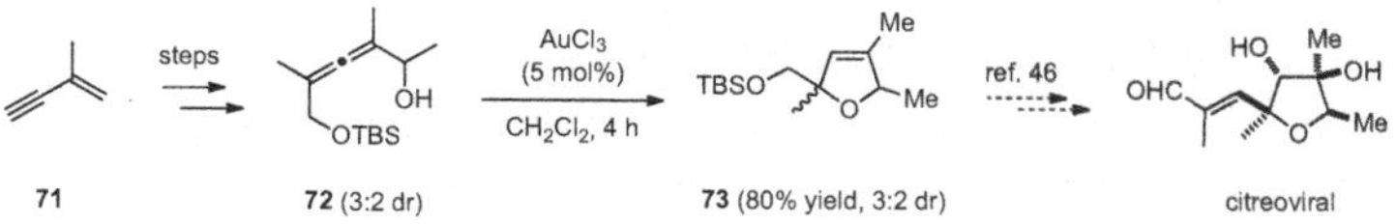

Scheme 20 The formal synthesis of citrovital

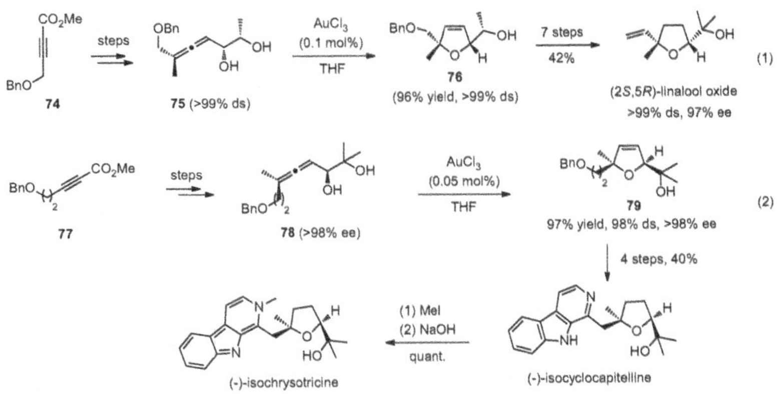

Scheme 21 The total synthesis of linalool oxide, isochrysotricine, and isocyclocapitelline

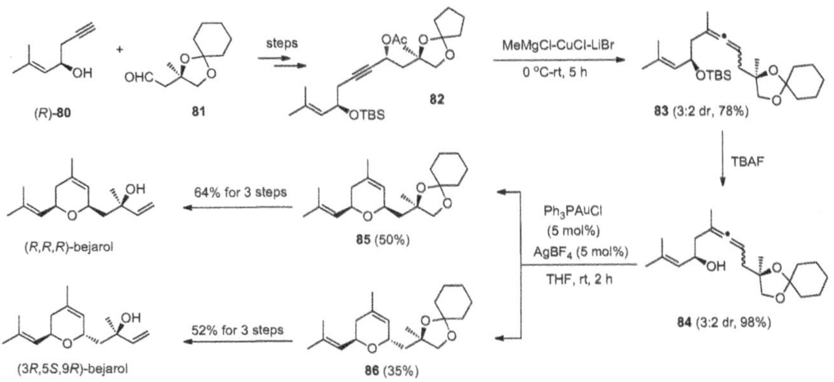

Scheme 22 The total synthesis of (*R*,*R*,*R*)-bejarol and (3*R*,5*S*,9*R*)-bejarol

was further applied to the enantioselective total synthesis of (−)-isocyclocapitelline and (−)-isochrysotricine [85], widely used in the traditional Chinese and Vietnamese medicine [86]. In this case (Eq. 2), high efficiency was maintained when the catalyst loading to was lowered down to 0.05 mol%, giving **79** in 97% yield, 98% ds, and >98% ee. From this precursor, both (−)-isocyclocapitelline and (−)-isochrysotricine were accessible.

On the other hand, the gold-catalyzed 6-*endo-trig* cyclization of β-hydroxyallenes readily gave pyran moiety, as shown in the first total synthesis of (*R*,*R*,*R*)-bejarol and its (3*R*,5*S*,9*R*)-isomer by the Krause group (Scheme 22) [87]. Starting from alcohol (*R*)-**80** and aldehyde **81**, propargyl acetate **82** was obtained and transformed to allene **83** in a good yield of 78%, albeit with poor 3:2 dr value. After removing TBS group, the following gold-catalyzed 6-*endo-trig* cyclization gave a mixture of dihydropyrans **85** and **86**, which were then converted to the target molecules in short sequences.

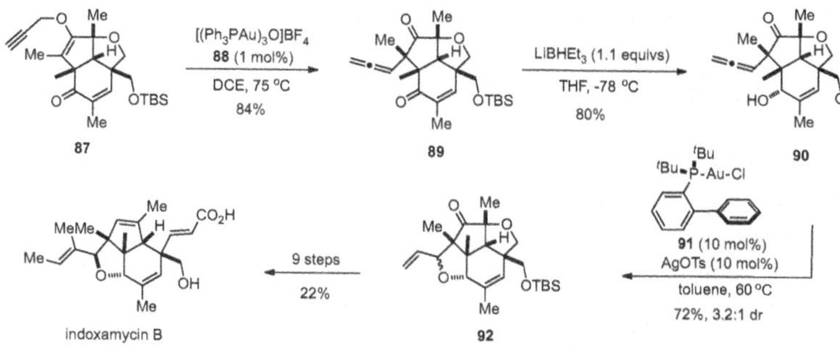

Scheme 23 The total synthesis of indoxamycin B

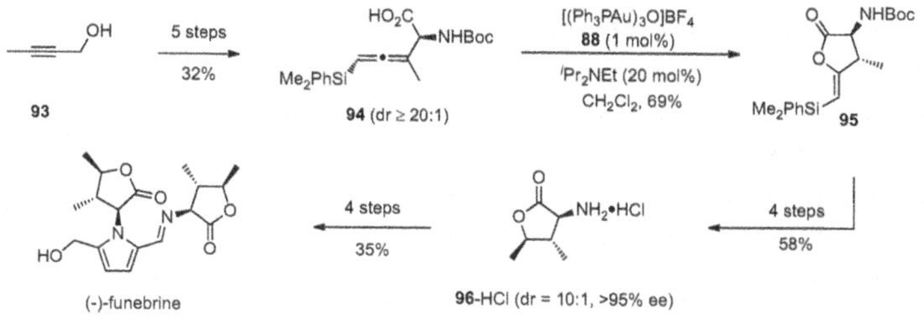

Scheme 24 The total synthesis of (−)-funebrine

In addition, γ-hydroxyallene was also viable for ring formation, as evidenced by the impressive creation of embedded tetrahydrofuran ring of (±)-indoxamycin B by Carreira et al. (Scheme 23) [88]. Because of the failure to forge allene **89** via Saucy–Marbet rearrangement starting from **87** at elevated temperature, they switched to the trinuclear [Au(I)]-oxo complex [(Ph$_3$PAu)$_3$O]BF$_4$ **88** to facilitate the rearrangement process, which allowed for the expected resultant **89** as the only diastereomer in 84% yield. The chemo- and diastereoselective reduction of compound **89** to γ-hydroxyallene **90** was realized by using LiBHEt$_3$ as the reducing reagent. The subsequent [Au(I)]-catalyzed 5-*exo-trig* cyclization yielded tetracycle **92** in 72% yield and 3.2:1 dr. With this important intermediate in hand, an extra linear nine steps led to the formation the desired natural product indoxamycin B.

Apart from hydroxy group, carboxy group can also participate in the hydrofunctionalization of allenes under [Au(I)] catalysis. For instance, Sakaguchi and Ohfune applied the gold-catalyzed γ-butyrolactonization of allenylsilane as one of the crucial steps in the total synthesis of (−)-funebrine (Scheme 24) [89]. From the simple 2-butyn-1-ol **93**, allenylsilane **94** was prepared in five steps with an excellent diastereomeric ratio (≥20:1). The use of 3 mol% of [(Ph$_3$PAu)$_3$O]BF$_4$ catalyzed the γ-butyrolactonization of **94** well to generate the expected product **95** in 69% yield. Furthermore, by adding 20 mol% of ^{i}Pr$_2$NEt as additive, the catalyst loading could be lowered into 1 mol%. The subsequent eight steps finalized the naturally occurring (−)-funebrine.

2.1.5 Cascade Cyclization Reactions Based on Activating Alkyne

While gold-catalyzed direct addition of oxygen-based nucleophiles to alkynes or allenes allows a facile access to oxoheterocycles, it is also possible to develop cascade cyclizations based on pre-organized multifunctional reactants, which are initiated by the activation of alkyne with gold catalyst. An attractive feature of this strategy is to provide the possibility of efficient buildup of complex molecular skeleton in one-pot procedure.

For example, based on a chiral 1,6-enynes with a carbonyl group at the alkenyl side chains, Echavarren et al. disclosed an [Au(I)]-catalyzed novel [2+2+2] alkyne/alkene/carbonyl cycloaddition cascade to build up a complex [3.1.0] bicyclic system for the total synthesis of natural product pubinernoid B and (+)-orientalol F (Scheme 25) [90]. The enantioenriched enyne (*S*)-**98** was prepared from racemic epoxyfarnesol **97** via Sharpless epoxidation and some routine transformations. Noteworthy, the TES protecting of **99** was important to enable the desired cyclization; otherwise, the multiple functionalities of the corresponding free propargylic alcohol might lead to possible side reactions such as Meyer–Schuster rearrangement or nucleophilic attack by gold catalysis. In the presence of 3 mol% of [IPrAuNCPh]SbF_6 **100**, the tertiary propargyl stereocenter induced the stereoselective enyne cyclization through gold carbenoid intermediate **I**, wherein the OTES group and gold carbene was anti-oriented. The following intramolecular attack proceeded faster than the 1,5-migration of the OTES, to generate an oxonium cation **II** with three additional stereocenters. Finally, a Prins-type ring closure gave access to **101** with 65% yield as the core skeleton of pubinernoid B and (+)-orientalol F.

This approach was also applied by the groups of Ma [91] and Echavarren [92] for the construction of englerins A and B, respectively. As demonstrated in Scheme 26, both groups initiated their synthetic route with cyclization precursors **103** and **106**,

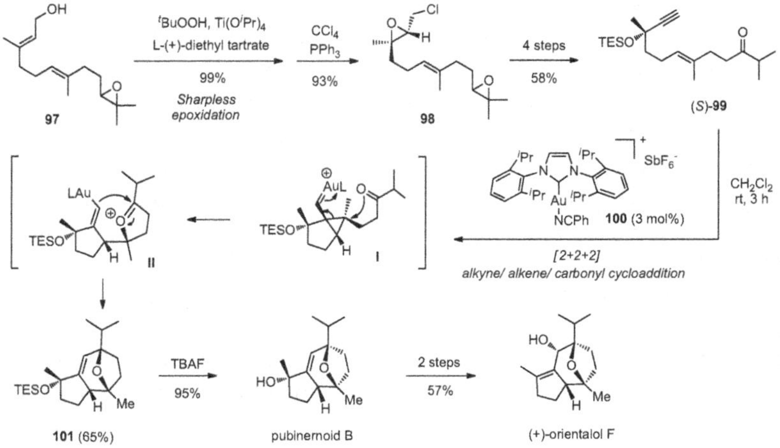

Scheme 25 The total synthesis of pubinernoid B and (+)-orientalol F

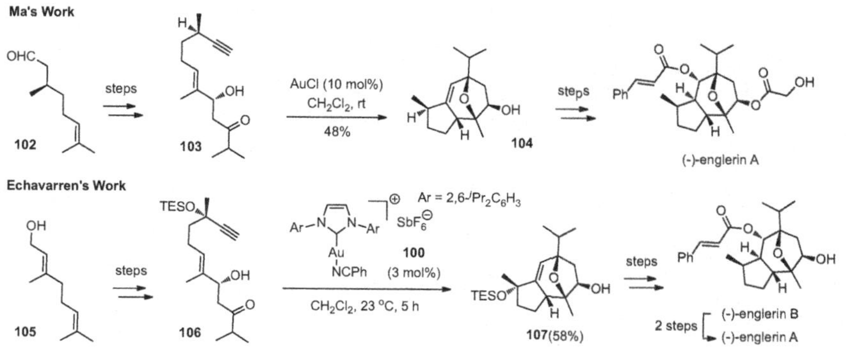

Scheme 26 The total synthesis of englerin A and B

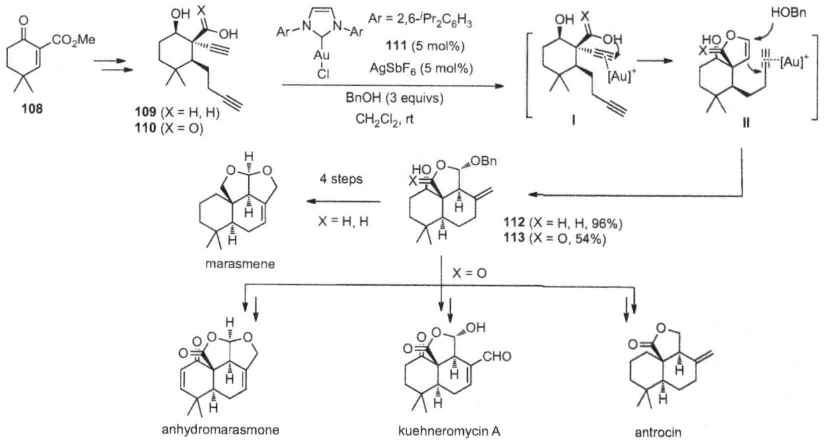

Scheme 27 The total synthesis of four kinds of sesquiterpenoids

which merely had tiny difference in the substituent at the chiral center. The core structure of englerins could be elaborated with similar catalytic system, while Ma and co-workers used 10 mol% of AuCl, and Echavarren et al. employed 3 mol% of [IPrAuNCPh]SbF_6 **100**.

Yang, Li, Luo et al. also exploited a gold-catalyzed tandem reaction of 1,7-diynes **109** or **110** with both internal and external oxygen-based nucleophile to construct the core polycyclic structure **112** or **113** for the total synthesis of drimane-type sesquiterpenoids (Scheme 27) [93]. The 1,7-diynes were readily accessed from cyclohexenone **108**. The use of 5 mol% IPrAuCl/$AgSbF_6$ triggered the cyclization of homopropargylic alcohol **109** (or 3-butynoic acid **110**) through intermediate **I** to form olefin **II**, which was then attacked by an external BnOH prior to a ring closure reaction to give the corresponding adduct **112** in 96% yield (or **113** in 54% yield). This highly chemo-, regio-, and diastereoselective process paved the

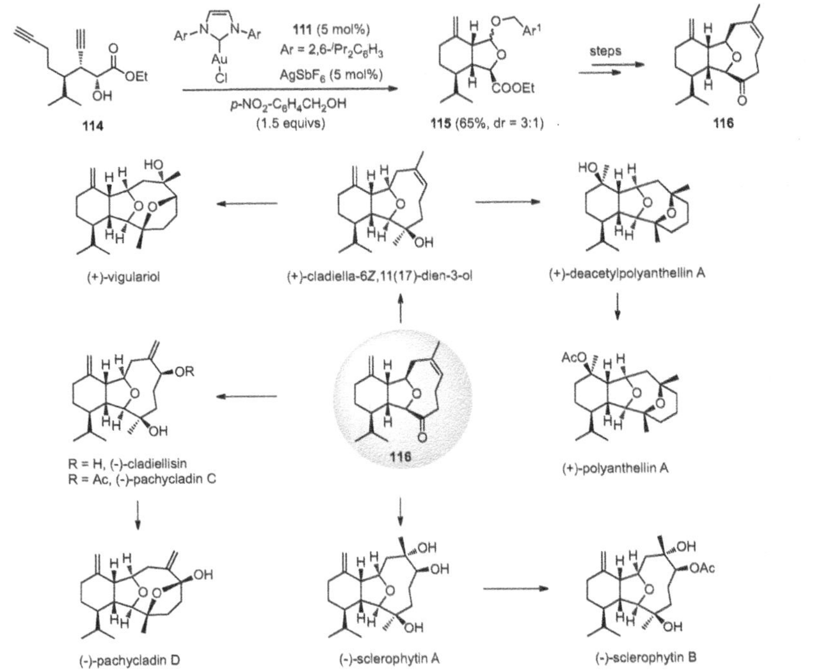

Scheme 28 The synthesis of the core structure of cladiellins

way for the total synthesis of four natural products, marasmene, anhydromarasmone, kuehneromycin A, and antrocin.

The cladiellins belong to a large family of the diterpenoids isolated from marine invertebrates of *Octocorallia* species, with over 100 examples and a wide range of biological activities [94–96]. Among these interests in this family of natural products, Yang and Luo et al. noticed that nine members of the family shared the same tricyclic species **116** as the crucial architectural features of the nature products (Scheme 28). In view of this, a gold-catalyzed cascade reaction of 1,7-diynes **114** was once again employed, to furnish the 6-5-bicyclic ring system of this key intermediate [97]. It was found that in the presence of 5 mol% $IPrAuCl/AgSbF_6$, with *p*-nitrobenzyl alcohol as the nucleophile, the annulation product **115** was achieved in 65% yield with 3:1 dr. Importantly, scale-up of this process proceeded smoothly as well. Then the mixture of the diastereomers **115** was transformed to the core synthon **116** by an array of functional group manipulations, which contributed to the total synthesis of the nine natural products indicated below.

2.1.6 Miscellaneous Reactions

In addition to the formation of oxygen-containing heterocycles via the addition of oxygen-based nucleophiles to alkynes or allenes, the cyclization of monoallylic

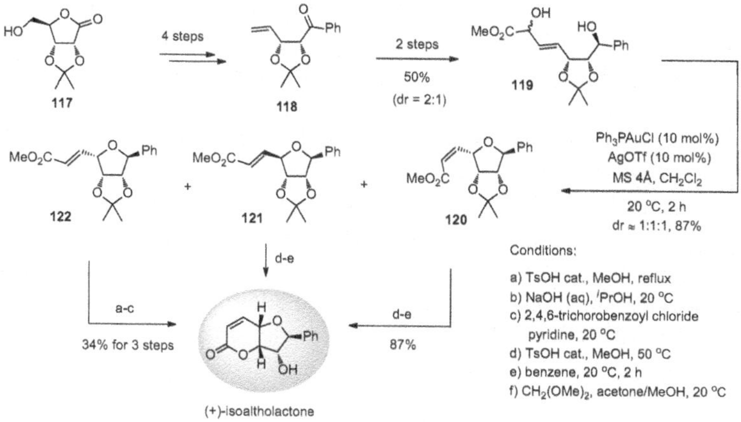

Scheme 29 The total synthesis of (+)-isoaltholactone

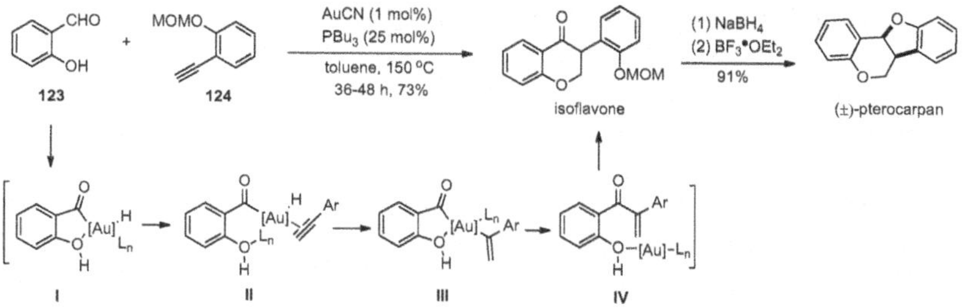

Scheme 30 The total synthesis of isoflavone and (±)-pterocarpan

diols is also a potentially viable method. This is exemplified by Robertson's concise synthesis of (+)-isoaltholactone [98], which was isolated from *Goniothalamus malayanus*, *G. montanus*, and *G. tapis*, and may have potential use in the procurement of abortion and for undefined postnatal treatments [99]. As shown in Scheme 29, the key precursor **119** for cyclization was obtained as an epimeric mixture (2:1 dr) after six steps from D-ribonolactone acetonide **117**. Nevertheless, the use of 10 mol% of in situ generated $Ph_3PAuOTf$ catalyzed cyclization to afford expected products **120**, **121**, and **122** in 87% combined yield with ca. 1:1:1 dr. In this step, the diastereoselectivity of **119** (2:1) was conserved into the *E*/*Z* ratio of the alkene moiety of adducts (**121** + **122**:**120** ≈ 2:1). The moderate dr did not affect the follow-up synthesis, since all diastereomers could be converted into the desired target (+)-isoaltholactone, via three steps from separable **122** or a one-pot process from the inseparable mixture of **120** and **121**.

A remarkable [Au(I)]-catalyzed annulation of salicylaldehydes and aryl acetylenes is also developed by Li et al. for the facile synthesis of isoflavones (Scheme 30) [100], which exist in human diet soybeans and soy-derived products and are phytoestrogens with weak oestrogenic activity [101–103]. This method was applied to the synthesis of isoflavone and (±)-pterocarpan [104]. Considerable effort was made to optimize the condition for the crucial annulation step of **123** with **124**, and it turned out that AuCN and PBu_3 could catalyze this reaction to give isoflavone in 73% yield. Subsequent one-pot reduction and acidic cyclization provided (±)-pterocarpan in 91% yield. Moreover, this protocol allowed the convenient preparation of various natural product derivatives. Probably, this annulation proceeded via an initial oxidative addition of the [Au(I)] catalyst into the aldehyde C–H bond, and the thus formed acyl [Au(III)] hydride complex **I** interacted with phenylacetylene to give intermediate **II**, followed by a hydrometalation forming intermediate **III**, which underwent a conjugated addition to afford isoflavone, and regenerated the [Au(I)] catalyst.

2.2 *Nitrogen-Containing Heterocycle Formation*

2.2.1 Intramolecular Hydroamination of Alkynes

Similar to the cyclization using oxygen-based nucleophiles, the Au-catalyzed intramolecular addition of amino groups to unsaturated C–C bonds resulted in nitrogen-containing heterocycles. An early example of the corresponding hydroamination of alkynes was reported by Utimoto and co-workers in the synthesis of solenopsin A (Scheme 31) [105, 106]. A 6-*exo-dig* cyclization based on 5-alkynylamine **125** underwent smoothly under the catalysis of [Au(III)], giving the enamine intermediate **II**, which was prone to generate the more stable tautomer **126** in 90% yield. After reduction, solenopsin A was attained. In this case, the gold(III)-catalyzed alkyne hydroamination emerged as a useful methodology to prepare piperidine alkaloid.

This methodology was later applied by Gouault et al. to the total synthesis of dendrobate alkaloids, including (+)-241D, isosolenopsin, isosolenopsin A, and (−)-epimyrtine (Scheme 32) [107, 108]. Starting from optically active *N*-Boc-D-alanine **127**, chiral amino ynone **128** were readily prepared via three steps. Under the catalysis cationic [Au(I)] complexes derived from Ph_3PAuCl and $AgSbF_6$ (5 mol%, each), the regiospecific 6-*endo-dig* cyclization of the electron-deficient alkyne moiety of **128** provided the anticipated pyridinone **129** in good yields (78–83%), which could be converted to these natural products by facile late-stage modifications.

The [Au(I)]-catalyzed hydroamination of alkyne was further employed by Fujii and Ohno to construct the tetrahydroisoquinoline core of (−)-quinocarcin

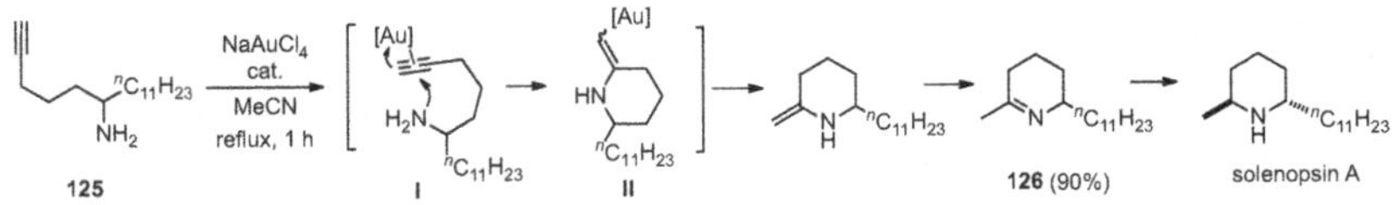

Scheme 31 The total synthesis of solenopsin A

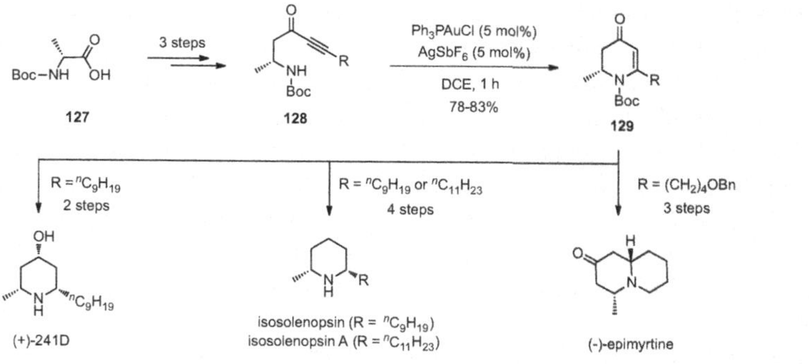

Scheme 32 The application of [Au(I)]-catalyzed intramolecular aza-Michael addition

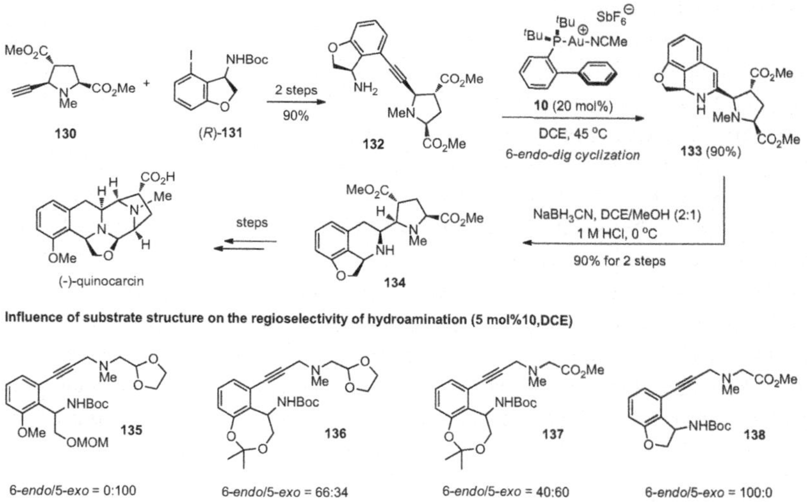

Scheme 33 The [Au(I)]-catalyzed hydroamination in the total synthesis of (−)-quinocarcin

(Scheme 33) [109, 110], which has drawn the attention from both synthetic and biological chemists for its polycyclic architecture and potent broad-spectrum antitumor activity [111, 112]. The Sonogashira coupling of **130** and **131**, followed by deprotection of the Boc group, delivered the cyclization precursor **132** in 90% overall yield. The desired 6-*endo-dig* hydroamination of alkyne proceeded smoothly with 20 mol% of cationic gold catalyst **10**, furnishing the enamine intermediate **133**, which was instable and was reduced immediately in one-pot operation to afford the key tetrahydroisoquinoline **134** in 90% isolated yield. It was found that the regioselectivity of this cyclization was largely dependent on the substrate structure. When substrate **135** was used for condition optimization, only undesired 5-*exo-dig* cyclization was observed no matter using CuBr, Ph_3PAuCl, JohnPhosAuCl, or **10** as the catalyst, probably because the nucleophilic attack preferably took place at the more cationic carbon atom bearing an aryl group.

Accordingly, the modification of substrates to control regioselectivity was tried. It was found that the inherent bias for the formation of 5-*exo-dig* products was overridden by using seven-membered acetonide-type substrates **136** and **137**, and 6-*endo-dig* products could be formed. Moreover, when dihydrobenzofuran-type substrate **138** was engaged, the desired 6-*endo* cyclization product was exclusively formed. A rational explanation for the excellent regioselectivity might attribute to the ring strain of the tricyclic core in corresponding product. Nevertheless, in the total synthesis, the Boc group of **132** must be removed, as the unfavorable repulsion between bulky Boc group and the methyl ester had negative influence on the reactivity.

In 2009, in the synthesis of an antitumor drug candidate nitidine [113], a gold(I)-catalyzed tandem reaction initiated by hydroamination of *o*-alkynyl benzylic amide derivative was developed by Takemoto et al. to construct 1,2-dihydroisoquinoline framework (Scheme 34) [114, 115]. In the presence of 5 mol% of cationic gold (I) catalyst **91** and five equivs of MeOH, alkynyl carbamate **141** bearing an acetal group cyclized to provide tetracyclic heterocycle **143** in 98% yield. Without the use of MeOH, both tetracyclic heterocycle **153** (47%) and hydroamination adduct **152** (50%) were obtained. A plausible mechanism for the tandem reaction was initiated by the gold(I)-catalyzed hydroamination of alkyne with 6-*endo-dig* cyclization to afford key intermediate **I**, followed by an enamide cyclization and aromatization to afford **143**, since the isolated side product **142** could be converted to **143** in 58% yield under the same cyclization condition. On the other hand, the vinyl gold moiety **I** might also be first hydrolyzed to undergo the following cyclization.

In 2013, an [Au(I)]-catalyzed cascade strategy was developed by Tokuyama et al. to forge the 5-indolizinone substructure of (−)-rhazinilam isolated from various *Apocynaceae* species. This natural product, along with the analogy (−)-rhazinicine [116], was recognized as lead compounds for new antitumor agents

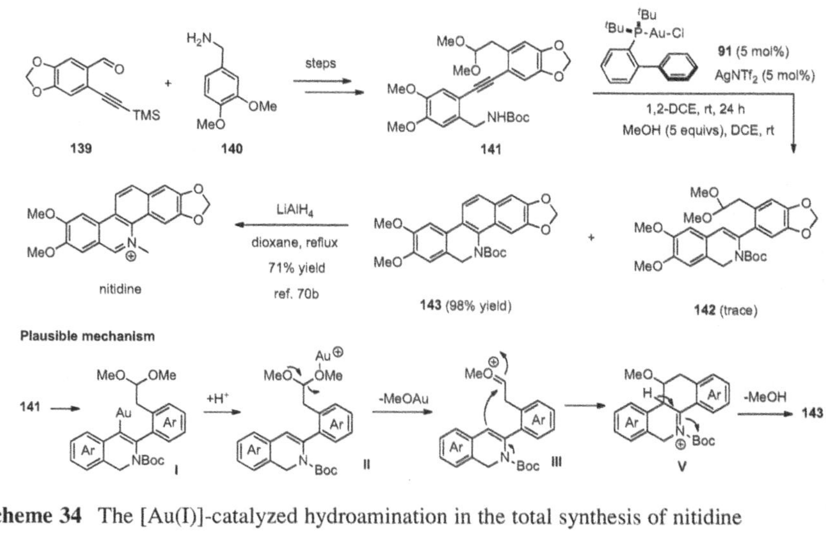

Scheme 34 The [Au(I)]-catalyzed hydroamination in the total synthesis of nitidine

[117]. As shown in Scheme 35, upon treatment with 30 mol% of $Ph_3PAuNTf_2$ and $KHSO_4$ under microwave irradiation, alkynyl amide **147** bearing an acetal group readily underwent a 6-*exo-dig* ring closure/pyrrole formation step, to provide 5-indolizinone **148** in 65% yield, which could be transformed to (−)-rhazinilam and (−)-rhazinicine with a few more steps.

During the total synthesis of communesin B, Crawley and Funk reported a rare example of [Au(I)]-catalyzed 7-*exo-dig*-type annulation reaction, to construct the bridge cycle (Scheme 36) [118]. After the coupling of amine **149** and bromide **150** to afford substrate **151**, a fluoride-promoted ring opening afforded the multifunctional alkyne–amine precursor **152**. In the presence of 1 mol% of $Ph_3PAuOTf$, the cyclization of **152** proceeded smoothly to give polycyclic compound **153** in 89% yield. Despite the synthesis of communesin B from **154** was not realized, this methodology nicely showed the potential of Au-catalyzed 7-*exo-dig* cyclization in constructing polycyclic scaffolds.

The gold-catalyzed indole synthesis from *o*-alkynyl aniline derivatives has also found wide application in the synthesis of indole-containing natural products. For

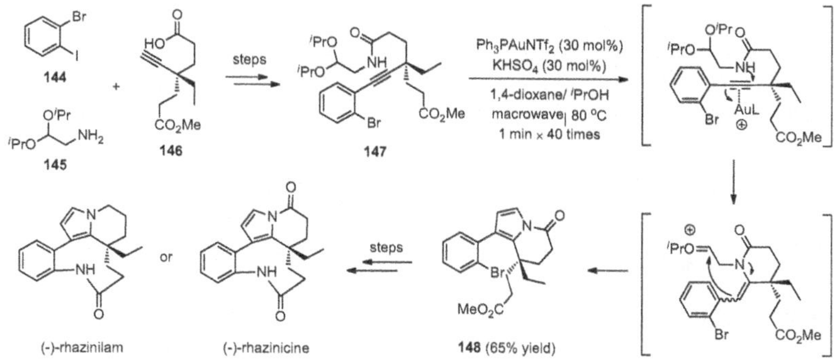

Scheme 35 The [Au(I)]-catalysis in the total synthesis of (−)-rhazinilam and (−)-rhazinicine

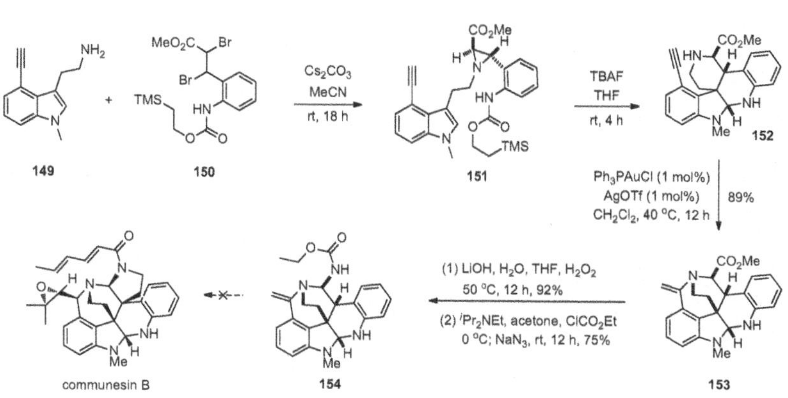

Scheme 36 The [Au(I)]-catalyzed 7-*exo-dig*-type annulation reaction

example, in the first enantioselective total synthesis of (−)-mersicarpine (Scheme 37) [119]. Fukuyama et al. found that the use of 5 mol% of $NaAuCl_4 \cdot 2H_2O$ could catalyze the desired 5-*endo-dig* hydroamination of *o*-alkynyl aniline **156**, providing 2-functionalized indole **157** in 78% yield. Noticeably, the coexistence of a hydroxy group at the side chain did not affect the indole formation at all. With this intermediate in hand, (−)-mersicarpine was accessed over five more steps.

In the total synthesis of (+)-terreusinone (Scheme 38) [120], which might be used to protect organisms from harmful effects of solar UV irradiation [121], the Sperry group connected the [Au(I)]-catalyzed indole synthesis from *o*-alkynyl anilines with a Larock indolization to furnish the core structure, as they failed to construct fused indole-skeleton **163** via double Larock indolization. After indole **161** was prepared by a one-pot Larock indolization–Sonogashira coupling procedure, the other indole moiety was formed via the hydroamination of **162** under the catalysis of 12 mol% catalyst **10** in toluene at 60°C, affording adduct **163** in 85% yield. Finally, (+)-terreusinone was accomplished with Frémy's salt **164** under buffered conditions.

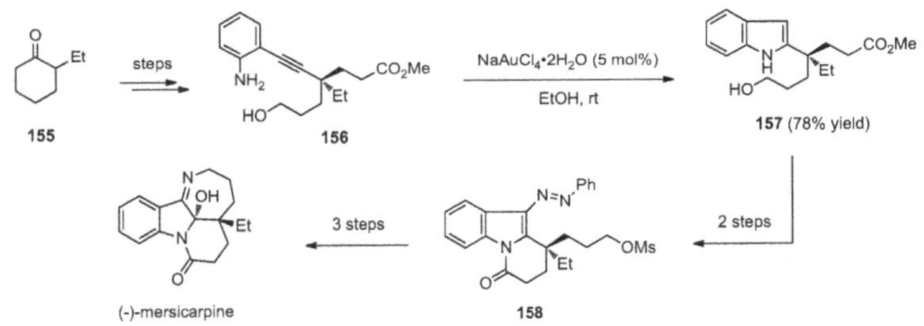

Scheme 37 [Au(III)]-catalyzed indole synthesis in the total synthesis of (−)-mersicarpine

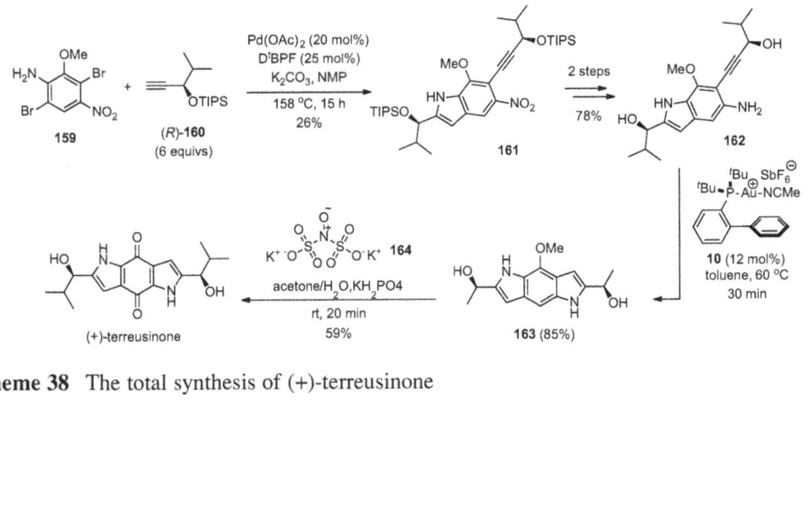

Scheme 38 The total synthesis of (+)-terreusinone

2.2.2 Miscellaneous Reactions

The gold-catalyzed hydroamination of other unsaturated compounds is applied to the total synthesis of natural products. For example, a gold-mediated synthesis of piperidine fragment from functionalized allenes **165** was incorporated in the synthesis of (−)-swainsonine (Scheme 39) [122], which is known to exhibit glycosidase inhibitory properties and have therapeutic potential in a range of conditions [123, 124]. Interestingly, the regioselectivity of this protocol was greatly influenced by reaction condition. Undesired *O*-attack product **166** was obtained when $AuCl_3$ was used as catalyst, no matter using CH_2Cl_2 or MeCN as the solvent, probably due to the O–Si bond cleavage induced by the trace amount of HCl in system. Thus, $CaCO_3$ was added to neutralize the acid, but *N*-attack product **167** was still obtained in only 13% yield when using MeCN as the solvent (condition 1, Scheme 39). Nevertheless, the yield of **167** was dramatically increased to 99% when a mixed solvent of CH_2Cl_2 and MeCN was used, with the presence of $CaCO_3$, to solubilize and stabilize the catalyst (condition 2). The single diastereoisomer **167** was taken forward to complete the total synthesis of swainsonine [125].

A gold-mediated tandem hydration/elimination/intramolecular aza-Michael addition sequence was also developed to construct the piperidine-ketone moiety of (+)-andrachcinidine [126] by Floreancig et al. (Scheme 40) [127]. By mixing the enantioenriched homopropargylic ethers **169** (94% ee) with 5 mol% of Ph_3PAuCl and 10 mol% of $AgSbF_6$ in the presence of H_2O, the desired product **170** was obtained in 89% yield. It was proposed that the initial hydration of alkyne moiety

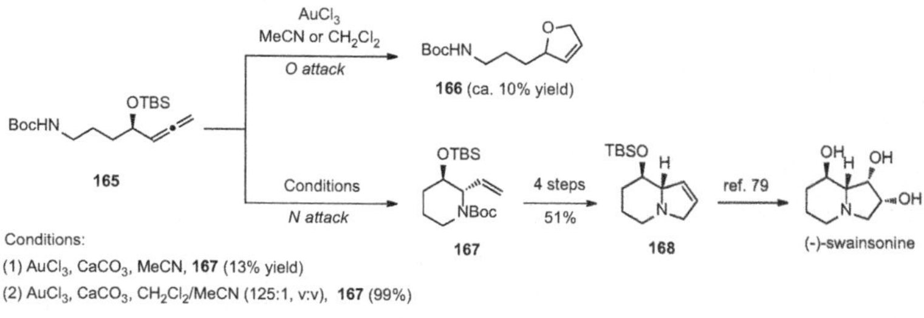

Scheme 39 The total synthesis of (−)-swainsonine

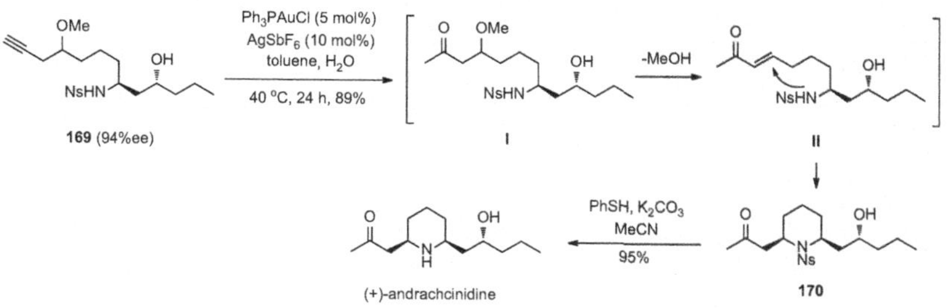

Scheme 40 A tandem hydration/elimination/aza-Michael addition to (+)-andrachcinidine

worked well to give β-methoxyketone intermediate **I**, followed by an elimination of the methoxy group to afford enone **II**, which underwent an intramolecular aza-Michael addition to provide the desired piperidine **170**. After removal of the Ns protecting group with PhSH in basic condition, the target (+)-andrachcinidine was obtained in 95% yield.

Recently, He and Xiang reported the use of nitriles as nitrogen source to react with α-oxo gold carbenes generated via alkyne oxidation to synthesize the oxazole-containing natural products, pimprinine and WS-30581B (Scheme 41) [128]. The coupling of alkyne **171** with MeCN or ^{n}BuCN readily afforded 2,5-disubstituted oxazoles **173** under the catalysis of 5 mol% of $Ph_3PAuNTf_2$, in the presence of stoichiometric amount of 8-methylquinoline N-oxide **172** as oxidant. The key intermediates in this reaction are the highly reactive α-oxo gold carbene **II**, which is formed from alkyne and quinoline N-oxides via an addition–elimination process.

The formation of a reactive α-imino gold carbene using the combination of alkyne and azide is also attractive strategy to the construction of nitrogen-containing heterocycles [129, 130]. In this context, Zhang et al. developed a gold-catalyzed one-step construction of 2,3-dihydro-1*H*-pyrrolizines from electron-deficient alkynes tethered with an azido group (Scheme 42) [131]. Probably, a 5-*exo-dig* cyclization took place to produce α-imino gold carbene

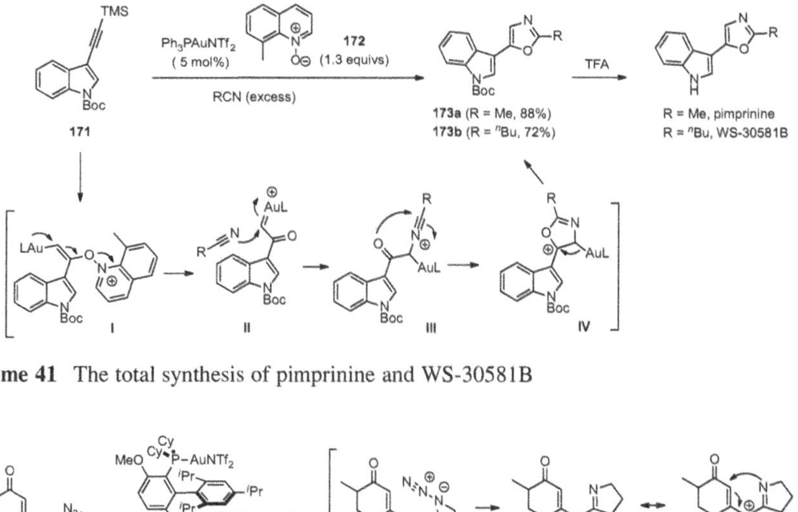

Scheme 41 The total synthesis of pimprinine and WS-30581B

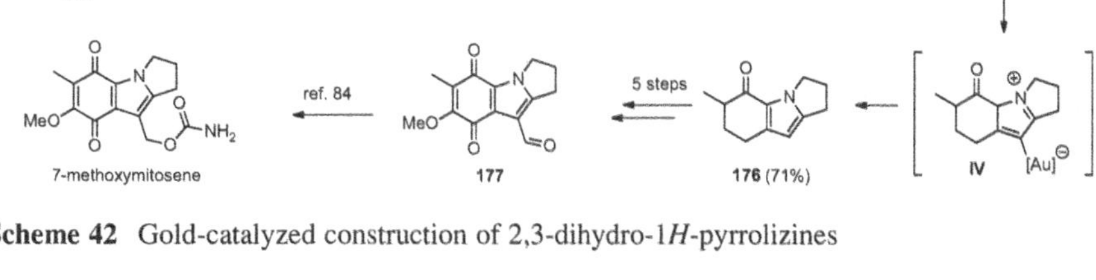

Scheme 42 Gold-catalyzed construction of 2,3-dihydro-1*H*-pyrrolizines

intermediate **II**, which was mesomeric to a destabilized 1-azapentadienium ion **III**, and underwent an electrocyclic ring closure to deliver the pyrrole derivative. Accordingly, from pre-organized material **174**, the desired pyrrole derivative **176** was obtained in 71% yield under the catalysis of BrettPhosAuNTf$_2$ **175**, which could be transformed to compound **177**, key intermediate to 7-methoxymitosene of antitumor activities [132].

2.3 Catalytic Asymmetric Gold Catalysis in Natural Product Synthesis

The aforementioned gold-catalyzed cyclization reactions are all based on the use of optically active substrates, if chiral products involved. Although limited, chiral gold complex-catalyzed enantioselective heterocycle-forming reactions have been applied to the synthesis of natural products. In this context, the pioneering work of Hayashi and Ito in the asymmetric gold catalysis, namely, the aldol reaction of isocyanoacetate **179** and aldehydes to oxazolines [133], constituted a facile access to optically active β-hydroxyamino acids widely present in many natural products. For example, a highly diastereo- and enantioselective synthesis of *threo*- and *erythro*-sphingosines, sphingolipid bases in mammals, had been developed by the same group (Scheme 43) [134].

Starting from (*E*)-2-hexadecenal **178** and isocyanoacetate **179**, only 1 mol% of gold complex derived from chiral ferrocenylphosphine **180** allowed the synthesis of the desired adduct **181** in 93% ee and 80% yield when using CH$_2$Cl$_2$ as the solvent at 25°C. Hence, *threo*-sphingosine was accessible by two steps in 85% yield, and the *erythro* isomer could be obtained by two more steps. The proposed mechanism

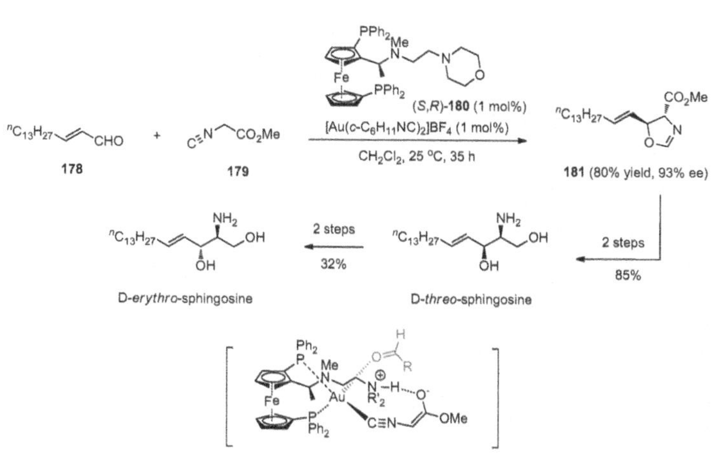

Scheme 43 The total synthesis of D-*erythro*-sphingosine and D-*threo*-sphingosine

for this type of aldol reaction catalyzed by ferrocenylphosphine-derived chiral gold complex was shown in Scheme 43. While [Au(I)] cation bound tightly to the two phosphorus atoms, the two nitrogen atoms in the side chain were in free state, and the terminal tertiary amine moiety served as Brønsted base to deprotonatively activate isocyanoacetate. Besides, the conjugate acid of terminal dialkylamino group stabilized the enolate of isocyanoacetate to organize a more favorable arrangement of both substrates to ensure high stereoselectivity. Since the reactive site of the enolate intermediate was far from the auxiliary chirality, it was necessary to use a longer chain on the catalyst backbone for stereocontrol.

Later in 1994, Hughes et al. applied Hayashi's method to the synthesis of 3-hydroxylysine [135], a useful intermediate in the asymmetric synthesis of potent naturally occurring protein kinase C inhibitor, balanol (Scheme 44) [136]. From 4-phthalimidobutanal **182** and methyl isocyanoacetate **179**, the dihydrooxazole formation worked well under the catalysis of 5 mol% of Hayashi's chiral gold (I) catalyst to give **184** in 90% yield and 19:1 dr. After a single recrystallization, **184** was obtained as a single diastereomer in 73% yield, which was then hydrolyzed in refluxing HCl (6.0 M), to produce (−)-3-hydroxylsine in 75% yield and 99% ee.

In the total synthesis of (−)-α-kainic acid reported by Bachi and Melman [137], enal **186** bearing a thioether moiety worked well under Hayashi's condition to afford the key intermediate oxazoline **188** in 80% yield without raising the catalyst loading (Scheme 45).

Togni et al. observed a double stereodifferentiation effect in the ferrocenylphosphine **191** derived chiral gold(I) catalyst-mediated aldol reaction of chiral aldehyde **189** with ethyl isocyanoacetate **190** (Scheme 46) [138]. While (*S*,*R*)-**191**

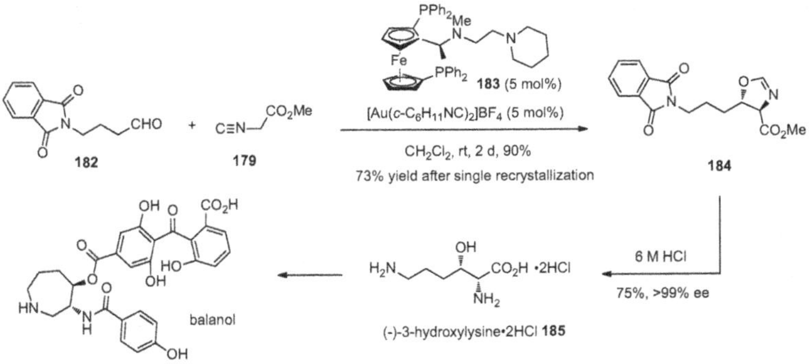

Scheme 44 The total synthesis (−)-3-hydroxylysine

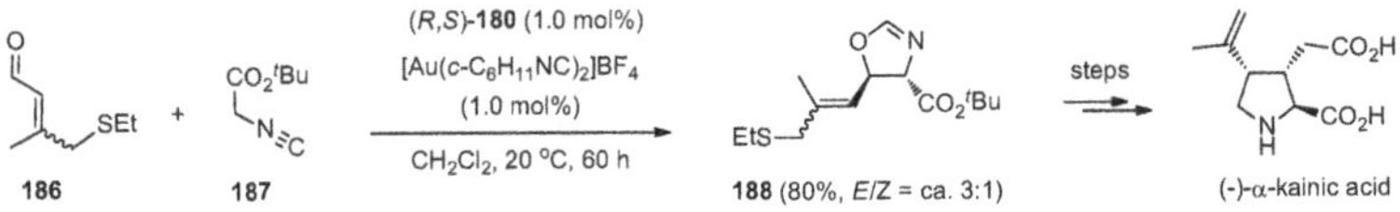

Scheme 45 The total synthesis of (−)-α-kainic acid

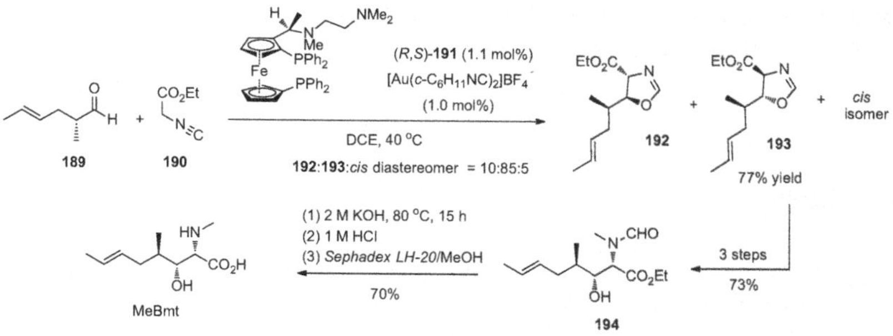

Scheme 46 The total synthesis of MeBmt

gave undesired diastereomer **192** as the major, the use of opposite enantiomer (*R*,*S*)-**191** afforded the desired diastereomer **193** dominantly. The oxazoline **193** could be converted to compound **194** in three steps with a 73% overall yield, and subsequent transformations of **194** completed the synthesis of MeBmt, an unusual amino acid in the immunosuppressive undecapeptide cyclosporine [139].

3 Conclusion

The chemo- and regioselective formation of heterocycles plays an important role in the natural product synthesis and drug development, owing to the wide occurrence of structure diverse heterocycle moieties in natural products, drugs, and bioactive compounds. In particular, selective cyclization initiated by the direct C–X bond formation represents a very important yet challenging task, as the strong coordinating ability of heteroatoms often has negative influence on the performances of metal catalysts. Very helpfully, as the studies highlighted in this chapter, the good compatibility of soft gold catalysts with various types of oxygen- or nitrogen-based nucleophiles allows the facile construction of cyclic, bicyclic, polycyclic, or spirocyclic heterocycles via the activation of soft alkyne, allene, or alkene functionalities. The powerfulness of this golden tool has been vividly demonstrated by the facile synthesis of a variety of different heterocycle-containing complex natural products, and, impressively, good selectivity is achieved in many cases. On the other hand, golden synthetic opportunities are still open, which include the application of gold-catalyzed enantioselective C–X bond formation in the synthesis of related natural products, the exploration of gold-catalyzed C–S bond formation reactions for the synthesis of natural products with S-containing heterocycles (gold catalysis is not always incompatible with S-containing substances, see for an example [140]), and the merging of gold catalysis with other catalysts [141] for the exploitation of novel reactions for the construction of molecular complexity from simple staring materials to make the total synthesis of natural products easier. It can be anticipated that with the discovery of new catalytic properties of gold catalysts and the exploitation of new gold catalysts, new types of gold-catalyzed cyclizations will be developed to facilitate natural products.

References

1. Gorin DJ, Toste FD (2007) Nature 446:395–403
2. Leyva-Pérez A, Corma A (2012) Angew Chem Int Ed 51:614–635
3. Hashmi ASK (2007) Chem Rev 107:3180–3211
4. Li Z, Brouwer C, He C (2008) Chem Rev 108:3239–3265
5. Arcadi A (2008) Chem Rev 108:3266–3325
6. Corma A, Leyva-Pérez A, Sabater MJ (2011) Chem Rev 111:1657–1712
7. Hashmi ASK, Rudolph M (2011) Chem Commun 47:6536–6544
8. Fürstner A, Davies PW (2007) Angew Chem Int Ed 46:3410–3449
9. Brenzovich WE (2012) Angew Chem Int Ed 51:8933–8935
10. Fürstner A (2009) Chem Soc Rev 38:3208–3221
11. Skouta R, Li CJ (2008) Tetrahedron 64:4917–4938
12. Jiménez-Núñez E, Echavarren AM (2007) Chem Commun 333–346
13. Gorin DJ, Sherry BD, Toste FD (2008) Chem Rev 108:3351–3378
14. Hashmi ASK, Frost TM, Bats JW (2000) J Am Chem Soc 122:11553–11554
15. Hashmi ASK, Schwarz L, Choi JH, Frost TM (2000) Angew Chem Int Ed 39:2285–2288
16. Kennedy-Smith JJ, Staben ST, Toste FD (2004) J Am Chem Soc 126:4526–4527
17. Sherry BD, Toste FD (2004) J Am Chem Soc 126:15978–15979
18. Nieto-Oberhuber C, Muñoz MP, Buñuel E, Nevado C, Cárdenas DJ, Echavarren AM (2004) Angew Chem Int Ed 43:2402–2406
19. Zhang L, Kozmin SA (2004) J Am Chem Soc 126:11806–11807
20. Hashmi ASK, Rudolph M (2008) Chem Soc Rev 37:1766–1775
21. Rudolph M, Hashmi ASK (2012) Chem Soc Rev 41:2448–2462
22. Zhang Y, Luo T, Yang Z (2014) Nat Prod Rep 31:489–503
23. Zeldin RM, Toste FD (2011) Chem Sci 2:1706–1709
24. Teller H, Fürstner A (2011) Chem Eur J 17:7764–7767
25. Lu Z, Li Y, Deng J, Li A (2013) Nat Chem 5:679–684
26. González AZ, Benitez D, Tkatchouk E, Goddard WA III, Toste FD (2011) J Am Chem Soc 133:5500–5507
27. Liu Z, Wasmuth AS, Nelson SG (2006) J Am Chem Soc 128:10352–10353
28. Cui L, Peng Y, Zhang L (2009) J Am Chem Soc 131:8394–8395
29. Zhang L, Cui L (2010) Sci China: Chem 53:113–118
30. Shan ZH, Liu J, Xu LM, Tang YF, Chen JH, Yang Z (2012) Org Lett 14:3712–3715
31. Trost BM, Dong G (2008) Nature 456:485–488
32. Trost BM, Dong G (2010) J Am Chem Soc 132:16403–16416
33. Pettit GR, Gao F, Blumberg PM, Herald CL, Coll JC, Kamano Y, Lewin NE, Schmidt JM, Chapuis JC (1996) J Nat Prod 59:286–289
34. Trost BM, Yang H, Thiel OR, Frontier AJ, Brindle CS (2007) J Am Chem Soc 129:2206–2207
35. Bihelovic F, Saicic RN (2012) Angew Chem Int Ed 51:5687–5691
36. Bihelovic F, Karadzic I, Matovic R, Saicic RN (2013) Org Biomol Chem 11:5413–5424
37. Bister B, Bischoff D, Ströbele M, Riedlinger J, Reicke A, Wolter F, Bull AT, Zähner H, Fiedler HP, Süssmuth RD (2004) Angew Chem Int Ed 43:2574–2576
38. Benson S, Collin MP, Arlt A, Gabor B, Goddard R, Fürstner A (2011) Angew Chem Int Ed 50:8739–8744

39. Pettit GR, Inoue M, Kamano Y, Herald DL, Arm C, Dufresne C, Christie ND, Schmidt JM, Doubek DL, Krupa TS (1988) J Am Chem Soc 110:2006–2007
40. Serrano M (1997) Exp Cell Res 237:7–13
41. Fortner KC, Kato D, Tanaka Y, Shair MD (2010) J Am Chem Soc 132:275–280
42. Fernández A, Levine ZG, Baumann M, Sulzer-Mossé S, Sparr C, Schläger S, Metzger A, Baxendale IR, Ley SV (2013) Synlett 24:514–518
43. Higa T, Tanaka J, Kitamura A, Koyama T, Takahashi M, Uchida T (1994) Pure Appl Chem 66:2227–2230
44. Christen AA, Peaden RN (1981) Plant Dis 65:319–321
45. Lin Y, Wu X, Feng S, Jiang G, Luo J, Zhou S, Vrijmoed LLP, Jones EBG, Krohn K, Steingrover K, Zsila F (2001) J Org Chem 66:6252–6256
46. Panda B, Sarkar TK (2013) J Org Chem 78:2413–2421
47. Sarkar D, Venkateswaran RV (2011) Tetrahedron 67:4559–4568
48. Feng Y, Jiang X, De Brabander JK (2012) J Am Chem Soc 134:17083–17093
49. Trost BM, O'Boyle BM, Hund D (2009) J Am Chem Soc 131:15061–15074
50. Takahashi K, Yoshihara T, Kurosawa K (2005) J Antibiot 58:420–424
51. Tachibana K, Scheuer PJ, Tsukitani Y, Kikuchi H, Van Engen D, Clardy J, Gopichand Y, Schimtz FJ (1981) J Am Chem Soc 103:2469–2471
52. Fang C, Pang Y, Forsyth CJ (2010) Org Lett 12:4528–4531
53. Isobe M, Ichikawa Y, Goto T (1986) Tetrahedron Lett 27:963–966
54. Isobe M, Ichikawa Y, Bai DL, Masaki H, Goto T (1987) Tetrahedron 43:4767–4776
55. Ley SV, Humphries AC, Eick H, Downham R, Ross AR, Boyce RJ, Pavey JBJ, Pietruszka J (1998) J Chem Soc Perkin Trans 1 (23):3907–3911
56. Kona CN, Ramana CV (2014) Tetrahedron 70:3653–3656
57. Tlais SF, Dudley GB (2010) Org Lett 12:4698–4701
58. Tlais SF, Dudley GB (2011) Beilstein J Org Chem 7:570–577
59. Li X, Yao Y, Zheng Y, Sattler I, Lin W (2007) Arch Phram Res 30:812–815
60. Li Y, Zhou F, Forsyth CJ (2007) Angew Chem Int Ed 46:279–282
61. Hoffmeister L, Fukuda T, Pototschnig G, Fürstner A (2015) Chem Eur J 21:4529–4533
62. Kazlauskas R, Murphy PT, Wells RJ, Blackman AJ (1982) Aust J Chem 35:113–120
63. Chaładaj W, Corbet M, Fürstner A (2012) Angew Chem Int Ed 51:6929–6933
64. Stout EP, Hasemeyer AP, Lane AL, Davenport TM, Engel S, Hay ME, Fairchild CR, Prudhomme J, Le Roch K, Aalbersberg W, Kubanek J (2009) Org Lett 11:225–228
65. Lee JS, Shin J, Shin HJ, Lee HS, Lee YJ, Lee HS, Won H (2014) Eur J Org Chem 2014:4472–4476
66. Chen Y, Wang L, Sun N, Xie X, Zhou X, Chen H, Li Y, Liu Y (2014) Chem Eur J 20:12015–12019
67. Wang C, Chen Y, Xie X, Liu J, Liu Y (2012) J Org Chem 77:1915–1921
68. Kono K, Tanaka M, Ono Y, Hosoya T, Ogita T, Kohama T (2001) J Antibiot 4:415–420
69. Zhu J, Germain AR, Porco JA Jr (2004) Angew Chem Int Ed 43:1239–1243
70. Ma S (2005) Chem Rev 105:2829–2872
71. Pan F, Fu C, Ma S (2004) Chin J Org Chem 24:1168–1190
72. Ma S (2003) Acc Chem Res 36:701–712
73. Ma S (2009) Acc Chem Res 42:1679–1688
74. Hashmi ASK (2000) Angew Chem Int Ed 39:3590–3593
75. Hoffmann-Röder A, Krause N (2001) Org Lett 3:2537–2538
76. Krause N, Hoffmann-Röder A, Canisius J (2002) Synthesis 1759–1774
77. Marshall JA, Pinney KG (1993) J Org Chem 58:7180–7184
78. Naves YR, Bachmann P (1945) Helv Chim Acta 28:1227–1230
79. Kawakami M, Ganguly SN, Banerjee J, Kobayashi A (1995) J Agric Food Chem 43:200–207
80. Wüst M, Mosandl A (1999) Eur Food Res Technol 209:3–11
81. Wang D, Ando K, Morita K, Kubota K, Kobayashi A (1994) Biosci Biotechnol Biochem 58:2050–2053

82. Ito Y, Sugimoto A, Kakuda T, Kubota K (2002) J Agric Food Chem 50:4878–4884
83. Bentley R (2006) Chem Rev 106:4099–4112
84. Volz F, Wadman SH, Hoffmann-Röder A, Krause N (2009) Tetrahedron 65:1902–1910
85. Volz F, Krause N (2007) Org Biomol Chem 5:1519–1521
86. Phuong NM, Sung TV, Porzel A, Schmidt J, Merzweiler K, Adam G (1999) Phytochemistry 52:1725–1729
87. Sawama Y, Sawama Y, Krause N (2008) Org Biomol Chem 6:3573–3579
88. Jeker OF, Carreira EM (2012) Angew Chem Int Ed 51:3474–3477
89. Okada T, Sakaguchi K, Shinada T, Ohfune Y (2011) Tetrahedron Lett 52:5744–5746
90. Jiménez-Núñez E, Molawi K, Echavarren AM (2009) Chem Commun 7327–7329
91. Zhou Q, Chen X, Ma D (2010) Angew Chem Int Ed 49:3513–3516
92. Molawi K, Delpont N, Echavarren AM (2010) Angew Chem Int Ed 49:3517–3519
93. Shi H, Fang L, Tan C, Shi L, Zhang W, Li CC, Luo T, Yang Z (2011) J Am Chem Soc 133:14944–14947
94. Bernardeli P, Paquette LA (1998) Heterocycles 49:531–556
95. Ellis JM, Crimmins MT (2008) Chem Rev 108:5278–5298
96. Welford AJ, Collins I (2011) J Nat Prod 74:2318–2328
97. Yue G, Zhang Y, Fang L, Li CC, Luo T, Yang Z (2014) Angew Chem Int Ed 53:1837–1840
98. Unsworth WP, Stevens K, Lamont SG, Robertson J (2011) Chem Commun 47:7659–7661
99. Colegate SM, Din LB, Latiff A, Salleh KM, Samsudin MW, Skelton BW, Tadano KI, White AH, Zakaria Z (1990) Phytochemistry 29:1701–1704
100. Skouta R, Li CJ (2007) Angew Chem Int Ed 46:1117–1119
101. Máximo P, Lourenço A (1998) Phytochemistry 48:359–362
102. Holder CL, Churchwell MI, Doerge DR (1999) J Agric Food Chem 47:3764–3770
103. Hendrich S, Lee KW, Xu X, Wang HJ, Murphy PA (1994) J Nutr 124:1789S–1792S
104. Skouta R, Li CJ (2007) Tetrahedron Lett 48:8343–8346
105. Fukuda Y, Utimoto K, Nozaki H (1987) Heterocycles 25:297–300
106. Fukuda Y, Utimoto K (1991) Synthesis 975–978
107. Gouault N, Le Roch M, de Pinto GC, David M (2012) Org Biomol Chem 10:5541–5546
108. Trinh TTH, Nguyen KH, de Amaral PA, Gouault N (2013) Beilstein J Org Chem 9:2042–2047
109. Chiba H, Oishi S, Fujii N, Ohno H (2012) Angew Chem Int Ed 51:9169–9172
110. Chiba H, Sakai Y, Ohara A, Oishi S, Fujii N, Ohno H (2013) Chem Eur J 19:8875–8883
111. Siengalewicz P, Rinner U, Mulzer J (2008) Chem Soc Rev 37:2676–2690
112. Kahsai AW, Zhu S, Wardrop DJ, Lane WS, Fenteany G (2006) Chem Biol 13:973–983
113. Taira Z, Matsumoto M, Ishida S, Icikawa T, Sakiya Y (1994) Chem Pharm Bull 42:1556–1561
114. Enomoto T, Girard AL, Yasui Y, Takemoto Y (2009) J Org Chem 74:9158–9164
115. Hanaoka M, Yamagishi H, Marutani M, Mukai C (1987) Chem Pharm Bull 35:2348–2354
116. Sugimoto K, Toyoshima K, Nonaka S, Kotaki K, Ueda H, Tokuyama H (2013) Angew Chem Int Ed 52:7168–7171
117. David B, Sévenet T, Morgat M, Guénard D, Moisand A, Tollon Y, Thoison O, Wright M (1994) Cell Motil Cytoskeleton 28:317–326
118. Crawley SL, Funk RL (2006) Org Lett 8:3995–3998
119. Nakajima R, Ogino T, Yokoshima S, Fukuyama T (2010) J Am Chem Soc 132:1236–1237
120. Wang C, Sperry J (2011) Org Lett 13:6444–6447
121. Gao Q, Garcia-Pichel F (2011) Nat Rev Microbiol 9:791–802
122. Bates RW, Dewey MR (2009) Org Lett 11:3706–3708
123. Oredipe OA, Furbert-Harris PM, Green WR, White SL, Olden K, Laniyan I, Parish-Gause D, Vaughn T, Griffin WM, Sridhar R (2003) Pharm Res 47:69–74
124. Oredipe OA, Furbert-harris PM, Laniyan I, Green WR, Griffin WM, Sridhar R (2003) Cell Mol Biol 49:1089–1099
125. Buschmann N, Rückert A, Blechert S (2002) J Org Chem 67:4325–4329

126. Mill S, Hootelé C (2000) J Nat Prod 63:762–764
127. Jung HH, Floreancig PE (2007) J Org Chem 72:7359–7366
128. Wu Y, Peng S, Ouyang Y, Qian P, He W, Xiang J (2012) Acta Chim Sin 70:1367–1370
129. Gorin DJ, Davis NR, Toste FD (2005) J Am Chem Soc 127:11260–11261
130. Hiroya K, Matsumoto S, Ashikawa M, Ogiwara K, Sakamoto T (2006) Org Lett 8:5349–5352
131. Yan ZY, Xiao Y, Zhang L (2012) Angew Chem Int Ed 51:8624–8627
132. Allen GR, Poletto JF, Weiss MJ (1965) J Org Chem 30:2897–2904
133. Ito Y, Sawamura M, Hayashi T (1986) J Am Chem Soc 108:6405–6406
134. Ito Y, Sawamura M, Hayashi T (1988) Tetrahedron Lett 29:239–240
135. Hughes PF, Smith SH, Olson JT (1994) J Org Chem 59:5799–5802
136. Kulanthaivel P, Hallock YF, Boros C, Hamilton SM, Janzen WP, Ballas LM, Loomis CR, Jiang JB, Katz B, Steiner JR, Clardy J (1993) J Am Chem Soc 115:6452–6453
137. Bachi MD, Melman A (1997) J Org Chem 62:1896–1898
138. Togni A, Pastor SD, Rihs G (1989) Helv Chim Acta 72:1471–1478
139. Wenger RM (1985) Angew Chem Int Ed Engl 24:77–85
140. Li J, Ji K, Zheng R, Nelson J, Zhang L (2014) Chem Commun 50:4130–4133
141. Zhou J (2014) Multicatalyst system in asymmetric catalysis. Wiley, New York

Index

The manufacturer's authorised representative in the EU is Springer Nature Customer Service Centre GmbH, Europaplatz 3, 69115 Heidelberg, Germany. If you have any concerns regarding our products, please contact ProductSafety@springernature.com

Printed and bound by CPI Group (UK) Ltd, Croydon, CR0 4YY
30/09/2026
02225228-0010